The Variation Method in Quantum Chemistry

This is Volume 33 of
PHYSICAL CHEMISTRY
A series of monographs
Edited by ERNEST M. LOEBL, *Polytechnic Institute of New York*

A complete list of the books in this series appears at the end of the volume.

The Variation Method in Quantum Chemistry

SAUL T. EPSTEIN

Theoretical Chemistry Institute and Department of Physics
University of Wisconsin
Madison, Wisconsin

1974

ACADEMIC PRESS New York San Francisco London

A Subsidiary of Harcourt Brace Jovanovich, Publishers

To my mother and father, secure in the knowledge that they would have liked it sight unseen.

ACADEMIC PRESS, INC.
111 Fifth Avenue, New York, New York 10003

United Kingdom Edition published by
ACADEMIC PRESS, INC. (LONDON) LTD.
24/28 Oval Road, London NW1

Library of Congress Cataloging in Publication Data

Epstein, Saul T
The variation method in quantum chemistry.

(Physical chemistry, a series of monographs,)
Bibliography: p.
1. Quantum chemistry. 2. Calculus of variations.
I. Title. II. Series.
QD462.E67 541′.28 73-18960
ISBN 0-12-240550-1

PRINTED IN THE UNITED STATES OF AMERICA

Contents

Preface ix

Chapter I **General Theory of the Variation Method**

1. Some Background 1
2. The Variation Principle 4
3. The Variation Method: Introduction 8
4. The Variation Method: Details 12
5. The Variation Method and Moments of the Schrödinger Equation . . 19
References 24

Chapter II **Applications of the Variation Method**

6. The Linear Variation Method 26
7. Linear Spaces and Excited States 33
8. The Unrestricted Hartree–Fock Approximation 37
9. The Unrestricted Self-Consistent Field Approximation 52
References 55

Chapter III **The Generalized Brillouin Theorem**

10. Derivation and Applications of the Theorem 57
References 67

Chapter IV **Special Theorems Satisfied by Optimal Trial Functions**

11. Introduction 69
12. Reality 70
13. Unitary Invariance 73
14. Symmetry 83
15. Generalized Hellmann–Feynman Theorems 87

Chapter V Perturbation Theory and the Variation Method: General Theory

16. Hypervirial Theorems: General 92
17. Momentum Theorems 96
18. Force Theorems 98
19. Torque Theorems 101
20. Virial Theorems 104
21. Orthogonality and Related Theorems 115
References 118

Chapter VI Perturbation Theory and the Variation Method: Applications

22. The Variation Principle and Perturbation Theory 120
23. Perturbation Analysis of the Variation Method: Introduction 121
24. Perturbation Analysis of the Variation Method: General Formalism . . 124
25. Variation Methods within the Variation Method 135
26. The ν^{2n+1} Theorem and Interchange Theorems 144
References 154
27. Perturbation Analysis of the Linear Variation Method 156
28. Perturbation Analysis of USCF and UHF: One-Particle Perturbations . 165
29. The $\mathscr{Z}^{-1}$ Expansion 176
References 185

Chapter VII The Hylleraas Variation Method

30. Perturbation Analysis of the Variation Principle and the Hylleraas Variation Method 187
31. The Second-Order Hylleraas Variation Method: Details 192
32. The Linear Hylleraas Variation Method 199
33. Improvable Upper Bounds to Second-Order Energies for Excited States . 206
34. The Second-Order Hylleraas Variation Method with Two Perturbations 213
References 219

Chapter VIII Special Theorems Satisfied by Optimal First-Order Trial Functions

35. Derivation of Theorems 221
References 233

Chapter IX **Corrections to Approximate Calculations**

36. Mostly First-Order Corrections 234
References 243

Appendix A **The Max-Min Theorem** 244

Appendix B **Lagrange Multipliers** 247

Appendix C **Theorems Satisfied by Optimal Time-Dependent Variational Wave Functions** 250

Appendix D **Various Hypervirial Theorems in the Presence of Magnetic Fields** 260

Appendix E **Proof That (33-18) Is an Improvable Bound** 265

Author Index 268

Subject Index 273

Preface

What follows is very far from being a treatise on the variation method. Rather it represents an attempt to bring together into a coherent whole those aspects of the variation method as applied to bound states problems in quantum mechanics which have been of especial interest to me. Hopefully the results will be of some interest to others as well.

A fair appraisal of the contents of the book can be gained simply by inspecting the table of contents. As will be gathered, most of the text is devoted to generalities, general points of view, general theorems, etc. To balance this to some extent, we do describe in detail, in Sections 6, 8, and 9, three standard variational procedures which then serve in Sections 27–30 and also here and there in other sections as concrete examples to which we can apply the general theorems. Also we give a number of references to papers in the literature where other examples may be found.

Since, as will be seen, the level of discussion, both as regards physics and mathematics, is fairly elementary (and certainly should be readily accessible to anyone with two semesters of quantum mechanics), we have omitted many steps, while at the same time making frequent use of phrases like "it is obvious that," "it is easily shown," or "one readily finds," with the expectation that the reader will be alerted by them to work out the omitted steps. Also, at the ends of sections we have appended selections of problems, some of them asking the reader more explicitly to work out details, others indicating extensions and generalizations of results in the text.

Although my general interest in the variation method certainly predated my arrival at the University of Wisconsin, the great intensification of that interest in recent years owes much to the stimulation provided by my association with Professor Joseph O. Hirschfelder, and with the Theoretical Chemistry Institute generally. In particular, in addition to acknowledging my great debt to Professor Hirschfelder, I would like also to express my thanks to Professors W. Byers Brown, P. R. Certain, and P. D. Robinson for many discussions of these matters through the years. Also special thanks are due to my wife, Jean Hoopes Epstein, who has assisted in every phase of the preparation of this manuscript. Finally thanks are due to the National Aeronautics and Space Administration and the National Science Foundation which have supported the Institute.

Chapter I / General Theory of the Variation Method

1. Some Background

Principles of "least this" and "maximum that" have for a long time fascinated scientists and philosophers alike.[1] Moreover, even principles of "stationary this or that," though they may have somewhat less philosophical appeal, nonetheless continue to interest scientists because on the one hand they usually provide a very compact way of stating the mathematical essentials of a theory, and on the other provide a useful avenue to the development of new theories.

Thus it is not surprising that in his first paper on wave mechanics [3], Schrödinger presented his equation in the form of a variation principle, and only later [4] revealed some of the background which led to his writing down of the equation in the first place.

An obvious reason for the philosophical fascination of minimal (or maximal) principles is that they seem to make physical laws less arbitrary, more rational, suggest a purpose, etc. Indeed, it is tempting to take a variation principle quite literally and then imagine that the variations are actually taking place; that at each moment or at each point, the system, whatever it is, is sampling all sorts of possible behavior, with the actual behavior then being selected on the basis that it will make the least change in something, that is, it will make some quantity stationary. Moreover, Feynman [5], following on some earlier work of Dirac, has shown that in a very real sense quantum mechanics can be "derived" from classical mechanics by omitting the final selection processes and assuming that all behavior, not just the classical one, is possible.

In a not dissimilar vein, Ruedenberg and co-workers [6] have argued that one can get real insight into the nature of chemical binding by imagining that as a molecule forms, it actually does, so to speak, try one wave function,

[1] See, for example, Yourgrau and Mandelstam [1] and Born [2].

then another, relaxing a bit here, tightening a bit there, before finding the most suitable one.

Although these general theoretical and philosophical considerations concerning variation principles are of great importance, probably of more practical significance are the associated variation methods for approximating the solutions of actual physical problems. The use of such methods has a long history in science,[2] and in particular it is not much of an exaggeration to say that in applied quantum mechanics, most approximation procedures are either direct applications of the variation method or can profitably be related to it in one way or another.

In this book, we will be concerned with the variation method for approximating the energy eigenvalues and energy eigenfunctions of a bound quantum mechanical system. Thus we consider situations in which at least the low lying energy levels are discrete, and we concentrate our attention on these discrete levels. More particularly, our standard example will be that of N electrons interacting with one another and with various fixed nuclei via Coulomb interactions, and possibly also with external static electric and magnetic fields. Thus the Hamiltonian for the system, including various nuclear interactions as well, is

$$H = \sum_{s=1}^{N} \left\{ \frac{[\mathbf{p}_s + (1/c)\mathscr{A}(\mathbf{r}_s)]^2}{2} - \sum_a \frac{\mathscr{Z}_a}{|\mathbf{r}_s - \mathbf{R}_a|} - \Phi(\mathbf{r}_s) \right\}$$
$$+ \sum_{s=1}^{N} \sum_{\substack{t=1 \\ s>t}}^{N} \frac{1}{|\mathbf{r}_s - \mathbf{r}_t|} + \sum_{a>b}\sum \frac{\mathscr{Z}_a \mathscr{Z}_b}{|\mathbf{R}_a - \mathbf{R}_b|} + \sum_a \mathscr{Z}_a \Phi(\mathbf{R}_a) \qquad (1)$$

where we have used atomic units [mass of electron = −charge of electron = (Planck's constant)/2π = 1]. In Eq. (1), c is the velocity of light, $\mathscr{A}$ is the vector potential, and Φ is the scalar potential; the electric and magnetic fields $\mathscr{E}$ and $\mathscr{B}$ are then given by

$$\mathscr{E} = -\boldsymbol{\nabla}\Phi \qquad (2)$$

$$\mathscr{B} = \boldsymbol{\nabla} \times \mathscr{A} \qquad (3)$$

Also, $\mathbf{r}_s$ is the coordinate of the sth electron, $\mathbf{p}_s$ is the associated momentum operator, $\mathbf{R}_a$ is the coordinate of the ath nucleus, and $\mathscr{Z}_a$ is its charge.

The Hamiltonian (1) is, of course, the Hamiltonian for a molecule in the so-called "clamped nuclei approximation" with complete neglect of all spin (electronic and nuclear) and relativistic effects. As such, it provides

[2] See, for example, Mikhlin [7], Courant [8], Finlayson [9], and Birkhoff [10].

a conceptually simple but physically meaningful example to which we can apply the results of our general considerations. Nevertheless, it should be clear to the reader, without our making the point each time, that many of our general conclusions are quite independent of the detailed form of H. Also, although our language will be appropriate to the coordinate representation, most results will be representation independent, holding equally well in momentum space, for example, with appropriate interpretation of the symbols.

Our general notation will be quite a standard one. The scalar product of two wave functions ψ_1 and ψ_2 will be denoted by

$$(\psi_1, \psi_2) \tag{4}$$

with the "integration," for the N-electron example, being over both space and spin. It has the standard properties:

$$(\psi_1, \psi_2) = (\psi_2, \psi_1)^*, \quad (\psi_1, \alpha\psi_2) = \alpha(\psi_1, \psi_2), \quad (\alpha\psi_1, \psi_2) = \alpha^*(\psi_1, \psi_2) \tag{5}$$

where α is a number and the asterisk denotes complex conjugate. Also, we will assume that

$$(\psi, \psi) = 0 \tag{6}$$

implies that

$$\psi \equiv 0 \tag{7}$$

(that is, we will not worry about sets of measure zero). Finally, the statement that an operator θ is Hermitian means that

$$(\psi_1, \theta\psi_2) = (\theta\psi_1, \psi_2) \tag{8}$$

In our discussions, we will have frequent use for matrix and vector notation. A particular vector will usually be distinguished by a subscript in parentheses; thus $V_{(j)}$ with components $V_{(j)k}$ or $V_{k(j)}$, and the scalar product of two vectors will be denoted by

$$W_{(k)}^{+} \cdot V_{(j)} \equiv \sum_l W_{(k)l}^{*} V_{(j)l} \tag{9}$$

If this scalar product multiplies another vector X, we will, however, usually write the result as $(W_{(k)}^{+} \cdot V_{(j)})X$ rather than simply $W_{(k)}^{+} \cdot V_{(j)}X$. Finally, as usual, a Hermitian matrix is one for which

$$M_{lj} = M_{jl}^{*} \tag{10}$$

and which therefore has the property

$$V_{(j)}^{+} \cdot MV_{(i)} = (MV_{(j)})^{+} \cdot V_{(i)} \tag{11}$$

while a symmetric matrix is one for which

$$M_{lj} = M_{jl} \tag{12}$$

PROBLEMS

1. What is the numerical value of c in atomic units?

2. If in atomic units $\mathscr{E} = \mathscr{B} = 1$, what are the field strengths in more practical units? Are they large or small by laboratory standards?

3. Presumably you are familiar with the standard properties of Hermitian operators and matrices and of their eigenfunctions and eigenvalues—that they can be diagonalized by a unitary transformation, the orthogonality of the eigenfunctions, reality of the eigenvalues, etc. If not, please consult almost any graduate level quantum mechanics text. Also, you should be familiar with the fact that the average value of any Hermitian operator is greater than or equal to its smallest eigenvalue.

4. A Hermitian matrix M is said to be positive definite if all its eigenvalues are positive. Show that a necessary and sufficient condition for this is that $X^{+} \cdot MX > 0$ for all X.

5. Show that a positive-definite Hermitian matrix always has an inverse (which is also positive definite), and a positive square root.

2. The Variation Principle

Given any function ψ for which the requisite integrals exist (we will refer to such functions as *trial functions*), we can calculate the real number

$$E = (\psi, H\psi)/(\psi, \psi) \tag{1}$$

Since E would be the average energy of the system if the system were in the state described by the function ψ, we have that E cannot be less than the smallest possible energy, that is, E cannot be smaller than the smallest eigenvalue of H.

To learn more, we consider another trial function ψ' and Δ, the difference between ψ and ψ'; thus

$$\psi = \psi' + \Delta \tag{2}$$

Then with the definition

$$E' \equiv (\psi', H\psi')/(\psi', \psi') \tag{3}$$

one finds upon inserting (2) into (1) that

$$E = \frac{E'(\psi', \psi') + (\psi', H\Delta) + (\Delta, H\psi') + (\Delta, H\Delta)}{(\psi, \psi)}$$

or, since

$$(\psi', \psi') + (\psi', \Delta) + (\Delta, \psi') + (\Delta, \Delta) = (\psi, \psi) \tag{4}$$

we have

$$E = E' + \frac{(\psi', (H - E')\Delta) + (\Delta, (H - E')\psi') + (\Delta, (H - E')\Delta)}{(\psi, \psi)}$$

Finally, using the hermiticity of $(H - E')$, we rewrite this as

$$E = E' + \frac{((H - E')\psi', \Delta) + (\Delta, (H - E')\psi') + (\Delta, (H - E')\Delta)}{(\psi, \psi)} \tag{5}$$

We will now draw several important conclusions from this result.

Result 1. Suppose that

$$(H - E')\psi' = 0 \tag{6}$$

that is, suppose that ψ' and E' are an eigenfunction and the corresponding eigenvalue of H. Then (5) becomes

$$E = E' + \frac{(\Delta, (H - E')\Delta)}{(\psi, \psi)} \tag{7}$$

from which we see that E differs from E' by terms which are of at least second order in the difference between ψ and ψ'. Therefore if we think of ψ as being a continuously variable quantity, then it follows that *the eigenvalues of H are stationary points of E as a functional of ψ.*

Result 2. We will now show that E has no other stationary points. Thus suppose that E' is a stationary point of E as a functional of ψ. Then the first-order terms on the right hand side of (5) must vanish for all Δ and therefore the first-order terms in

$$((H - E')\psi', \Delta) + (\Delta, (H - E')\psi')$$

must vanish for all Δ. In particular this must be true for

$$\Delta = \alpha(H - E')\psi' \tag{8}$$

where α is a continuously variable number. Thus we must have

$$(\alpha + \alpha^*)((H - E')\psi', (H - E')\psi') = 0$$

which can be satisfied only if

$$(H - E')\psi' = 0$$

Therefore *if E' is a stationary point, then E' is an eigenvalue and the corresponding ψ is an eigenfunction.* The characterization of the eigenvalues and eigenfunctions of H provided by Results 1 and 2 constitutes a statement of the *variation principle.*

Result 3. Suppose now that E' is the smallest eigenvalue of H. Then $(H - E')$ has only nonnegative eigenvalues and thus its average value is always nonnegative. In particular, then, this means that however large Δ may be, still

$$(\Delta, (H - E')\Delta) > 0$$

and hence from (7) we see, consistent with our initial observation, that *the smallest eigenvalue of H* is not just a stationary point, *it is the absolute minimum of E as a functional of ψ.* Conversely, *E is always an upper bound to the smallest eigenvalue.*

On the other hand, if E' is a higher eigenvalue, then by choosing Δ to be a linear combination of the lower eigenfunctions (still higher eigenfunctions—and this can include continuum states if there are any), we can evidently make $(\Delta, (H - E')\Delta)$ less than (greater than) zero. Thus the intermediate eigenvalues of H are only saddle points of E as a functional of ψ, and even locally are neither maxima nor minima.

Result 4. That the smallest eigenvalue is the absolute minimum of E is a very striking result. However, it does not in general serve to characterize the energies of the actual ground states of atoms or molecules since, because of the requirements of the Pauli principle, these ground states are usually not the lowest eigenstates of the appropriate Hamiltonians; for example, the ground state of the lithium atom is $(1s)^2 2s$ and not $(1s)^3$.

Happily, however, there is a similar theorem which is applicable to the physical ground states and to the various excited states as well: often the eigenstates of H can be classified into types according to their symmetries or, more generally, according to the eigenvalues of other operators K which commute with H. If then E' is the smallest eigenvalue of H for states of a certain type (for example, for states satisfying the Pauli principle), it follows that $(H - E')$ will have a nonnegative expectation value with respect to functions of that type because such functions will be orthogonal to all eigenfunctions of H associated with smaller eigenvalues.

If now we confine attention to ψ of that same type, then $\Delta = \psi - \psi'$ will also be of that type and therefore however large Δ may be, still

$$(\Delta, (H - E')\Delta) \geq 0$$

Thus *the smallest eigenvalue of H for states of a given type is the absolute minimum of E as a functional of trial functions of that type*, and conversely, if one uses only trial functions of that type, then *E is always an upper bound to that lowest eigenvalue.*

Result 5. We now observe that if H commutes with K, it will follow that if ψ' is of a definite type, then the Δ, which played the decisive role in Result 2, namely $(H - E')\psi'$, will also be of that type. Thus we may generalize the result found there as follows: If H commutes with K, if ψ' is of a certain type, and if E' is a stationary point with respect to all variations of that type, then E' is an eigenvalue and ψ' is an eigenfunction. In summary, combining this last result with Result 1, *the variation principle applies separately to the eigenfunctions and eigenvalues of each type.*

Result 6. A further generalization of Result 4 is evidently the following. Let E' be an arbitrary eigenvalue of H and confine attention to ψ which are orthogonal to all eigenfunctions of H whose associated eigenvalues are less than E'. Then we will have

$$(\Delta, (H - E')\Delta) \geq 0$$

and thus *an arbitrary eigenvalue of H is an absolute minimum of E as a functional of trial functions orthogonal to eigenfunctions associated with smaller eigenvalues.*

Before concluding this section, it is appropriate to make some remarks about the existence of the integrals in (1) and hence about the functions which can be trial functions. For one thing, of course, as befits a bound state function, they must be normalizable so that the denominator exists. Further, if we are working in configuration space, then, for the usual Hamiltonians of atomic and molecular physics, for example for (1-1), the existence of $(\psi, H\psi)$ requires that ψ be twice differentiable. However, if one uses the formally equivalent $(\nabla\psi, \cdot\nabla\psi)$ instead of $(\psi, -\nabla^2\psi)$ it can be shown [11] that the results which we have found in this section continue to hold even if ψ is only once differentiable.[3] Further, the former form is usually much more convenient to deal with numerically. Also, it was in this form that Schrödinger [3] gave his original variational formulation of quantum mechanics, and finally, as Schrödinger has emphasized [13], it is an especially useful one for transforming the Schrödinger equation from Cartesian coordinates to arbitrary coordinates (it is easier to transform $\nabla\psi$ than to transform $\nabla^2\psi$). Nevertheless, in spite of all these virtues of $(\nabla\psi, \cdot\nabla\psi)$, we will continue to use the expression (1) because it is much easier to deal with formally, and because in most applications the ψ *are* twice differentiable.

PROBLEMS

1. If ψ and ψ' are physically distinct (for example, orthogonal to one another), can $E = E'$ in (7)?

2. It may seem odd that out of all possible Δ's the special choice (8) already sufficed to derive the general result. However, show that the first-order contributions of any physically distinct additions to this Δ would vanish identically.

3. Devise a projection operator (presumably you are familiar with the general notion of projection operators) which is such that if one knows the lower eigenvalues of H, then one can produce trial functions which are orthogonal to the lower eigenfunctions. (See, for example, Löwdin [14].)

3. The Variation Method: Introduction

The results of the previous section are, aside from their theoretical interest, of great practical importance because they suggest a soundly based method

[3] For variation principles which allow ψ to be discontinuous, see, for example, Rudge [12].

for approximating the eigenvalues and eigenfunctions of H. According to the variation principle, we can find the eigenvalues and eigenfunctions of H by calculating E for all ψ and then looking for stationary points. In practice this is usually impossible—one cannot examine *all* ψ. However, what one *can* do is to examine a restricted class of trial functions, a class no larger than one can handle, and then take the stationary points of E and the corresponding ψ *within this restricted class* as approximations to the eigenfunctions and eigenvalues of H.

This procedure is known as the *variation method.* We will call the ψ which yield stationary values "optimal trial functions" and denote them by $\hat{\psi}$, possibly with a subscript. The corresponding E we will denote by $\hat{E}$, again possibly with a subscript.

We said that this is a soundly based method. To support that assertion, consider first the lowest state of a given type. Then Result 4 of Section 2 tells us that we have a good approximation scheme in that it is capable of systematic improvement. Namely if we enlarge the class of trial functions (assumed to be of appropriate type), then the smallest of the $\hat{E}$, since it is the absolute minimum of E within the set, can only decrease (or at any rate not increase), whence, from Result 4, it follows that we will have a better approximation to the eigenvalue. In short, we have a guaranteed and improvable upper bound. (This, of course, ignores, as we shall, the possibility of pathological situations involving some sort of discontinuous behavior in which the absolute minimum of E is not a minimum in the calculus sense and hence is not the smallest $\hat{E}$.) Note also that we have a quadratic convergence to the eigenvalue since, from Result 1, the error in the eigenvalue is of second order in the error of the eigenfunction.

For the higher states of a given type, the situation at this point is not so clear. Result 6 of Section 2 is of little practical use since one usually cannot guarantee the required orthogonality. We can of course say that if we enlarge the class of trial functions, we will make the higher $\hat{E}$ "more stationary," but this may or may not represent a numerical improvement. However, in a later section we will discuss a practical way of choosing trial functions (the linear variation method) which yields systematically improvable upper bounds also to higher eigenvalues.

Even from these brief remarks it should be clear that the variational approximation to eigenvalues *is* a soundly based one.[4] It is harder to make

[4] Schaefer [15] describes the details and discusses the results of many variational calculations for atoms and molecules. See also Burden and Wilson [15a]. A survey of high precision results for the helium atom is given by Stewart [16]. A survey of both experimental and theoretical results for two-electron diatomics has been given by Herzberg [17].

a definite statement about the quality of the eigenfunction approximation, in part because there are so many figures of merit which one might use—the overlap between the approximation and the eigenfunction, the accuracy of particular expectation values, the energy variance $(\hat{\psi}, (H - \hat{E})^2\hat{\psi})/(\hat{\psi}, \hat{\psi})$, etc. Also, although there is a considerable literature on the subject (more precisely, it is concerned with getting bounds on such quantities) it has thus far found little quantitative application except to those systems with very few electrons, like He.[5] We will therefore not attempt a quantitative discussion of these many possibilities except to note that it is well documented in the literature that in general there need be no correlation between accuracy in the approximation to the energy and accuracy in the values of other average values.[6]

In a general way, however, one usually says, and expects, that the approximation to the eigenvalue furnished by $\hat{E}$ is better than the approximation to the eigenfunction furnished by $\hat{\psi}$ because, as we have already noted, the error in the former is of second order in the error in the latter. In this connection, though, it should be kept in mind that second-order quantities are guaranteed to be smaller than first-order quantities only if the order parameter is sufficiently small. Thus αx^2, where α is a number, is formally of second order in x for all α but numerically it is smaller than the first-order quantity x only for $x < 1/\alpha$.

Moreover, it should be recognized that energy eigenvalues as such are usually of direct interest only for very light systems. Rather, one is usually interested in comparatively small energy *differences*—excitation energies, ionization energies, bond energies, changes in the molecular energy with the nuclear configuration, and so forth—rather than the total energies. Thus the accuracy of the total energy may not be of immediate concern. Moreover, even if each theoretical number is a guaranteed upper bound, the difference need not be, so that, so to speak, one does not know where one is when calculating differences. Further, since improving the individual upper bounds will not necessarily improve the difference, it is not particularly clear what it is best to do. Indeed, one may well question the use of the variation method at all in approximating energy differences and there continues to be much discussion and use of methods which yield energy differences directly though still with no guaranteed bounds.[7]

One solution to the problem would be to supplement the upper bound

[5] See, for example, the recent review by Weinhold [18].

[6] See, for example, Bishop and Macias [19].

[7] The paper by Simons [20] contains an extensive bibliography.

by a lower bound since the difference between an upper bound and a lower bound is again an upper bound. Now in fact lower bound formulas do exist and there is a large literature on the subject. However, as yet they have not been found to be particularly useful, that is, have not given very tight lower bounds, except for the very simplest systems, notably helium.[8] Rather lower bounds are generally, for the same effort, worse than upper bounds.

Another approach to getting a bound on a difference is to take one number from experiment. The difficulty here is that most calculations are a priori of low or modest accuracy as far as the total energy is concerned and the differencing will simply reveal that fact, i.e., the bound will be poor. A less bothersome point is that even in calculations of presumably high accuracy one is almost certainly using an H which is an approximation to the "true" Hamiltonian and hence further corrections (which usually have the effect that the bound is no longer guaranteed) must be applied to $\hat{E}$ (or to the experimental number) in order that the two numbers refer to the same physical (or mathematical) problem.

As we said, most calculations are of low or modest accuracy as far as individual $\hat{E}$ are concerned. Here it is very important to keep in mind that contrary to the tone of our earlier remarks, differencing can often *improve* the situation, even very dramatically. For example, in the normal states of an atom, most of the total energy is provided by the inner shells. However, the inner shells are scarcely disturbed by the excitation of valence electrons, and therefore it is not surprising that even if the individual $\hat{E}$ are very poor because of an inadequate treatment of the inner shells, still when one takes differences, the results are quite accurate or at least not grossly wrong. A similar situation prevails for the tightly bound electrons in molecules. Further, in dealing with molecules, if one can differentiate between what is going on in the individual atoms and what is specific to the formation of the molecule,[9] then one might well expect, and one does find, that even large inaccuracies in the treatment of the former will cancel out when one takes differences to compute molecular binding energies, etc.[10]

Returning to the question of "where one is," even if one is interested in $\hat{E}$, and even if one knows that it is an upper bound, still, without an at least equally good lower bound, then just from the theory itself, one has no real idea as to how far from the actual eigenvalue one is. However, if one has worked hard to reduce $\hat{E}$ (that is, steadily enlarged the set of

[8] See, for example, Pekeris [21] and Wang and Weinhold [22].

[9] See, for example, Wahl and Das [23].

[10] For a spectacular example, see Schaefer *et al.* [24]; see also Bertoncini and Wahl [25, 26].

trial functions) and if the value has stabilized to a certain number of decimals, then one will usually claim accuracy to that number of decimals (this could of course be made rigorous if one could give a quantitative proof of convergence), though one may well be badly in error. Similar remarks of course also apply to the calculation of expectation values generally.

4. The Variation Method: Details

First we introduce the basic notation that we will be using from now on. As implied by our description of the variation method as a search for stationary points, in general we imagine the set of trial functions to be a continuous one, labeled by variational parameters which may be arbitrary numbers, real and/or complex, and/or arbitrary functions of one or more variables. In general discussions, we will usually denote these parameters by A_i or collectively by A, though in particular examples we may use other symbols. Thus we write

$$\psi = \psi(A_1, A_2, \ldots) \equiv \psi(A)$$

where we have omitted specific mention of particle coordinates in the argument of ψ. Indeed, as a rule, we will usually omit mention of whatever variables are not needed at any particular moment. The optimal values of the A_i will be denoted by $\hat{A}_i$; thus

$$\hat{\psi} = \psi(\hat{A})$$

Starting with any one member of the set $\psi(A)$, we can generate all the others by replacing the A_i by $A_i + \delta A_i$ and letting the δA_i take on all possible values consistent with whatever they are. Then from Taylor's theorem, we have

$$\psi(A + \delta A) = \psi(A) + \sum_i \frac{\partial \psi(A)}{\partial A_i} \delta A_i + \frac{1}{2} \sum_i \sum_j \frac{\partial^2 \psi(A)}{\partial A_i \, \partial A_j} \delta A_i \, \delta A_j + \cdots \tag{1}$$

whence it follows that

$$\sum_i \frac{\partial \psi(A)}{\partial A_i} \delta A_i \equiv \delta \psi(A) \tag{2}$$

the first differential of $\psi(A)$, or as we will call it, *the variation of* $\psi(A)$ is the first order change in ψ produced by changing the A_i. Similarly one-half of

$$\sum_i \sum_j \frac{\partial^2 \psi(A)}{\partial A_i \, \partial A_j} \delta A_i \, \delta A_j \equiv \delta_2 \psi(A) \tag{3}$$

the second differential of ψ or the second variation of ψ, is the second-order change in ψ, etc. Note that only if $\delta_2\psi$, $\delta_3\psi$, ... all vanish is $\delta\psi$ the total change in ψ (analogous to the quantity Δ in Section 2); however, δA_i *is* the total change in A_i.

One final point concerning notation. In writing δA_i for change in A_i, we have more or less followed tradition. However, this notation is deceiving in that it suggests that the possible values of the change depend on the value from which one starts, whereas in fact this is usually *not* the case, and certainly we will assume that it is not in our formal discussion. For example, if A_i is a real number, then whatever particular value we may start with, we may add to it *any* other real number. (Apparent exceptions to this can usually be overcome by redefining the parameters. Thus if $\psi = \exp(-Ar^2)$, with A a real number, then A must be positive. However, if we replace A by e^A, then A can be any real number.) It is to emphasize this point that we will therefore be superficially inconsistent in our notation and denote a change in $\hat{A}_i$ by δA_i again rather than by $\delta \hat{A}_i$. Also note that it is implied that the δA form at least a *real* linear space in that if δA_i and $\delta' A_i$ are possible changes in A_i, then so is $\alpha\, \delta A_i + \beta\, \delta' A_i$ for any real numbers α and β. (This observation will play a large role in our discussion starting in Section 16.)

Having settled on our notation, a straightforward way to implement the variation method is the following. The first step is to calculate δE, the first differential of E (the variation of E) produced by changing the variational parameters from A to $A + \delta A$. Then to determine the $\hat{A}$, one sets $\delta\hat{E} = 0$ and looks for solutions, if any, of the resulting equations. Finally, one uses the $\hat{A}$ to calculate $\hat{E}$ and $\hat{\psi}$.

To connect up with more familiar things, suppose that the A_i are simply real numbers. Then obviously

$$\delta E = \sum_i \frac{\partial E}{\partial A_i}\, \delta A_i$$

whence $\delta\hat{E} = 0$ yields

$$\sum_i \frac{\partial \hat{E}}{\partial \hat{A}_i}\, \delta A_i = 0 \tag{4}$$

We will now show that if the A_i are independent, then (4) yields the familiar conditions for a stationary point

$$\frac{\partial \hat{E}}{\partial \hat{A}_i} = 0 \qquad \text{all} \quad i \tag{5}$$

The proof is simple. Equations (5) certainly suffice to guarantee (4). In addition, if the A_i are independent, Eqs. (5) are also necessary since "independent" means that we can change each A_i separately. Thus we can choose all δA_i in (4) equal to zero except one which evidently leads to (5).

Thus in this simple case, the procedure which we have described is precisely equivalent to the standard calculus prescription. We should, however, point out that in many practical calculations which must be largely numerical rather than analytical in character, the solving of equations like (5) is often partially or completely bypassed in favor of some direct numerical search procedure to locate the stationary points of E.[11]

We will now describe a second procedure, one which, as we will see, is very convenient for theoretical discussions, and hence the one which we will mostly use. If we write (2-1) as

$$(\psi, (H - E)\psi) = 0$$

then clearly δE is determined by

$$(\delta\psi, (H - E)\psi) + (\psi, (H - E)\delta\psi) - \delta E(\psi, \psi) = 0 \tag{6}$$

Now the $\hat{\varphi}$ are those ψ that make $\delta E = 0$ for all variations possible within the set. Thus we must have

$$(\delta\hat{\varphi}, (H - \hat{E})\hat{\varphi}) + (\hat{\varphi}, (H - \hat{E})\delta\hat{\varphi}) = 0 \qquad \text{“all”} \quad \delta\hat{\varphi} \tag{7}$$

where the quotation marks are to remind us that we are requiring that Eq. (7) hold only for variations within the set. In more detail, (7) becomes, using (2),

$$\sum_i \left(\frac{\partial\hat{\varphi}}{\partial\hat{A}_i}\,\delta A_i, (H - \hat{E})\hat{\varphi}\right) + \left(\hat{\varphi}, (H - \hat{E})\,\frac{\partial\hat{\varphi}}{\partial\hat{A}_i}\,\delta A_i\right) = 0 \qquad \text{all} \quad \delta A_i \tag{8}$$

where now the quotation marks are not needed since the limitations of the set are implicit in the nature and number of the A_i.

Equation (7) together with

$$(\hat{\varphi}, (H - \hat{E})\hat{\varphi}) = 0 \tag{9}$$

are then the equations to be used to determine the $\hat{\varphi}$ and $\hat{E}$, and hence are equations which characterize the variation method. An obvious next step would be to eliminate $\hat{E}$ from (7) by means of (9), solve the resultant

[11] See, for example, Kari and Sutcliffe [27]; also Koutecký and Bonačić-Koutecký [28] and Bloomer and Bruner [29].

equations for the $\hat{A}$, and then return to (9) to determine $\hat{E}$, a procedure which is obviously essentially the same as the straightforward one which we outlined earlier.

However, we will now show that for theoretical purposes, and often for practical purposes as well, one can dispense with (9), and take (7) alone as *the* equation characterizing the variation method. To see this, first suppose that the set of trial functions has no fixed overall scale, or more technically that it is invariant to multiplication by a real constant. That is, if ψ is a member of the set, then so is $B\psi$ for any real constant B. In such cases, then it must be that for some choice of the δA, $\psi(\hat{A} + \delta A)$ will equal $B(\delta A)\psi(\hat{A})$, where $B(\delta A)$ is a real number, with of course $B(0) = 1$. Therefore for these δA, $\delta\hat{\psi}$ will evidently have the form $\delta\hat{\psi} = \delta B\hat{\psi}$. If now we use this $\delta\hat{\psi}$ in (7), then, cancelling a factor of δB, we have (9), which proves the point in this case.

On the other hand, it is sometimes convenient to fix the scale of the ψ's, in particular to require that they be normalized. In such a case, then as far as actual calculations are concerned, one must adopt the straightforward approach. However, for our present purposes, the point is that conceptually we can trivially replace this set of trial functions by another set which has no fixed scale but which yields the same physical results, so that for theoretical purposes we can again simply use (7). Namely let us imagine enlarging the set of trial functions by multiplying each by an arbitrary number which is then to be used as a new variational parameter. Since the new set has no fixed overall scale, (7) alone now suffices, but since the E's obviously do not involve the new parameter (it cancels out between numerator and denominator), the final results, the $\hat{E}$ and the $\hat{\psi}$, are not changed in any way. Thus this conceptual increase in flexibility of the set of trial functions has no physical consequences but, as we have shown, does allow us the theoretical economy of using (7) alone to characterize the variation method.

Once one has located a stationary point, a natural question to ask is, is it a minimum, a maximum, or just a saddle point? The way to answer this question is to look at the second differential of $\hat{E}$. To determine $\delta_2\hat{E}$, we simply take the first differential of Eq. (6) and then put $\psi = \hat{\psi}$ and $E = \hat{E}$ and use the fact that $\delta\hat{E} = 0$ to find

$$\delta_2\hat{E}(\hat{\psi}, \hat{\psi}) = (\delta_2\hat{\psi}, (H - \hat{E})\hat{\psi}) + 2(\delta\hat{\psi}, (H - \hat{E})\delta\hat{\psi}) + (\hat{\psi}, (H - \hat{E})\delta_2\hat{\psi}) \quad (10)$$

If we now substitute from (2) and (3), then we have $\delta_2\hat{E}$ expressed as a quadratic form in the δA_i and the δA_i^*, and then the question to be answered is, is the form positive (local minimum), negative (local maximum) or indefinite (saddle point)?

When all the A_i are numbers—a common case in practice—it is possible to reduce the problem of determining the sign of $\delta_2\hat{E}$ to the problem of determining the signs of the eigenvalues of a certain Hermitian matrix [30]. To do this, we first use (2) and (3) and the fact that the δA_i are numbers to rewrite (10) as

$$\begin{aligned}\delta_2\hat{E}(\hat{\psi}, \hat{\psi}) = \sum_i \sum_j \Big\{ &\delta A_i{}^* \, \delta A_j{}^* \left(\frac{\partial^2 \hat{\psi}}{\partial \hat{A}_i \, \partial \hat{A}_j}, (H - \hat{E})\hat{\psi} \right) \\ &+ 2\delta A_i{}^* \, \delta A_j \left(\frac{\partial \hat{\psi}}{\partial \hat{A}_i}, (H - \hat{E}) \frac{\partial \hat{\psi}}{\partial \hat{A}_j} \right) \\ &+ \delta A_i \, \delta A_j \left(\hat{\psi}, (H - \hat{E}) \frac{\partial^2 \hat{\psi}}{\partial \hat{A}_i \, \partial \hat{A}_j} \right) \Big\} \end{aligned} \tag{11}$$

which we further rewrite as

$$\delta_2\hat{E}(\hat{\psi}, \hat{\psi}) = \sum_i \sum_j \{ \delta A_i{}^* \, \delta A_j{}^* \, \eta_{ij} + 2\delta A_i{}^* \, \delta A_j \, \gamma_{ij} + \delta A_i \, \delta A_j \, \eta^*_{ij} \} \tag{12}$$

where

$$\begin{aligned} \eta_{ij} &= \left(\frac{\partial^2 \hat{\psi}}{\partial \hat{A}_i \, \partial \hat{A}_j}, (H - \hat{E})\hat{\psi} \right) = \eta_{ji}, \\ \gamma_{ij} &= \left(\frac{\partial \hat{\psi}}{\partial \hat{A}_i} \, (H - \hat{E}) \, \frac{\partial \hat{\psi}}{\partial \hat{A}_j} \right) = \gamma^*_{ji} \end{aligned} \tag{13}$$

As it stands, the right hand side of (12) is not in familiar Hermitian form, since it involves not only $\delta A_i{}^* \, \delta A_j$ but also possibly $\delta A_i{}^* \, \delta A_j{}^*$ and $\delta A_i \, \delta A_j$. To put it in a more familiar form, we proceed as follows. We introduce the "two-element" vector

$$\begin{pmatrix} \delta A \\ \delta A^* \end{pmatrix} \equiv \delta \mathbb{A} \tag{14}$$

whose rows are themselves vectors with components δA_i and $\delta A_i{}^*$, respectively. Similarly we introduce the "2×2" matrix

$$\begin{pmatrix} \gamma & \eta \\ \eta^* & \gamma^* \end{pmatrix} \equiv \Lambda \tag{15}$$

where γ is itself a matrix whose components are the γ_{ij}, γ^* is itself a matrix where components are $(\gamma^*)_{ij} \equiv \gamma^*_{ij}$, etc. One then readily finds, using (13), that (12) can be written as

$$\delta_2\hat{E}(\hat{\psi}, \hat{\psi}) = (\delta A^* \; \delta A) \begin{pmatrix} \gamma & \eta \\ \eta^* & \gamma^* \end{pmatrix} \begin{pmatrix} \delta A \\ \delta A^* \end{pmatrix} \tag{16}$$

or

$$\delta_2\hat{E}(\hat{\psi}, \hat{\psi}) = \delta \mathbb{A}^+ \cdot \Lambda \, \delta \mathbb{A} \tag{17}$$

Furthermore, since from (13) γ is a Hermitian matrix and η a symmetric matrix, Λ is obviously Hermitian, so that the right hand side of (17) is now in standard form and the question of the sign of $\delta_2\hat{E}$ is reduced to the question of the signs of the eigenvalues of Λ. That is, if we solve

$$\Lambda Y_{(i)} = \lambda_i Y_{(i)} \tag{18}$$

for the eigenvectors and eigenvalues of Λ, the former being chosen orthonormal

$$Y_{(i)}^{+} \cdot Y_{(j)} = \delta_{ij} \tag{19}$$

then since we can write

$$\delta A = \sum_i Y_{(i)}(Y_{(i)}^{+} \cdot \delta A) \tag{20}$$

(17) becomes, using (1-11), (18), and (19),

$$\delta_2\hat{E}(\hat{\psi}, \hat{\psi}) = \sum \lambda_i | Y_{(i)}^{+} \cdot \delta A |^2 \tag{21}$$

Therefore $\delta_2\hat{E}$ is nonnegative (nonpositive) if all the λ are nonnegative (nonpositive), and is of indefinite sign otherwise. Note also that to prove that a stationary point is only a saddle point, it is sufficient to prove that the largest λ_i is positive while the smallest is negative, and that one way to do this is to use the fact that for any vector Y

$$(Y^{+} \cdot \Lambda Y)/(Y^{+} \cdot Y) \tag{22}$$

is an upper bound to the smallest eigenvalue of Λ and a lower bound to the largest. Therefore if one can find two Y's, one for which (22) is positive and one for which (22) is negative, the point will be proven.

Actually this discussion is not yet quite complete, since we have implicitly assumed that δA could be chosen to be a multiple of any of the $Y_{(i)}$, or more precisely, if λ_i is degenerate, could be chosen to be at least one of the degenerate $Y_{(i)}$. However, this is not immediately obvious since δA has a rather special structure, its two components being the complex conjugates of one another. To complete the analysis, let

$$Z \equiv \begin{pmatrix} x \\ y \end{pmatrix} \tag{23}$$

be an eigenvector of Λ with eigenvalue λ,

$$\Lambda Z = \lambda Z \tag{24}$$

Then, taking the complex conjugate of (24) and using the facts that $\lambda^* = \lambda$ and that Λ^* is just Λ with its two rows interchanged, one readily verifies that

$$Z' \equiv \begin{pmatrix} y^* \\ x^* \end{pmatrix} \tag{25}$$

is also an eigenvector with eigenvalue λ. Therefore since $(Z + Z')$ and $i(Z - Z')$, at least one of which will be nonzero, both have the special form, it follows that our assumption was justified. Namely if $\lambda_{(j)}$ is nondegenerate, then, to within an irrelevant phase factor, $Y_{(j)}$ must be of the special form, while if λ_j is degenerate, at least one of the $Y_{(j)}$ may be freely assumed to have the form.

If all quantities are explicitly real, or if all the A_i are linear parameters so that η vanishes, then there is no need to introduce the two-component formalism since (12) is then already a Hermitian quadratic form. Thus if we introduce the $y_{(i)}$, the orthonormal eigenvectors of $\gamma + \eta$, then we will have

$$\delta_2\hat{E}(\hat{\psi}, \hat{\psi}) = 2 \sum_i \lambda_i \mid y_{(i)}^{+} \cdot \delta A \mid^2 \tag{26}$$

where the λ_i are the associated eigenvalues. (We continue to denote them by λ_i since, under the given conditions, it is easy to see that they are also the eigenvalues of Λ.)

PROBLEMS

1. Presumably you already have some acquaintance with the variation method and have done some simple calculations as part of a course in quantum mechanics. If not, or if you need refreshing, please consult the appropriate sections of any graduate level quantum mechanics text.

2. Be sure that you can derive (10). Following it, we spoke of *local* minima and *local* maxima. Why did we introduce the qualification local?

3. Prove that (22) is an upper bound to the smallest eigenvalue of Λ and a lower bound to the largest.

4. In connection with the discussion following Eq. (23) show that the special form is maintained under Schmidt orthogonalization. From this, argue that there is therefore no loss in generality in assuming that *all* the $Y_{(i)}$ have the special form.

5. Often in practice, though the δA are complex, still η and γ are real. Denoting the eigenvalues and eigenvectors of $(\gamma \pm \eta)$ by $\lambda_i^{(\pm)}$ and $y_i^{(\pm)}$ respectively, show that in such a case

$$\begin{pmatrix} y_i^{(+)} \\ y_i^{(+)} \end{pmatrix} \quad \text{and} \quad \begin{pmatrix} y_i^{(-)} \\ -y_i^{(-)} \end{pmatrix}$$

are the eigenvectors of Λ, and that the eigenvalues are the $\lambda_i^{(+)}$ and $\lambda_i^{(-)}$. Show, in accord with (26), that when δA *is* real only the $y_i^{(+)}$ contribute to (21).

5. The Variation Method and Moments of the Schrödinger Equation

Using the hermiticity of $(H - \hat{E})$, our basic equation, Eq. (4-7), can be written as

$$(\delta\hat{\psi}, (H - \hat{E})\hat{\psi}) + ((H - \hat{E})\hat{\psi}, \delta\hat{\psi}) = 0, \qquad \text{“all”} \quad \delta\hat{\psi} \tag{1}$$

or

$$\operatorname{Re}(\delta\hat{\psi}, (H - \hat{E})\hat{\psi}) = 0, \qquad \text{“all”} \quad \delta\hat{\psi} \tag{2}$$

However, there are circumstances, and they frequently occur in practice, in which one can replace (2) by the formally simpler and seemingly stronger conditions

$$(\delta\hat{\psi}, (H - \hat{E})\hat{\psi}) = 0 \qquad \text{“all”} \quad \delta\hat{\psi} \tag{3}$$

First of all, this is obviously the case if H and all of the trial functions are explicitly real. We will now show that it is also true for any H if the A_i are independent and if, in addition, to go to the other extreme, there are no a priori reality restrictions whatsoever on them.

If the A_i are independent, we can change them one at a time, whence we have from (4-8)

$$\left(\frac{\partial\hat{\psi}}{\partial\hat{A}_j}\,\delta A_j, (H - \hat{E})\hat{\psi}\right) + \left((H - \hat{E})\hat{\psi}, \frac{\partial\hat{\psi}}{\partial A_j}\,\delta A_j\right) = 0 \qquad \text{all} \quad \delta A_j \tag{4}$$

If there are no a priori reality restrictions on A_j then we are able to change its real and imaginary parts (which we naturally assume to be the same sort of thing—each an arbitrary number, or each an arbitrary function of one variable, or whatever) separately. Thus we have

$$\left(\frac{\partial\hat{\psi}}{\partial\hat{A}_j}\,\delta A_{j\mathrm{R}}, (H - \hat{E})\hat{\psi}\right) + \left((H - \hat{E})\hat{\psi}, \frac{\partial\hat{\psi}}{\partial\hat{A}_j}\,\delta A_{j\mathrm{R}}\right) = 0, \qquad \text{all} \quad \delta A_{j\mathrm{R}} \tag{5}$$

and

$$-\left(\frac{\partial \hat{\varphi}}{\partial \hat{A}_j}\,\delta A_{j\mathrm{I}},\,(H-\hat{E})\hat{\varphi}\right)+\left((H-\hat{E})\hat{\varphi},\,\frac{\partial \hat{\varphi}}{\partial \hat{A}_j}\,\delta A_{j\mathrm{I}}\right)=0, \qquad \text{all} \quad \delta A_{j\mathrm{I}} \tag{6}$$

where $\delta A_{j\mathrm{R}}$ and $\delta A_{j\mathrm{I}}$ are the real and imaginary parts of δA_j.

Now (6) is to be true for all $\delta A_{j\mathrm{I}}$ and therefore in particular, for a given $\delta A_{j\mathrm{R}}$, (6) must be true with $\delta A_{j\mathrm{I}} = \delta A_{j\mathrm{R}}$. Making this substitution in (6) and comparing with (5) then immediately yields

$$\left(\frac{\partial \hat{\varphi}}{\partial \hat{A}_j}\,\delta A_{j\mathrm{R}},\,(H-\hat{E})\hat{\varphi}\right)=0, \qquad \text{all} \quad \delta A_{j\mathrm{R}} \tag{7}$$

Similarly for a given $\delta A_{j\mathrm{I}}$, (5) must be satisfied with $\delta A_{j\mathrm{R}} = \delta A_{j\mathrm{I}}$, which then leads to

$$\left(\frac{\partial \hat{\varphi}}{\partial \hat{A}_j}\,\delta A_{j\mathrm{I}},\,(H-\hat{E})\hat{\varphi}\right)=0, \qquad \text{all} \quad \delta A_{j\mathrm{I}} \tag{8}$$

If now we multiply (8) by $-i$ and add to (7), then the result is (3).

Having gone through this in detail, it is now useful to note the following quick derivation: If the A_j are independent, and if there are no a priori reality restrictions, and if $\delta\hat{\varphi}$ is a possible variation of $\hat{\varphi}$ within the set, then so is $\delta'\hat{\varphi} \equiv i\,\delta\hat{\varphi}$. (Proof: Choose $\delta' A_{j\mathrm{R}} = -\delta A_{j\mathrm{I}}$, $\delta' A_{j\mathrm{I}} = \delta A_{j\mathrm{R}}$.) Therefore (1) must also be satisfied if we replace $\delta\hat{\varphi}$ by $i\,\delta\hat{\varphi}$. Doing this then yields

$$-(\delta\hat{\varphi},\,(H-\hat{E})\hat{\varphi}) + ((H-\hat{E})\hat{\varphi},\,\delta\hat{\varphi}) = 0, \qquad \text{“all”} \quad \delta\hat{\varphi} \tag{9}$$

Comparison with (1) then yields (3).

Another way of expressing our result is the following: The first term on the left hand side in (1) can be thought of as the result of varying $\hat{\varphi}^*$ and the second as the result of varying $\hat{\varphi}$. Therefore one may say that under the given conditions, one can vary $\hat{\varphi}$ and $\hat{\varphi}^*$ independently.

One case in which the preceding theorem would fail would be if A_2, say, were in fact A_1^* since then, because we obviously cannot change A_2 and A_1 independently, our proof will not go through (or, alternatively, if we replace them by two independent real parameters, the proof will not go through because we would then have reality constraints). However, there is a theorem with a similar sound to it which still does hold, namely that in $\hat{E}$, we can vary $\hat{A}_1$ and $\hat{A}_1^*$ $(= \hat{A}_2)$ separately.[12] We will now derive this

[12] See, for example, Bhabha [31].

theorem for the simple case in which $\hat{A}_1$ is a number. Then supposing as before that A_{1R} and A_{1I} can be arbitrary, we have

$$\frac{\partial \hat{E}}{\partial \hat{A}_{1R}} = 0, \qquad \frac{\partial \hat{E}}{\partial \hat{A}_{1I}} = 0 \tag{10}$$

However,

$$\frac{\partial \hat{E}}{\partial \hat{A}_{1R}} = \frac{\partial \hat{E}}{\partial \hat{A}_1} + \frac{\partial \hat{E}}{\partial \hat{A}_1{}^*} \tag{11}$$

and

$$\frac{\partial \hat{E}}{\partial \hat{A}_{1I}} = i\frac{\partial \hat{E}}{\partial \hat{A}_1} - i\frac{\partial \hat{E}}{\partial \hat{A}_1{}^*} \tag{12}$$

which when inserted into (10) clearly yield

$$\frac{\partial \hat{E}}{\partial \hat{A}_1} = 0, \qquad \frac{\partial \hat{E}}{\partial \hat{A}_1{}^*} = 0 \tag{13}$$

which proves the point. To recover our earlier result, however, we evidently need further that $\hat{\psi}$ involve only $\hat{A}_1$ and not $\hat{A}_1{}^*$, and hence that $\hat{\psi}^*$ involve only $\hat{A}_1{}^*$ and not $\hat{A}_1$.

When (3) applies, it provides an interesting and suggestive interpretation of the variation method. In a general way, given a function F, quantities of the form

$$(G,\ F) \tag{14}$$

for various choices of G can be thought of as "moments" of F. Thus we can say that when (3) applies, the variation method approximates making $(H - \hat{E})\hat{\psi}$ equal to zero by requiring the vanishing of a restricted set of moments of $(H - \hat{E})\hat{\psi}$. One can also express the same idea in more familiar language as follows: The statement that $(H - \hat{E})\hat{\psi}$ is zero is equivalent to the statement that $(H - \hat{E})\hat{\psi}$ is orthogonal to all functions. Therefore when (3) applies, we may say that the variation method approximates having $(H - \hat{E})\hat{\psi}$ equal to zero by making $(H - \hat{E})\hat{\psi}$ orthogonal to a restricted set of functions.

The approximation of requiring that certain moments of $(H - \hat{E})\hat{\psi}$ vanish is certainly one which one might come upon, and indeed one which people *have* come upon, without reference to the variation method.[13] In particular, consider the *linear variation method* (which we will discuss in more detail in succeeding sections) in which the set of trial functions con-

[13] Many references are given by Finlayson [9]; see especially Chapter 1. See also Bangudu *et al.* [32].

sists of functions of the form

$$\psi = \sum_{l=1}^{M} A_l \phi_l \tag{15}$$

where the ϕ_l (the "basis set") are a given set of functions, and where the A_l are arbitrary numerical parameters.

If no reality conditions are imposed on the A_l, then (3) applies so that with

$$\hat{\psi} = \sum_{l=1}^{M} \hat{A}_l \phi_l, \qquad \delta\hat{\psi} = \sum_{l=1}^{M} \delta A_l \phi_l \tag{16}$$

we have

$$\sum_{k=1}^{M} \sum_{l=1}^{M} \delta A_k{}^* \, (\phi_k, (H - \hat{E})\phi_l)\hat{A}_l = 0 \qquad \text{all} \quad \delta A_k$$

and therefore since the A_k are independent

$$\sum_{l=1}^{M} (\phi_k, (H - \hat{E})\phi_l)\hat{A}_l = 0, \qquad k = 1, \ldots, M \tag{17}$$

Now the point we want to make is that one can arrive at these same equations, and people often do, by first writing down the "Schrödinger equation" (the reason for the quotes will be discussed in a moment):

$$(H - \hat{E}) \sum_{l=1}^{M} \hat{A}_l \phi_l = 0$$

and then simply taking the scalar product with each ϕ_k in turn.

Moreover, this sort of approach to the derivation of Eqs. (11) suggests other possibilities. Since, however, the use of the caret has special reference to the variation method, let us consider the more neutral "equation"

$$(H - \tilde{E}) \sum_{l=1}^{M} \tilde{a}_l \phi_l = 0 \tag{18}$$

Then we note that although the procedure of "taking the scalar..." provides one way of trying to determine $\tilde{E}$ and the $\tilde{a}_l$, there are other possibilities. For example, one might try to satisfy (18) identically at M selected points, or more generally one might try multiplying through by quite another set of M functions θ_k and integrate to find

$$\sum_{l=1}^{M} (\theta_k, (H - \tilde{E})\phi_l)\tilde{a}_l = 0, \qquad k = 1, \ldots, M \tag{19}$$

(Evidently this reduces to the previous suggestion if the θ_k are Dirac delta functions.)

Given these many possibilities, new questions naturally arise: Are they equivalent? And is one superior to the other? First, as to the equivalence: In general the different procedures (different choices for the set of θ_k) will lead to different answers. The point is simply that (18) as it stands is almost certainly an inconsistent equation—there are no $\tilde{a}_l$ and $\tilde{E}$ which satisfy it (hence our use of quotation marks), or more precisely, it is a consistent equation only if there happens to be an eigenfunction of H which can be written as a linear combination of the ϕ_l. Since in practice in a complicated problem this is unlikely, we may take it that the "equation" (18) is not consistent, and hence it follows that different methods of "solution" will in general lead to different results.

Now as to the advantages of one method over another. As we have seen, the variation method leads to (17) and therefore, as we know, this endows it with the virtue that the lowest $\hat{E}$ is a guaranteed and improvable upper bound to the lowest eigenvalue of H of appropriate type. Moreover, as we shall see in the next section, it is even more virtuous: The M values of $\hat{E}$ which are solutions of (17) are, in order, guaranteed and improvable upper bounds to the M lowest eigenvalues of H of appropriate type. Thus there is considerable reason to choose (17).

However, recently there has been a revival of interest in the use of equations of the form

$$(\theta_k, (H - \tilde{E})\tilde{\psi}) = 0, \qquad k = 1, \ldots \tag{20}$$

where $\tilde{\psi}$ may be of the form (15), but may also be of a much more complicated structure, and where the θ_k may either be quite distinct from $\tilde{\psi}$, or may involve some of the arbitrary parameters and/or functions in $\tilde{\psi}$ which are to be determined from Eqs. (20). In any case, the reason for the interest is quite simply that with the forms of ψ which are in use (or which one would like to use) in the applications of (1) to atoms and molecules, the integrals involved are often quite difficult (or impossible in practice), whereas with a $\tilde{\psi}$ of similar form and with a suitable choice of the θ_k, the integrals in (15) are quite tractable.

We will not discuss such methods further here but instead will refer the interested reader to the literature [33–36]. Also, methods like it have been much used under various names in applied mathematics. It should be emphasized, however, that such methods do not in general yield guaranteed bounds, and hence additional flexibility does not necessarily imply improved numerical accuracy. Indeed, they need not even yield real $\tilde{E}$.[14]

[14] See, for example, Hegyi *et al.* [33].

PROBLEMS

1. Define E by $(\phi, (H - E)\psi) = 0$, where ϕ and ψ are trial functions. Show that the eigenvalues of H are stationary points of E. Does E have any obvious minimal properties? By an obvious extension of the variation method and a suitable choice of trial functions ϕ and ψ, derive (19) [37].

2. Referring to Eq. (19), show that if, say, the set ϕ_k contains an eigenfunction, then one of the $\tilde{E}$ will equal the corresponding eigenvalue. Referring to Problem 1, what will the associated optimal ϕ and ψ be?

3. Use trial functions

$$\psi = A\psi' + B\psi'', \qquad \phi = A(\psi' + \psi'') + B(\psi' - \psi'')$$

where A and B are numerical parameters and ψ' and ψ'' are eigenfunctions. Show that in general the optimal energies will not be eigenvalues. What is the essential difference between these trial functions and those implied in Problem 2?

References

1. W. Yourgrau and S. Mandelstam, "Variational Principles in Dynamics and Quantum Theory," 3rd ed. Saunders, Philadelphia, Pennsylvania, 1968.
2. M. Born, "Physics in My Generation," 2nd ed. rev., Chapter 4. Springer-Verlag, Berlin and New York, 1969.
3. E. Schrödinger, *Ann. Phys. (Leipzig)* **79**, 361 (1926).
4. E. Schrödinger, *Ann. Phys. (Leipzig)* **79**, 489 (1926).
5. R. P. Feynman, *Rev. Mod. Phys.* **20**, 367 (1948).
6. K. Ruedenberg, *Rev. Mod. Phys.* **34**, 326 (1962); M. J. Feinberg and K. Ruedenberg, *J. Chem. Phys.* **54**, 1495 (1971).
7. S. G. Mikhlin, "Variational Methods in Mathematical Physics." Pergamon, Oxford, 1964.
8. R. Courant, *Bull. Amer. Math. Soc.* **49**, 1 (1963).
9. B. A. Finlayson, "The Method of Weighted Residuals and Variational Principles." Academic Press, New York, 1972.
10. G. Birkhoff, "The Numerical Solution of Elliptic Equations." SIAM, Philadelphia, Pennsylvania, 1972.
11. E. Courant and D. Hilbert, *Methods Math. Phys.* **1**, bottom of p. 457 (1943).
12. M. R. H. Rudge, *Proc. Roy. Soc. Ser. A* **328**, 429 (1972), and references therein.
13. E. Schrödinger, *Ann. Phys. (Leipzig)* **79**, 734 (1926).
14. P.-O. Löwdin, *Phys. Rev.* **97**, 1509 (1955).
15. H. F. Schaefer III, "The Electronic Structure of Atoms and Molecules." Addison-Wesley, Reading, Massachusetts, 1972.

15a. F. R. Burden and R. M. Wilson, *Advan. Phys.* **21**, 825 (1972).
16. A. L. Stewart, *Advan. Phys.* **12**, 299 (1963).
17. G. Herzberg, *Science* **177**, 123 (1972).
18. F. Weinhold, *Advan. Quantum Chem.* **6**, 299 (1972).
19. D. M. Bishop and A. Macias, *J. Chem. Phys.* **55**, 647 (1971).
20. J. Simons, *J. Chem. Phys.* **57**, 3787 (1972).
21. C. Pekeris, *Phys. Rev.* **126**, 1471 (1962).
22. P. S. C. Wang and F. Weinhold, *J. Chem. Phys.* **57**, 1738 (1972).
23. A. C. Wahl and G. Das, *Advan. Quantum Chem.* **5**, 261 (1970).
24. H. F. Schaefer III, D. R. McLaughlin, F. E. Harris, and B. J. Alder, *Phys. Rev. Lett.* **25**, 988 (1970).
25. B. J. Bertoncini and A. C. Wahl, *Phys. Rev. Lett.* **25**, 991 (1970).
26. B. J. Bertoncini and A. C. Wahl, *J. Chem. Phys.* **58**, 1259 (1973), and references therein.
27. R. Kari and B. T. Sutcliffe, *Chem. Phys. Lett.* **7**, 149 (1970), and references therein.
28. J. Koutecký and V. Bonačić-Koutecký, *Chem. Phys. Lett.* **15**, 558 (1972).
29. W. L. Bloomer and B. L. Bruner, *J. Chem. Phys.* **58**, 3735 (1973).
30. D. J. Thouless, *Nucl. Phys.* **21**, 255 (1960); **22**, 78 (1961).
31. H. J. Bhabha, *Rev. Mod. Phys.* **21**, 451 (1949).
32. E. A. Bangudu, K. Jankowski, and D. A. Dion, *Chem. Phys. Lett.* **19**, 418 (1973), and references therein.
33. M. G. Hegyi, M. Mezei, and T. Szondy, *Theor. Chim. Acta* **21**, 168 (1971), and references there to earlier work.
34. S. F. Boys and N. C. Handy, *Proc. Roy. Soc. Ser. A* **311**, 309 (1969), and references there to earlier work.
35. J. B. Delos and S. M. Blinder, *J. Chem. Phys.* **47**, 2784 (1967).
36. S. M. Rothstein, J. E. Welch, and H. J. Silverstone, *J. Chem. Phys.* **51**, 2932 (1969), and references there to earlier work.
37. N. C. Handy and S. T. Epstein, *J. Chem. Phys.* **53**, 1392 (1970), appendix and references therein.

Chapter II / Applications of the Variation Method

6. The Linear Variation Method

Let us now return to Eq. (5-17), which we repeat here:

$$\sum_{l=1}^{M} (\phi_k, (H - \hat{E})\phi_l)\hat{A}_l = 0, \qquad k = 1, \ldots, M \tag{1}$$

This is a set of linear homogeneous equations to determine the $\hat{A}_k$. It has nontrivial solutions (that is, not all $\hat{A}_k = 0$) only for certain values of $\hat{E}$, those for which the determinant of coefficients (the "secular determinant") vanishes:

$$|(\phi_k, (H - \hat{E})\phi_l)| = 0 \tag{2}$$

More precisely, Eq. (2) is the condition for nontrivial solutions only if, as we will henceforth assume, the ϕ_k are linearly independent since otherwise the determinant will vanish identically. However, this assumption evidently involves no loss of generality since if ϕ_M, say, is a linear combination of the others, then we can simply delete it from (5-15) without any loss of flexibility since the A_k are anyway arbitrary. In practical calculations, near-linear dependence of nonorthogonal basis sets can, however, often be a source of numerical difficulties.[1]

Equation (2), the "secular equation," is an Mth-order algebraic equation to determine $\hat{E}$; note that in accordance with the discussion in Section 4, we have not had to invoke (4-9) explicitly since the set of trial functions which we are using clearly has no fixed overall scale. We will denote the roots of (2) by $\hat{E}_k$, with $k = 1, 2, \ldots, M$ and $\hat{E}_1 \leq \hat{E}_2 \leq \cdots$. Similarly, we will denote the corresponding $\hat{\psi}$ by $\hat{\psi}_k$ and the associated $\hat{A}_i$ by $\hat{A}_{(k)i}$.

The set of trial functions (5-15) has the special property of forming a linear space (a subset of Hilbert space) since any linear combination of such trial functions is again a member of the set. In contrast, the set of

[1] See, for example, Wallis *et al.* [1].

functions $\exp(-Ax^2)$, with A a numerical parameter, does not form a linear space since, for example, $\exp(-x^2) + \exp(-7x^2)$ is not again of the form $\exp(-Ax^2)$. There are other interesting sets of trial functions which form linear spaces. Thus the set of all functions of a given symmetry form a linear space. Also, there has been considerable interest in the so called "s limit" for heliumlike ions in which the set consists of all functions of the form $\psi(r_1, r_2)$ where r_1 and r_2 are the distances of the two electrons from the nucleus.[2]

If we introduce the $M \times M$ Hermitian Hamiltonian matrix $\mathscr{H}$ whose elements are the $(\phi_k, H\phi_l)$, the $M \times M$ Hermitian, positive-definite overlap matrix S whose elements are (ϕ_k, ϕ_l) and the M-element vector $\hat{A}$ whose elements are the $\hat{A}_l$, then we can write (1) in compact matrix notation as

$$(\mathscr{H} - \hat{E}S)\hat{A} = 0 \tag{3}$$

Thus we are led to a somewhat generalized matrix eigenvalue problem (it evidently becomes an ordinary matrix eigenvalue problem if, as one always may without changing the set of ψ's, one orthonormalizes the ϕ_k so that S becomes the unit matrix). More generally we will now show that whenever the set of trial functions forms a linear space, then although the $\hat{\psi}_k$ and $\hat{E}_k$ (we will use the same notation for the general linear case as for the special case of the linear variation method) are probably only approximations to the eigenfunctions and eigenvalues of H, they are *exact* eigenfunctions and eigenvalues of the "projected Hamiltonian"

$$\bar{H} \equiv \Pi H \Pi$$

where Π is the Hermitian projection operator onto the linear space spanned by the trial functions:

$$\begin{aligned} \Pi\phi &= \phi \quad \text{if } \phi \text{ is in the space} \\ \Pi\phi &= 0 \quad \text{if } \phi \text{ is orthogonal to the space} \end{aligned} \tag{4}$$

To see this, we first observe that any function orthogonal to the space is an eigenfunction of $\bar{H}$ with eigenvalue zero. Since any other eigenfunction of $\bar{H}$ can be assumed, without loss of generality, to be orthogonal to these, it follows that to find any other eigenfunction of $\bar{H}$, it is sufficient to search

[2] See, for example, Handler and Joy [2]. For other examples of this type see Katriel and Adam [3].

for the stationary points of

$$\frac{(\psi, \bar{H}\psi)}{(\psi, \psi)} = \frac{(\psi, \Pi H \Pi \psi)}{(\psi, \psi)} = \frac{(\Pi\psi, H\Pi\psi)}{(\psi, \psi)}$$

as ψ ranges through the space. However, if ψ is in the space, then $\Pi\psi = \psi$, whence we see that it is sufficient to search for the stationary points of

$$(\psi, H\psi)/(\psi, \psi)$$

as ψ ranges through the space, which of course is just what one does in the variation method.

The observation that for a linear space the $\hat{\varphi}_k$ and $\hat{E}_k$ are eigenfunctions and corresponding eigenvalues of the Hermitian operator $\bar{H}$ allows us to invoke the standard quantum mechanical theorems for discrete eigenfunctions of Hermitian operators. This immediately leads us to the following important results, which, among other things, show that the $\hat{\varphi}_k$ and $\hat{E}_k$ have some of the formal properties of the eigenfunctions and eigenvalues of H. These results can of course also be derived, but with more labor and less insight, directly from the variational equations, for example, from the equations (1) in the case of the linear variation method.

Result 1. If $\hat{E}_k$ is nondegenerate, then $\hat{\varphi}_k$ is unique up to a constant factor and is automatically orthogonal to all $\hat{\varphi}_l$, $l \neq k$. We will restrict the constant factor to a phase factor by imposing the further condition that $\hat{\varphi}_k$ be normalized. If $\hat{E}_k = \hat{E}_l$, $k \neq l$, that is, if there is degeneracy, then $\hat{\varphi}_k$ and $\hat{\varphi}_l$ are arbitrary to within a linear transformation. We will restrict that transformation to be unitary by imposing the further requirement that $\hat{\varphi}_k$ and $\hat{\varphi}_l$ be orthonormal. In short, either automatically or by free choice, we have

$$(\hat{\varphi}_k, \hat{\varphi}_l) = \delta_{kl} \tag{5}$$

Result 2. Since $\bar{H}\hat{\varphi}_l = \hat{E}_l\hat{\varphi}_l$, we have that

$$(\hat{\varphi}_k, \bar{H}\hat{\varphi}_l) = \hat{E}_l(\hat{\varphi}_k, \hat{\varphi}_l) = \hat{E}_l\, \delta_{kl} \tag{6}$$

or, since $\Pi\hat{\varphi}_k = \hat{\varphi}_k$ and $\Pi\hat{\varphi}_l = \hat{\varphi}_l$,

$$(\hat{\varphi}_k, H\hat{\varphi}_l) = \hat{E}_l\, \delta_{kl} \tag{7}$$

Thus, summarizing (5) and (7) in words: Like a set of exact eigenfunctions, the $\hat{\varphi}_k$ are an orthonormal set of functions which diagonalize H within the space which they span.

Result 3. If H commutes with a Hermitian or unitary operator K and if Π commutes with K, that is, if the space is invariant to the action of K, then $\bar{H}$ will also commute with K, whence, like an exact eigenfunction, if $\hat{E}_k$ is nondegenerate, then $\hat{\varphi}_k$ will automatically also be an eigenfunction of K.

Result 4. From the considerations of Section 2, but applied to $\bar{H}$ instead of to H, it should be clear that $\hat{E}_1$ is the minimum of E with respect to variations within the space, and that the higher $\hat{E}_k$ are saddle points, (and in the case of the linear variation method, that $\hat{E}_M$ is the maximum of E with respect to variations within the space). Since it is instructive, we will now also derive these results using (4-10). Because, however, the details depend on the method of parameterization, we will limit ourselves to two examples. First consider the linear variation method. Since the $\hat{\varphi}_k$ are M in number and obviously linearly independent (they are orthonormal), they form a basis for the space. Therefore we can write $\delta\hat{\varphi}_k = \sum_{l=1}^{M} \hat{\varphi}_l \, \delta A_l$. Also, we clearly have $\delta_2\hat{\varphi}_k = 0$, whence (4-16) yields

$$\delta_2 \hat{E}_k = 2 \sum_{j=1}^{M} \sum_{l=1}^{M} \delta A_j^* \, \delta A_l (\hat{\varphi}_j, (H - \hat{E}_k)\hat{\varphi}_l)$$

which from (5) and (7) becomes

$$\delta_2 \hat{E}_k = 2 \sum_{l=1}^{M} | \, \delta A_l \, |^2 (\hat{E}_l - \hat{E}_k)$$

and therefore, as we said, $\delta_2\hat{E}_1$ is certainly nonnegative, $\delta_2\hat{E}_M$ is certainly nonpositive, while the other $\delta_2\hat{E}_k$ are of indefinite sign.

As a second example, consider something like the s limit, where

$$\psi = A$$

Then $\delta\hat{\varphi}_k = \delta A$ and $\delta_2\hat{\varphi}_k = 0$, whence (4-10) yields

$$\delta_2 \hat{E}_k(\hat{\varphi}_k, \hat{\varphi}_k) = 2(\delta A, (H - \hat{E}_k) \, \delta A)$$

However, δA belongs to the space. [Since the space is linear, $\psi(A + \delta A) - \psi(A) = \delta A$ belongs to the space.] Therefore we can replace H by $\bar{H}$ and write

$$\delta_2 \hat{E}_k(\hat{\varphi}_k, \hat{\varphi}_k) = 2(\delta A, (\bar{H} - \hat{E}_k) \, \delta A)$$

from which the results immediately follow.

There is also a useful converse to Results 1 and 2: If we have a set of functions χ_k with properties (5) and (7), that is, if

$$(\chi_k, \chi_l) = \delta_{kl} \tag{8}$$

$$(\chi_k, H\chi_l) = \varepsilon_l \, \delta_{kl} \tag{9}$$

then if we use them as basis functions in a linear variation calculation, the optimal trial functions will be just the χ_k again and the $\hat{E}_k$ will be the ε_k. To prove this, let Γ be the projector onto the space of the χ_k. If we introduce the projected Hamiltonian $\tilde{H} \equiv \Gamma H \Gamma$, then evidently it will be sufficient to prove that $(\tilde{H} - \varepsilon_l)\chi_l = 0$. To prove this, we first note that $(\tilde{H} - \varepsilon_l)\chi_l = \Gamma(H - \varepsilon_l)\chi_l$. However, from (8) and (9), $(H - \varepsilon_l)\chi_l$ is orthogonal to the space since its scalar product with any χ_k vanishes. Therefore $\Gamma(H - \varepsilon_l)\chi_l = 0$, whence, as desired, we have

$$(\tilde{H} - \varepsilon_l)\chi_l = 0$$

Although the set of trial functions (5-15) and the "*s*-limit" functions each form a linear space, the fact that the space formed by the former is of finite dimensionality while that formed by the latter is infinite has important consequences in practice. The point is simply that one can fairly readily solve finite problems, particularly algebraic problems, to arbitrary accuracy and the same is true for ordinary differential equations (these optimistic statements should of course be qualified by appropriate remarks concerning "roundoff error," problem size, and machine capacity and speed). However, really infinite problems, typically involving the solution of nonseparable partial differential or integral equations or infinite sets of algebraic equations, are usually intractable and make it necessary to introduce further approximations, although recently partial differential equations in two variables, such as occur with the *s*-limit problem, have begun to come under direct attack.[3] Usually these further approximations consist simply in again using the variation method but now with the ψ a finitely parameterized, though not necessarily linear, subset of the functions in the infinite linear space. Of course, if the subset does form a linear space, then the resulting $\hat{\psi}$ and $\hat{E}$ will be eigenfunctions and eigenvalues of $\pi H \pi$, where π is the projector onto that subspace.

Since the linear variation method leads to a finite algebraic problem, it has been widely used and goes under various names: the Ritz method, the Rayleigh–Ritz method, the method of linear variational parameters,

[3] See, for example, Winter *et al.* [4] and Barraclough and Mooney [5].

etc. As the name Rayleigh suggests, its use predates quantum mechanics; it has been applied to all kinds of vibration problems, and quite generally wherever eigenvalue problems occur [6–10].

In atomic and molecular problems, one common application of the linear variation method is in the configuration interaction method (CI).[4] Here, with H usually the clamped nuclei Hamiltonian, the ϕ_k are Slater determinants or linear combinations of Slater determinants, made out of given spin orbitals (the spin orbitals often also involving nonlinear parameters—see end of Section 7). If one uses all the determinants of appropriate type which one can make from the given spin orbitals, then one speaks of complete CI; otherwise, one speaks of incomplete CI. In this connection, it is important to keep in mind that even with a modest number of spin orbitals, the complete CI problem, though finite, may become impractically large. For example, if one has 10 electrons and 20 spin orbitals, one can form

$$20!/(10!\,10!) = 184{,}756$$

Slater determinants! Of course, often for reasons of symmetry, not all of these need be used, but still the numbers can become enormous. Thus partial CI, involving a selection of (hopefully) the most important "configurations" ϕ_k becomes the practical alternative when one deals with even moderately complicated systems. Indeed almost any variation approach which ultimately is based on spin orbitals can be viewed as a partial CI with one or another special condition imposed to select the Slater determinants actually used.

PROBLEMS

1. Show that if as we assume the ϕ_k are linearly independent, then S is a positive-definite Hermitian matrix (recall Problem 4, Section 1). Show that if the ϕ_k are linearly dependent, then S will have zero eigenvalues, though no negative ones. Show that the number of zero eigenvalues will equal the number of linearly dependent ϕ_k.

2. Show directly from Eq. (3) that its solutions either automatically have, or can be assumed to have, the properties

$$\hat{A}^+_{(j)} \cdot S\hat{A}_{(k)} = \delta_{jk}; \qquad \hat{A}_{(j)} \cdot \mathscr{H}\hat{A}_{(k)} = \hat{E}_k\,\delta_{jk}$$

Use these results to rederive (5) and (7).

[4] See Schaefer [10].

3. Show that (3) is equivalent to an ordinary matrix eigenvalue problem for the Hermitian matrix $\mathscr{H}' \equiv S^{-1/2}\mathscr{H}S^{-1/2}$. Use this fact and the familiar properties of the solutions of such problems to derive the results of Problem 2.

4. Apropos of our discussion in the two paragraphs following Eq. (4-9), show that the set of trial functions $\phi_1 + A\phi_2$ yields the same results as the set $A_1\phi_1 + A_2\phi_2$.

5. Where in the discussion starting in the paragraph after Eq. (3) did we use the fact that Π is a *linear* operator? (If this played no role, then one could apply the same argument to any variation calculation with Π defined in some way as the projector onto whatever set of trial functions one is using.)

6. Carry out, in a general way, a linear variation calculation using the basis functions $\phi_1 = \chi_1$, $\phi_2 = \chi_1 + \varepsilon\chi_2$, where ε is a real number. Then let $\varepsilon \to 0$. Show that in general the results do *not* agree with those of the (trivial) variation calculation using the single basis function χ_1 [11]. Try to find a simple explanation of this phenomenon (hint: $2 \neq 1$).

7. Show that in Dirac bra-ket notation (if you are not familiar with this notation, see, for example, the textbook of Messiah [12], which contains a concise summary) Π for the linear variation method is given by

$$\Pi = \sum_{k=1}^{M} \sum_{l=1}^{M} |\phi_k\rangle (S^{-1})_{kl} \langle\phi_l| = \sum_{k=1}^{M} | \hat{\psi}_k \rangle\langle \hat{\psi}_k |$$

Show that the projector onto a single function ϕ is

$$\pi = \frac{| \phi \rangle\langle \phi |}{\langle\phi|\phi\rangle}$$

8. By choice of suitable vectors in (4-22), show that in the linear variation method λ_M is positive if $\hat{E}_M - \hat{E}$ is positive, and that λ_1 is negative if $\hat{E}_1 - \hat{E}$ is negative. Also, from the fact that $\lambda_i = Y_{(i)}^{+} \cdot \Lambda Y_{(i)}$, show that if all the $\hat{E}_i - \hat{E}$ are positive (negative), then all the λ_i are positive (negative). Use these results to rediscuss the nature of the stationary points in the linear variation method, this time without making a special choice of basis for $\delta\hat{\psi}_k$.

7. Linear Spaces and Excited States

We have twice mentioned that all the $\hat{E}_k$ furnished by the linear variation method have bounding properties. We now want to prove this [13]. More generally, we will show that whenever the set of trial functions form a linear space, then the successive $\hat{E}_k$ are upper bounds to the corresponding successive eigenvalues of H.

We first note that from (6-5) and (6-7), the average energy in a state described by $\psi = \sum_{k=1}^{M} b_k \hat{\varphi}_k$, namely

$$E = \frac{\sum_{k=1}^{M} \sum_{l=1}^{M} b_k^* b_l (\hat{\varphi}_k, H\hat{\varphi}_l)}{\sum_{k=1}^{M} \sum_{l=1}^{M} b_k^* b_l (\hat{\varphi}_k, \hat{\varphi}_l)}$$

can be written as

$$E = \frac{\sum_{k=1}^{M} | b_k |^2 \hat{E}_k}{\sum_{k=1}^{M} | b_k |^2} \tag{1}$$

from which it is clear that it is not greater than $\hat{E}_{k'}$, where k' is the largest k for which $b_k \neq 0$. Further, we note that there is at least one linear combination of $\hat{\varphi}_1, \hat{\varphi}_2, \ldots, \hat{\varphi}_N$ which is orthogonal to each of the lowest $(N - 1)$ eigenfunctions of H. From what we have just proven, the average energy for this function will be less than or equal to $\hat{E}_N$, while from Result 6 of Section 2, it is greater than or equal to E_N', the Nth smallest eigenvalue of H. Thus we have, as announced, that

$$E_N' \leq \hat{E}_N, \qquad N = 1, 2, \ldots, M \tag{2}$$

and it should be clear that if both H and $\bar{H}$ commute with K, then (2) applies to the eigenvalues of each type separately.

The preceding discussion of course implicity assumed that H has at least M bound states below any continua. If H has only $M' < M$ such bound states, then for the $\hat{E}_{M'+1}, \ldots, \hat{E}_M$ we can conclude only that they, like $\hat{E}_{M'}$, are upper bounds to $E'_{M'}$. However, in what follows we will not consider this possibility explicitly. Also, we will not worry about such interesting things as bound states and quasibound states imbedded in continua; rather we will simply refer the reader to the literature.[5]

We will now show further that the bounds (2) are improvable bounds in that if we are dealing with a finite space, then enlarging the space will

[5] For a recent review, see Taylor [14].

improve or at any rate not worsen them. Thus as already mentioned in Section 3, the linear variation method provides a soundly based method for approximating the higher eigenvalues of H.

We start with a basis set of M functions. Let us note this explicitly by writing $\hat{E}_k(M)$ instead of $\hat{E}_k$. Thus in particular (6-7) will be written

$$(\hat{\varphi}_k, H\hat{\varphi}_l) = \hat{E}_k(M)\, \delta_{kl} \tag{3}$$

Suppose now that we add one more function ϕ to our basis set. We assume, without loss of generality, that ϕ is normalized and orthogonal to all the ϕ_k, and hence orthogonal to all the $\hat{\varphi}_k$:

$$(\phi, \phi) = 1, \qquad (\phi, \hat{\varphi}_k) = 0 \tag{4}$$

and of course we continue to have

$$(\hat{\varphi}_k, \hat{\varphi}_l) = \delta_{kl} \tag{5}$$

Let us write our new optimal trial function as

$$\hat{\varphi} = \sum_{l=1}^{M} \hat{B}_l \hat{\varphi}_l + B_{M+1}\phi \tag{6}$$

where, for convenience, we have used the $\hat{\varphi}_k$ instead of the ϕ_k, and where we have denoted the new variational parameters by B_i.

If we now insert $\delta\hat{\varphi} = \sum_{l=1}^{M} \delta B_l\, \hat{\varphi}_l + \delta B_{M+1}\phi$, with the δB_k's arbitrary, into (5-3) and use (3)–(5), the following equations result:

$$(\hat{E}_k(M) - \hat{E})\hat{B}_k + (\hat{\varphi}_k, H\phi)\hat{B}_{M+1} = 0, \qquad k \leq M \tag{7}$$

and

$$\sum_{l=1}^{M} (\phi, H\hat{\varphi}_l)\hat{B}_l + [(\phi, H\phi) - \hat{E}]\hat{B}_{M+1} = 0 \tag{8}$$

From (7), we then have

$$\hat{B}_k = \frac{(\hat{\varphi}_k, H\phi)\hat{B}_{M+1}}{(\hat{E} - \hat{E}_k(M))}, \qquad k \leq M$$

which, when inserted into (8), yields an equation for $\hat{E}$:

$$\hat{E} - (\phi, H\phi) = \sum_{l=1}^{M} \frac{|\,(\phi, H\hat{\varphi}_l)\,|^2}{\hat{E} - \hat{E}_l(M)} \equiv \Omega \tag{9}$$

If the $\hat{E}_k(M)$ are all distinct and if none of the $(\phi, H\hat{\psi}_l)$ vanishes (we will shortly remove these restrictions), then Ω as a function of $\hat{E}$ clearly has the following properties: It has simple poles at $\hat{E} = \hat{E}_l(M)$, $l = 1, \ldots, M$. It is negative immediately to the left of the poles and positive to the right of the poles. It goes to zero through positive (negative) values when $\hat{E}$ tends to positive (negative) infinity.

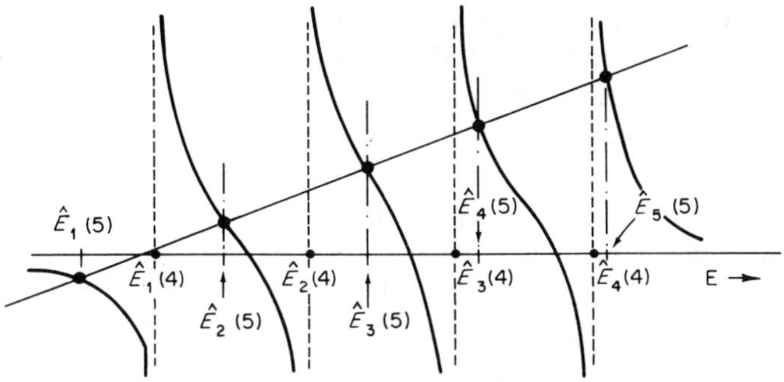

Fig. 7-1.

The solutions of (9)—let us denote them by $\hat{E}_k(M + 1)$—are then the intersections of Ω with the straight line $\hat{E} - (\phi, H\phi)$. For example, for $M = 4$, the situation looks as shown in Fig. 7-1. Therefore we have (in general) the "separation theorem"

$$\hat{E}_{k-1}(M) \leq \hat{E}_k(M + 1) \leq \hat{E}_k(M) \tag{10}$$

and hence in particular

$$\hat{E}_k(M + 1) \leq \hat{E}_k(M) \tag{11}$$

which is what we wanted to prove. (For a more elegant proof, see Appendix A.)

Turning now to the case of degeneracy among the $\hat{E}_k(M)$, the qualitative picture is not changed since one can view degeneracy as a limiting case of nondegeneracy. Consider, for example, what happens if, say, $\hat{E}_2(4)$ becomes degenerate with $\hat{E}_3(4)$ at the value ε. Then in the degenerate limit the branch of the Ω curve between $\hat{E} = \hat{E}_2(4)$ and $\hat{E} = \hat{E}_3(4)$ becomes a vertical line at $\hat{E} = \varepsilon$ and therefore $\hat{E}_3(5)$ will again equal ε, though there will be no more degeneracy. More generally, if there is an n-fold degeneracy among the $\hat{E}_k(M)$ at the value ε, then the $\hat{E}_k(M + 1)$ will have at least an $(n - 1)$-fold degeneracy also at the value ε. However, in any case, (10) still holds.

Finally, let us consider the possibility that one of the $(\phi, H\hat{\psi}_l)$, say $(\phi, H\hat{\psi}_1)$, vanishes. Then (7) yields

$$(\hat{E}_1(M) - \hat{E})\hat{B}_1 = 0$$

and, in addition, we see that $\hat{B}_1$ does not appear in (8). Thus one solution of the equations is $\hat{E} = \hat{E}_1(M)$, $\hat{\psi} = \hat{\psi}_1$, and therefore in general the vanishing of $(\phi, H\hat{\psi}_l)$ means that $\hat{E}_l(M)$ remains an $\hat{E}$ and $\hat{\psi}_l$ remains a $\hat{\psi}$ as one goes to the next stage. In any case, again Eq. (10) still holds.

Although it is primarily of theoretical interest, we will now compare the excited-state bound one gets using only the linear variation method to what one gets if, in addition, one imposes orthogonality to lower states as discussed in Result 6 of Section 2. As might be expected, the latter procedure, if it can be carried out, will generally yield a better bound [15].

Consider the first excited state. Then suppose that instead of simply using an M-dimensional basis set and trial functions of the form $\sum_{l=1}^{M} A_l\phi_l$, we further require that $\sum_{l=1}^{M} A_l(\psi', \phi_l) = 0$, where ψ' is the lowest eigenfunction of H (of a given type, if that is appropriate). Then we can use this requirement to determine one of the A_l for which $(\psi', \phi_l) \neq 0$, in terms of the others. Let this one be A_M. Thus

$$A_M = -\frac{1}{(\psi', \phi_M)} \sum_{l=1}^{M-1} A_l(\psi', \phi_l)$$

whence our trial functions can be written

$$\sum_{l=1}^{M-1} A_l\left[\phi_l - \phi_M \frac{(\psi', \phi_l)}{(\psi', \phi_M)}\right] = \sum_{l=1}^{M-1} A_l\chi_l$$

That is, this procedure corresponds to using the linear variation method with the $M - 1$ functions χ_l as the basis set. We now note that if we adjoin the function ϕ_M to the functions χ_l, we will effectively recover our original M-dimensional basis set and therefore it follows from (10) with $k = 2$ and M replaced by $M - 1$ that

$$\hat{E}_1(M-1) \leq \hat{E}_2(M)$$

On the other hand, we also know from Result 6 of Section 2 that if E_2' is the second lowest eigenvalue of H, then

$$E_2' \leq \hat{E}_1(M-1)$$

so we have

$$E_2' \leq \hat{E}_1(M-1) \leq \hat{E}_2(M) \tag{12}$$

which shows, as expected, that $\hat{E}_1(M-1)$, if we could calculate it, would be a better approximation to E_2' than is $\hat{E}_2(M)$. Another proof of this theorem and of its generalization to higher states is given in Appendix A.

The results which we have found in this section hold for any given choice of the ϕ_k. In practice, one often imbeds parameters in the ϕ_k and varies them as well. Usually these parameters are nonlinear parameters and the reason for introducing them is that they are very effective in that one nonlinear parameter can often do the work of many linear parameters. Thus a single parameter A can produce an optimal exponent in e^{-Ar}, whereas it will in general take several terms to do as well by linearly superposing, for example, e^{-r}, re^{-r}, r^2e^{-r}, etc. However, the difficulties of dealing with nonlinear parameters coupled with the increasing power of modern computing machinery often swings the balance in favor of more linear parameters, that is, more basis functions.

When nonlinear parameters are used, one usually chooses them so as to minimize each $\hat{E}_k$ separately. This in general will mean different parameter values in each $\hat{\psi}_k$. Therefore one price one pays in this approach is that probably (6-5) and (6-7) will no longer hold for $k \neq l$ since the final $\hat{\psi}_k$ will in general be eigenfunctions of quite different $\bar{H}$'s. Thus to this extent the $\hat{\psi}_k$ are less like eigenfunctions of H than before.

PROBLEMS

1. Discuss the case $(\phi, H\hat{\psi}_1) = 0$ graphically by means of a limiting process.

2. Discuss the degenerate case directly, that is, not by a limiting process, much as we discussed the case $(\phi, H\hat{\psi}_1) = 0$ directly in the text [16].

3. Use (11) to rederive $\hat{E}_k \geq E_k'$.

4. Let H be an operator whose highest eigenvalues are discrete. For example, a finite Hermitian matrix, or more trivially, the negative of (1-1). If E_T' is the largest eigenvalue, show that $E_{T-k}' \geq E_{M-k}$.

8. The Unrestricted Hartree–Fock Approximation

As we have seen, when the space of trial functions is a linear one, then formally one is on familiar ground in that the standard quantum mechanical results apply (though with respect to $\bar{H}$ rather than H). However, if one leaves the realm of linear spaces, then the situation can change dramatically and one must be prepared to encounter quite different behavior. For example, we have already noted the lack of orthogonality properties among different $\hat{\varphi}$.

As our major example of a nonlinear situation, we will take the unrestricted Hartree–Fock approximation (UHF), which is formally the simplest of a whole host of Hartree–Fock approximations, some of which we will refer to further as we proceed. To have something definite to talk about, the problem we will have in mind is that of an N-electron atom or molecule in the clamped-nuclei, nonrelativistic approximation. Thus the Hamiltonian is that of Eq. (1-1), or more schematically, to within an additive constant (the specifically nuclear terms)

$$H = \sum_{s=1}^{N} h(s) + \frac{1}{2} \sum_{\substack{s=1 \\ s \neq t}}^{N} \sum_{t=1}^{N} g(s, t) \tag{1}$$

where $h(s)$ includes the kinetic energy of the sth electron, its interaction with the nuclei, and its interaction with any external fields which may be present, and where $g(s, t) = g(t, s)$ denotes the mutual Coulomb interaction between the electrons.

Such a system is often well described in terms of an independent particle model (central field model, shell model, molecular orbital model) in which each electron moves more or less independently in a potential which is a combination of the nuclear and external potentials and some average potential produced by the other electrons. The wave function appropriate to such a model is then a single Slater determinant (or if symmetry considerations dictate, a sum of a few such determinants) made up from a set of spin orbitals describing the individual particle motions.

In the UHF approximation, one seeks those single Slater determinants which best, in the variational sense, approximate eigenfunctions of H. Thus the set of trial functions is the set of all Slater determinants

$$\psi = \begin{vmatrix} \chi_1(1) & \chi_1(2) & \chi_1(3) & \cdots & \chi_1(N) \\ \vdots & & & & \\ \chi_j(1) & & & & \\ \vdots & & & & \\ \chi_N(1) & & \cdots & & \chi_N(N) \end{vmatrix}$$

where the variational parameters A_i are the N spin orbitals χ_i. For conciseness, we will usually denote such determinants by

$$\psi = | \chi_1, \chi_2, \ldots, \chi_N |$$

This set of trial functions does not form a linear space since it can be shown [17] that in general a sum of Slater determinants is not itself a Slater determinant. As we have implied, one consequence of this is that the $\hat{\psi}$ belonging to different $\hat{E}$ usually have no simple formal relationships with one another. Therefore we will not introduce a further subscript to distinguish the different $\hat{\psi}$ and $\hat{E}$ since for the most part we will be considering only one $\hat{\psi}$ at a time. Nevertheless, there still is sufficient linearity, for example

$$| \chi_1, \chi_2, \ldots, \chi_N | + | \chi_1', \chi_2, \ldots, \chi_N | = | \chi_1 + \chi_1', \chi_2, \ldots, \chi_N |$$

so that in some circumstances one can assert that certain higher $\hat{E}$ values provided by the method furnish upper bounds to higher eigenvalues of H. We will not discuss the details here but will simply refer the interested reader to the literature [18].

Calculations with Slater determinants are much simplified if the χ_i are orthonormal, so we will make this assumption. However, we will now show that in fact this involves no loss in generality. Therefore even though in our specific calculations we will take the spin orbitals to be orthonormal, still for general theoretical purposes, we can continue to say that the set of trial functions for UHF is the set of *all* Slater determinants. First, however, we will state and prove a more general theorem. Consider any linear transformation of the spin orbitals

$$\varphi_i = \sum_{j=1}^{N} C_{ij}\chi_j \tag{2}$$

where the C_{ij} are any set of numbers. Then the theorem is that

$$| \varphi_1, \varphi_2, \ldots, \varphi_N | = | C | \, | \chi_1, \chi_2, \ldots, \chi_N | \tag{3}$$

where $| C |$ is the determinant of the matrix of the C_{ij}. To see this, we observe that in detail the left hand side of (3) is

$$\begin{vmatrix} \sum_{j=1}^{N} C_{1j}\chi_j(1) & \cdots & \sum_{j=1}^{N} C_{1j}\chi_j(N) \\ \vdots & & \\ \sum_{j=1}^{N} C_{Nj}\chi_j(1) & \cdots & \end{vmatrix}$$

which will be recognized as the determinant of the product of the two "matrices" C_{ij} and $\chi_j(s)$. Equation (3) then follows from the standard theorem that the determinant of a product of matrices is equal to the product of the determinants of the separate matrices.

Turning now to the question of assuming the χ_i orthonormal, we first note that there are many linear transformations of the type (3)[6]; for, example, the well-known Schmidt procedure, which, starting from a given linearly independent set of χ_i (and the χ_i must be linearly independent to start with or else $\psi \equiv 0$) will produce an orthonormal set of φ_i. From (3), then, we see that ψ is proportional to the Slater determinant formed from the φ_i, and therefore since the proportionality constant $| C |$ will simply cancel out in calculating E, we have the desired result that we can, without loss of generality, restrict attention to determinants formed from orthonormal spin orbitals.

Notice, however, that even with the requirement of orthonormality, the φ_i are certainly not unique since if we write

$$\varphi_i' = \sum_{j=1}^{N} C_{ij}\varphi_j$$

then the φ_i' will, if the C_{ij} form a unitary matrix, also be orthonormal, while from (3)

$$\psi' \equiv | \varphi_1', \varphi_2', \ldots, \varphi_N' | = | C | \psi$$

so that ψ' is physically equivalent to ψ. One important consequence of this observation is that we must expect that our variational equations will not determine the optimal spin orbitals uniquely. Instead, at any stationary point, they will be arbitrary up to a unitary transformation.

Actually we earlier encountered an example of nonuniqueness of the $\hat{A}_i$. Namely in the linear variation method only the ratios of the $\hat{A}_i$ were determined, while the overall scale could be chosen arbitrarily. Thus lack of uniqueness of the $\hat{A}_i$ is not uncommon. Further, before starting a calculation, it is well to try to be aware of potential ambiguities so as, for example, to be able to easily recognize equivalent solutions, and also so as to avoid spending time looking for general solutions of the variational equations when particular ones will do.

Alternatively, one may try to eliminate ambiguities by reparameterizing. Thus in the UHF approximation, it will turn out that the individual φ_i

[6] See, for example, Löwdin [19].

appear in E only in the form of the so called one-body density matrix

$$\varrho^{\mathrm{UHF}}(1, 1') \equiv \sum_{i=1}^{N} \varphi_i(1)\varphi_i{}^*(1') \tag{4}$$

which is, as we will show, invariant to unitary transformations. Therefore one can bypass the ambiguity by basing the whole theory of the UHF approximation on ϱ^{UHF} rather than on the φ_i[7] individually. However, we will not pursue that approach here.

To see that ϱ^{UHF} is invariant (as we said, we are not going to pursue the density matrix approach; however, we will need this result later) we write

$$\varphi_i = \sum_{j=1}^{N} C_{ij}\varphi_j{}'$$

Then

$$\varrho^{\mathrm{UHF}}(1, 1') = \sum_{k=1}^{N} \sum_{j=1}^{N} \sum_{i=1}^{N} C_{ik}^* C_{ij}\varphi_j{}'(1)\varphi_k'^*(1')$$

which in turn does become

$$\varrho^{\mathrm{UHF}}(1, 1') = \sum_{k=1}^{N} \varphi_k{}'(1)\varphi_k'^*(1') \equiv \varrho'^{\mathrm{UHF}}(1, 1') \tag{5}$$

if the transformation is unitary, that is, if

$$\sum_{l=1}^{N} C_{lk}^* C_{lj} = \delta_{kj}$$

Assuming then that the φ_i are orthonormal, we will now derive the equations which determine the $\hat{\varphi}_i$, the optimal spin orbitals. Since the overall scale of our trial functions is fixed (indeed $(\psi, \psi) = N!$), we must follow the straightforward procedure described at the outset of Section 4. Thus as the first step we find, using Slater's standard rules,[8] that

$$E = \sum_{j=1}^{N} (\varphi_j, h\varphi_j) + \tfrac{1}{2}\sum_{j=1}^{N} \sum_{k=1}^{N} [(\varphi_j\varphi_k, g\varphi_j\varphi_k) - (\varphi_j\varphi_k, g\varphi_k\varphi_j)] \tag{6}$$

where

$$(\varphi_j\varphi_k, g\varphi_l\varphi_M) \equiv \int d1 \int d2\, \varphi_j{}^*(1)\varphi_k{}^*(2)g(12)\varphi_l(1)\varphi_M(2) \tag{7}$$

the integral signs denoting an integral over space and a sum over spins.

[7] See, for example, McWeeny [20].

[8] See, for example, the book by Slater [21].

Next we calculate δE and then set $\delta\hat{E} = 0$. One then finds as the result

$$0 = \sum_{j=1}^{N} [(\delta\varphi_j, \hat{f}\hat{\varphi}_j) + (\hat{\varphi}_j, \hat{f}\delta\varphi_j)] \tag{8}$$

where the operator $\hat{f}$ is the so-called "Hartree–Fock single-particle Hamiltonian" and is defined by its action on an arbitrary spin orbital according to

$$\hat{f}(s)\chi(s) = h(s)\chi(s) + \sum_{j=1}^{N} [(\hat{\varphi}_j, g\hat{\varphi}_j)\chi(s) - (\hat{\varphi}_j, g\chi)\hat{\varphi}_j(s)] \tag{9}$$

where

$$(\hat{\varphi}_j, g\hat{\varphi}_j) \equiv \int d1\, \hat{\varphi}_j{}^*(1)g(1s)\hat{\varphi}_j(1) \tag{10}$$

and similarly

$$(\hat{\varphi}_j, g\chi) \equiv \int d1\, \hat{\varphi}_j{}^*(1)g(1s)\chi(1)$$

(Note that these latter quantities are not just numbers, they are functions of the coordinates s.) It is then easy to show that

$$(\chi', \hat{f}\chi) = (\hat{f}\chi', \chi) \tag{11}$$

that is, that $\hat{f}$ is Hermitian, so that (8) can be replaced by

$$0 = \sum_{j=1}^{N} [(\delta\varphi_j, \hat{f}\hat{\varphi}_j) + (\hat{f}\hat{\varphi}_j, \delta\varphi_j)] \tag{12}$$

Since the φ_i were assumed orthonormal, the $\delta\varphi_j$ are not independent. To take account of the orthonormality constraints and their complex conjugates

$$(\varphi_i, \varphi_j) = \delta_{ij}, \qquad i, j = 1, 2, \ldots, N \tag{13}$$

we introduce Lagrange multipliers ε_{ij} (see Appendix B) and replace (12) by

$$0 = \sum_{j=1}^{N} [(\delta\varphi_j, \hat{f}\hat{\varphi}_j) + (\hat{f}\hat{\varphi}_j, \delta\varphi_j)] - \sum_{i=1}^{N}\sum_{j=1}^{N} [(\delta\varphi_j, \hat{\varphi}_i) + (\hat{\varphi}_j, \delta\varphi_i)]\varepsilon_{ij} \tag{14}$$

or, rearranging a bit,

$$0 = \sum_{j=1}^{N} \left[\left(\delta\varphi_j, \hat{f}\hat{\varphi}_j - \sum_{i=1}^{N} \hat{\varphi}_i\varepsilon_{ij}\right) + \left(\hat{f}\hat{\varphi}_j - \sum_{i=1}^{N} \hat{\varphi}_i\varepsilon_{ji}^*, \delta\varphi_j\right)\right] \tag{15}$$

which is now to be true for all $\delta\varphi_j$ (i.e., for $\delta\varphi_j$ any one-electron function) without constraint and where (again from Appendix B) since $(\varphi_i, \varphi_j)^* = (\varphi_j, \varphi_i)$,

$$\varepsilon_{ij} = \varepsilon_{ji}^* \tag{16}$$

that is, the Lagrange multipliers should form a Hermitian matrix.

Assuming (16), then, (15) becomes

$$0 = \sum_{J=1}^{N} \left[\left(\delta\varphi_j, \hat{f}\hat{\varphi}_j - \sum_{i=1}^{N} \hat{\varphi}_i \varepsilon_{ij} \right) + \left(\hat{f}\hat{\varphi}_j - \sum_{i=1}^{N} \hat{\varphi}_i \varepsilon_{ij}, \delta\varphi_j \right) \right] \tag{17}$$

from which, following what is by now a familiar pattern, we are led to

$$\hat{f}\hat{\varphi}_j = \sum_i \hat{\varphi}_i \varepsilon_{ij} \tag{18}$$

In addition to satisfying (16), the ε_{ij} are of course to be chosen in such a way as to ensure the orthonormality of the $\hat{\varphi}_i$. We now observe that these two requirements are not independent since if the $\hat{\varphi}_i$ are orthonormal, then (16) will automatically be satisfied. Namely from (18) it then follows that

$$\varepsilon_{ij} = (\hat{\varphi}_i, \hat{f}\hat{\varphi}_j) = (\hat{f}\hat{\varphi}_j, \hat{\varphi}_i)^* = (\hat{\varphi}_j, \hat{f}\hat{\varphi}_i)^* = \varepsilon_{ji}^* \tag{19}$$

Further, we note that one choice of the Lagrange multipliers which will guarantee orthogonality is to put $\varepsilon_{ij} = 0$ for $i \neq j$, since with this choice the $\hat{\varphi}_i$ will all be eigenfunctions of the same Hermitian operator, namely $\hat{f}$. The diagonal elements ε_{ii} then remain to be chosen so that the $\hat{\varphi}_i$ are (or can be) normalized. Making this choice and using just a single subscript for the diagonal elements, we have the equations

$$\hat{f}\hat{\varphi}_i = \varepsilon_i \hat{\varphi}_i \tag{20}$$

We will refer to the solution of these equations as "canonical" UHF spin orbitals, and in general unless we specify otherwise, the symbol $\hat{\varphi}_j$ will refer to these. The ε_i are referred to as single-particle energies.

In view of our earlier discussion, it should come as no surprise that the general solution of (18) can be derived from the canonical solution by making an arbitrary unitary transformation. To see this, we note first that from (4) and (9) $\hat{f}$ involves the $\hat{\varphi}_i$ only through $\hat{\varrho}^{\mathrm{UHF}}$ and thus, from (5), is invariant to unitary transformations. Therefore if we write

$$\hat{\varphi}_i = \sum_{j=1}^{N} C_{ij} \hat{\varphi}_j'$$

where the C_{ij} form a unitary matrix, then one readily finds from (20) that

$$\hat{f}\hat{\varphi}_j' = \sum_{k=1}^{N} \hat{\varphi}_k' \varepsilon_{kj}$$

where

$$\varepsilon_{kj} = \sum_{i=1}^{N} C_{ij}^* \varepsilon_i C_{ik} = \varepsilon_{jk}^* \tag{21}$$

and thus we are back to the general equation (18). Conversely, given any solution of (18), we can find the unitary matrix which diagonalizes ε_{ij} and use it to pass from that solution of (18) to the solution of (20).

This discussion also shows that if the ε_i are nondegenerate, then the $\hat{\varphi}_i$ are unique (to within arbitrary phases) since then any nondiagonal unitary transformation will result in a nondiagonal ε_{ij}. On the other hand, if some of the ε_i are degenerate, then it follows that the $\hat{\varphi}_i$ are uniquely defined only to within unitary transformations within degenerate sets.

Although (20) obviously has a simpler mathematical structure than (18), it should be pointed out that the canonical spin orbitals are not necessarily the most useful ones physically. Other sets of spin orbitals derived from these by a unitary transformation may have more desirable properties, for example, they may be better localized. Also, it has been suggested that certain nonorthogonal sets may also be useful. However, we will not pursue these matters further here.[9]

The final step in the procedure, as outlined at the outset of Section 4, is to calculate $\hat{E}$. In particular, it is of interest to find the relation between $\hat{E}$ and the ε_i. From (6), we have

$$\hat{E} = \sum_{i=1}^{N} (\hat{\varphi}_i, h\hat{\varphi}_i) + \frac{1}{2} \sum_{i=1}^{N} \sum_{j=1}^{N} [(\hat{\varphi}_i\hat{\varphi}_j, g\hat{\varphi}_i\hat{\varphi}_j) - (\hat{\varphi}_i\hat{\varphi}_j, g\hat{\varphi}_j\hat{\varphi}_i)] \tag{22}$$

which we will rewrite as

$$\hat{E} = \sum_{i=1}^{N} (\hat{\varphi}_i, h\hat{\varphi}_i) + \sum_{i=1}^{N} \sum_{j=1}^{N} [(\hat{\varphi}_i\hat{\varphi}_j, g\hat{\varphi}_i\hat{\varphi}_j) - (\hat{\varphi}_i\hat{\varphi}_j, g\hat{\varphi}_j\hat{\varphi}_i)]$$
$$- \frac{1}{2} \sum_{i=1}^{N} \sum_{j=1}^{N} [(\hat{\varphi}_i\hat{\varphi}_j, g\hat{\varphi}_i\hat{\varphi}_j) - (\hat{\varphi}_i\hat{\varphi}_j, g\hat{\varphi}_j\hat{\varphi}_i)]$$

in which form the first line will be recognized as $\sum_i (\hat{\varphi}_i, \hat{f}\hat{\varphi}_i)$ and the second as the average electron–electron interaction energy, whence we have from

[9] For recent reviews, see England *et al.* [22] and Weinstein *et al.* [23].

(20) and the fact that the $\hat{\varphi}_i$ are orthonormal

$$\hat{E} = \sum_{i=1}^{N} \varepsilon_i - \frac{(\hat{\psi}, \frac{1}{2}\sum_{s=1}^{N}\sum_{t=1}^{N}{}_{s\neq t}\, g(s, t)\hat{\psi})}{(\hat{\psi}, \hat{\psi})} \tag{23}$$

In words, then, $\hat{E}$ differs from the sum of the single-particle energies by the average mutual potential energy of the electrons. Further, the physical origin of this difference is not hard to find. In detail, using (9), we can write (20) as

$$h\hat{\varphi}_i + \sum_{j=1}^{N} [(\hat{\varphi}_j, g\hat{\varphi}_j)\hat{\varphi}_i - (\hat{\varphi}_j, g\hat{\varphi}_i)\hat{\varphi}_j] = \varepsilon_i\hat{\varphi}_i \tag{24}$$

or

$$h\hat{\varphi}_i + \sum_{j=1, j\neq i}^{N} [(\hat{\varphi}_j, g\hat{\varphi}_j)\hat{\varphi}_i - (\hat{\varphi}_j, g\hat{\varphi}_i)\hat{\varphi}_j] = \varepsilon_i\hat{\varphi}_i$$

which is clearly in accord with the physical picture we gave at the outset of this section, the $\hat{\varphi}_i$ and the ε_i being determined by the nuclear and external fields plus the average effects of the other spin orbitals. However, this means that in summing the ε_i, we count these average effects twice; hence (23).

Thus, as far as the total energy is concerned, the ε_i do not have a simple physical meaning. However, in a certain approximation, they can be considered as ionization energies. Namely suppose we approximate the wave function of an $N - 1$ particle- system by a single determinant gotten from $\hat{\psi}$ by simply deleting one of the $\hat{\varphi}_j$, say $\hat{\varphi}_k$. Then E for the $(N - 1)$-particle system will differ from $\hat{E}$ in (22) by the removal of all the terms involving $\hat{\varphi}_k$, namely

$$\begin{aligned} \Delta E = (\hat{\varphi}_k, h\hat{\varphi}_k) &+ \tfrac{1}{2}\sum_{j=1}^{N} [(\hat{\varphi}_k\hat{\varphi}_j, g\hat{\varphi}_k\hat{\varphi}_j) - (\hat{\varphi}_k\hat{\varphi}_j, g\hat{\varphi}_j\hat{\varphi}_k)] \\ &+ \tfrac{1}{2}\sum_{j=1}^{N} [(\hat{\varphi}_j\hat{\varphi}_k, g\hat{\varphi}_j\hat{\varphi}_k) - (\hat{\varphi}_j\hat{\varphi}_k, g\hat{\varphi}_k\hat{\varphi}_j)] \end{aligned}$$

or

$$\Delta E = (\hat{\varphi}_k, h\hat{\varphi}_k) + \sum_{j=1}^{N} [(\hat{\varphi}_j\hat{\varphi}_k, g\hat{\varphi}_j\hat{\varphi}_k) - (\hat{\varphi}_j\hat{\varphi}_k, g\hat{\varphi}_k\hat{\varphi}_j)]$$

and hence finally

$$\Delta E = (\hat{\varphi}_k, \hat{f}\hat{\varphi}_k) = \varepsilon_k \tag{25}$$

Thus in this approximation, ε_k is the single-particle ionization energy, a result usually called Koopmans's theorem. Of course, this kind of wave function for the $(N - 1)$-particle system with, so to speak, "frozen spin orbitals" would not seem to be a very good approximation, especially if inner shell electrons are being ionized; presumably it would be better to do a complete UHF calculation for the $(N-1)$-particle system.[10] Nevertheless, the result does give some feeling for the empirical situation.[11]

In the preceding discussion, we removed a canonical spin orbital. It is of interest to point out that in a well-defined sense this is the optimal thing to do (this is really Koopmans's theorem) [28]. Namely let the $\hat{\varphi}_j'$ be just some orthonormal set of UHF spin orbitals, not necessarily the canonical set. Then we delete one of them to yield an approximate function for the $(N-1)$-particle system. We now fix the $\hat{\varphi}_j'$, i.e., fix the unitary transformation which relates them to the canonical spin orbitals, by requiring that they be such as to make the energy of the $(N-1)$-particle system stationary with respect to variation in that transformation. Moreover, since the energy of the N-particle system is fixed, this is equivalent to requiring that ΔE should be stationary, where clearly

$$\Delta E = (\hat{\varphi}_k', \hat{f}\hat{\varphi}_k')$$

since in the previous discussion we used the canonical nature of the spin orbitals only in the final step.

Now $\hat{f}$ is invariant to the unitary transformations in question, and so $\delta(\Delta E) = 0$ yields

$$(\delta\hat{\varphi}_k', \hat{f}\hat{\varphi}_k') + (\hat{\varphi}_k', \hat{f}\,\delta\hat{\varphi}_k') = 0$$

where

$$\hat{\varphi}_k' = \sum_{i=1}^{N} \hat{C}_i \hat{\varphi}_i \tag{26}$$

and where

$$\delta\hat{\varphi}_k' = \sum_{i=1}^{N} \delta C_i \hat{\varphi}_i \tag{27}$$

In addition we constrained $\hat{\varphi}_k'$ to be normalized. Introducing a Lagrange multiplier ε, then we have

$$(\delta\hat{\varphi}_k', (\hat{f} - \varepsilon)\hat{\varphi}_k') + ((\hat{f} - \varepsilon)\hat{\varphi}_k', \delta\hat{\varphi}_k') = 0$$

or since there are no reality constraints,

$$(\delta\hat{\varphi}_k', (\hat{f} - \varepsilon)\hat{\varphi}_k') = 0 \tag{28}$$

[10] For some other interpretations and uses of the ε_i, see Manne and Åberg [24], Shirley [25], and Brueckner *et al.* [26].

[11] See, for example, Slater [21, Section 15-6]. Also Siegbahn *et al.* [27].

Introducing (26) and (27), Eq. (28) becomes

$$\sum_{i=1}^{N} \delta C_i{}^* \, \hat{C}_i(\varepsilon_i - \varepsilon) = 0$$

which, since the δC_i can be arbitrary, then yields the final equations:

$$\hat{C}_i(\varepsilon_i - \varepsilon) = 0, \qquad i = 1, 2, \ldots, N$$

The solutions of these equations are obviously

$$\begin{aligned} \hat{C}_i &= 0, \qquad i \neq l \\ \hat{C}_l &= 1, \qquad l = 1, 2, \ldots, N \\ \varepsilon &= \varepsilon_l \end{aligned}$$

That is, as announced, $\hat{\varphi}_k{}'$, which we may call an optimal ionization spin orbital, should be a canonical spin orbital. The analogous question of optimal excitation spin orbitals has also been discussed in the literature [29].

In requiring that $\hat{\psi}$ be a single determinant, we have of course restricted the states which we can hope to describe accurately. Thus consider the ground states of isolated atoms. Then one does not expect to give a good description of states which one would usually think require the coupling of several determinants to produce the appropriate orbital angular momentum or spin. However, these cases aside, there are still surprises.

First, one expects that closed shell systems should be quite well described by single determinants and indeed one can show that for such systems (20) does admit closed shell solutions.[12] Thus for the helium atom, if one puts

$$\hat{\varphi}_1 = \hat{R}\uparrow, \qquad \hat{\varphi}_2 = \hat{R}\downarrow$$

where $\hat{R}$ is a radial function and $\uparrow$ and $\downarrow$ are the usual spin functions, then the two equations (20) becomes a single equation for $\hat{R}$ (and, moreover, an equation which has been shown quite rigorously to have solutions) [31].

Less trivially, if for neon one puts

$$\begin{aligned} &\hat{\varphi}_1 = \hat{R}\uparrow, &&\hat{\varphi}_2 = \hat{R}\downarrow, &&\hat{\varphi}_3 = \hat{R}'\uparrow, &&\hat{\varphi}_4 = \hat{R}'\downarrow \\ &\hat{\varphi}_5 = Y_{1-1}\hat{R}''\uparrow, &&\hat{\varphi}_6 = Y_{1-1}\hat{R}''\downarrow, &&\ldots, &&\hat{\varphi}_{10} = Y_{11}R''\downarrow \end{aligned}$$

where the Y_{1m} are spherical harmonics of order one, then the ten equations

[12] See, for example, Roothaan [30].

(20) reduce to three coupled equations for the three radial functions $\hat{R}$, $\hat{R}'$, and $\hat{R}''$. Thus far, then, things are as one expects. However, even though such closed shell solutions are possible, it is not necessarily true that they yield the determinant with the lowest E, and indeed there is a large literature devoted to this subject and related questions,[13] most of the investigations being concerned with the sign of $\delta_2\hat{E}$ since $\delta_2\hat{E} \geq 0$ is a necessary though not sufficient condition that $\hat{E}$ be an absolute minimum (finding useful sufficient conditions for the latter is really an unsolved problem).

Another class of cases in which one expects a single determinant to be a good approximation occurs if one has a half-filled shell outside of closed shells, each electron in the open shell having the same spin. The simplest example of this is the ground state of the lithium atom. However, if one tries

$$\hat{\varphi}_1 = \hat{R}\uparrow, \qquad \hat{\varphi}_2 = \hat{R}\downarrow, \qquad \hat{\varphi}_3 = \hat{R}'\uparrow$$

in (20), then one finds that the equations for $i = 1$ and 2 are inconsistent with one another. That is, UHF does not admit solutions of the form

$$\hat{\psi} = |\, \hat{R}\uparrow, \hat{R}\downarrow, \hat{R}'\uparrow \,|$$

Rather, it allows a solution of the form

$$\hat{\psi} = |\, \hat{R}\uparrow, \hat{R}''\downarrow, \hat{R}'\uparrow \,| \tag{29}$$

which, however, does not have total spin $\frac{1}{2}$, though it does have a projection of $\frac{1}{2}$. A similar situation exists when one has more electrons in the valence shell, and then not just with respect to spin, but with respect to orbital angular momentum as well.

There are several responses which have been made to this situation [34–36], while still more or less retaining the single-determinant approximation. For one thing, one may simply accept (29) as it is and not worry about its shortcomings (or even try to turn them into virtues).[14] Another approach involves taking just the spin-$\frac{1}{2}$ part of (29). Formally, one acts on (29) with an appropriate spin projection operator. Generalizing the latter method, one may also envision projection before variation, that is, using as the set of trial functions Slater determinants multiplied by projection operators.

[13] For discussions and extensive bibliography, see Paldus and Čižek [32] and Musher [33].

[14] Summary by Lunnel [34]. See also remarks on $\mathscr{X}^{-1}$ theorem in Section 10.

Historically, however, what was done first, and certainly in many ways it is the most obvious approach, was from the outset to restrict the trial functions to have the form

$$\psi = | R\uparrow, R\downarrow, R'\uparrow | \tag{30}$$

This is the so-called restricted Hartree–Fock approximation (RHF). Applied to closed shell situations, it evidently yields the same results as UHF since, as we have pointed out, UHF does admit closed shell solutions. Therefore since these are stationary points within the set of all Slater determinants, they are certainly stationary points within any restricted set. Thus, for example, if for neon one uses

$$\psi = | R\uparrow, R\downarrow, R'\uparrow, R'\downarrow, Y_{1-1}R''\uparrow, \ldots, Y_{11}R''\downarrow | \tag{31}$$

then the equations one derives for $\hat{R}$, $\hat{R}'$, and $\hat{R}''$ are precisely the equations which we mentioned earlier as being derived from (20). However, if one uses the set of functions (30) for Li, then one of course gets something different from UHF.

The fact that for closed shells the RHF solutions satisfy the UHF equations will be of considerable use to us in later discussions, allowing us some easy insights into certain properties of the RHF solutions. Indeed in a general way, and we had an example earlier in Section 4, even if it has no practical consequences, conceptually it is often useful to enlarge the set of trial functions as much as possible. In this same spirit, then, one can show, though we will not enter into any details (we leave them for the problems) that if one considers trial functions of the form

$$| \xi_1\uparrow, \xi_1\downarrow, \xi_2\uparrow, \xi_2\downarrow, \ldots, \xi_C\uparrow, \xi_C\downarrow, \xi_{C+1}\uparrow, \xi_{C+2}\uparrow, \ldots, \xi_{N-C}\uparrow | \tag{32}$$

where the ξ_i are orthonormal but otherwise arbitrary orbitals (we will refer to this as the SUHF approximation), then the RHF solutions for half-filled shells outside of closed shells do satisfy the SUHF equations. (Notice that the functions (32) are eigenfunctions of S_z with eigenvalue $(N-2C)/2$ and that they are also eigenfunctions of the total spin with the same eigenvalue.)

If we denote the doubly occupied orbitals in (32) by ξ_{d}, singly occupied orbitals by ξ_{o} and a general orbital by ξ_i, then one finds that the equations for the optimal orbitals take the form

$$\hat{f}_{\mathrm{o}}\hat{\xi}_{\mathrm{o}} = \sum_i \hat{\xi}_i\varepsilon_{i\mathrm{o}}, \qquad \hat{f}_{\mathrm{d}}\hat{\xi}_{\mathrm{d}} = \sum_i \hat{\xi}_i\varepsilon_{i\mathrm{d}} \tag{33}$$

$$\varepsilon_{ij} = \varepsilon_{ji}^{*} \tag{34}$$

subject to the constraints

$$(\hat{\xi}_i, \hat{\xi}_j) = \delta_{ij} \tag{35}$$

where $\hat{f}_o$ and $\hat{f}_d$ are Hermitian one-electron operators. However, since it turns out that $\hat{f}_o$ and $\hat{f}_d$ do not commute with one another, it is not possible to satisfy the constraints (35) for $i \neq j$ by choosing ε_{ij} diagonal. Nevertheless, some simplification is possible in that, as one readily shows, if the $\hat{\xi}_o$ and $\hat{\xi}_d$ are orthogonal to one another, then choosing $\varepsilon_{oo'}$ and $\varepsilon_{dd'}$ diagonal will ensure orthogonality among the $\hat{\xi}_o$ and $\hat{\xi}_d$ separately. Thus we can replace (33)–(35) by

$$\hat{f}_o\hat{\xi}_o = \varepsilon_o\hat{\xi}_o + \sum_d \hat{\xi}_d\varepsilon_{do}, \qquad \hat{f}_d\hat{\xi}_d = \varepsilon_d\hat{\xi}_d + \sum_o \hat{\xi}_o\varepsilon_{od} \tag{36}$$

$$\varepsilon_{do} = \varepsilon_{od}^* \tag{37}$$

$$(\hat{\xi}_o, \hat{\xi}_d) = 0, \qquad (\hat{\xi}_o, \hat{\xi}_o) = 1, \qquad (\hat{\xi}_d, \hat{\xi}_d) = 1 \tag{38}$$

In the case of UHF, we showed that the hermiticity of ε_{ij} and the orthogonality of the $\hat{\varphi}_i$ were intimately related. However, this is no longer the case here. Thus if (38) is satisfied, then (36) yields

$$\varepsilon_{do} = (\hat{\xi}_d, \hat{f}_o\hat{\xi}_o) \tag{39}$$

and

$$\varepsilon_{od} = (\hat{\xi}_o, \hat{f}_d\hat{\xi}_d) = (\hat{\xi}_d, \hat{f}_d\hat{\xi}_o)^* \tag{40}$$

and therefore (37) is not automatically satisfied but requires further that

$$(\hat{\xi}_o, (\hat{f}_o - \hat{f}_d)\hat{\xi}_d) = 0 \tag{41}$$

Though we have not given the precise formulas for $\hat{f}_o$ and $\hat{f}_d$, we have written Eq. (33) just as they appear from the variation method, the latter leading directly to

$$\sum_o \left(\delta\hat{\xi}_o, \left(\hat{f}_o\hat{\xi}_o - \sum_{i=1}^N \hat{\xi}_i\varepsilon_{io}\right)\right) + \sum_d \left(\delta\hat{\xi}_d, \left(\hat{f}_d\hat{\xi}_d - \sum_{i=1}^N \hat{\xi}_i\varepsilon_{id}\right)\right) \tag{42}$$

However, it should be noted that $\hat{f}_d$ differs by a factor of two from what one usually thinks of as a single-particle Hamiltonian since evidently the contribution of the one-particle part of H to $\hat{E}$ is

$$\sum_s (\hat{\xi}_o, h\hat{\xi}_o) + \sum_d 2(\hat{\xi}_d, h\hat{\xi}_d)$$

Therefore a more physical way of writing Eq. (36) is

$$\hat{f}_o \hat{\xi}_o = \varepsilon_o \hat{\xi}_o + \sum_d \hat{\xi}_d \, \varepsilon_{do}, \qquad \hat{f}_d{}' \hat{\xi}_d = \varepsilon_d{}' \hat{\xi}_d + \sum_o \hat{\xi}_o (\varepsilon_{od}/2) \tag{43}$$

where

$$\hat{f}_d{}' = \tfrac{1}{2} \hat{f}_d \tag{44}$$

and

$$\varepsilon_d{}' = \tfrac{1}{2} \varepsilon_d \tag{45}$$

Equations (20) or (43) or their RHF analogs are, as they stand, sets of three-dimensional coupled nonlinear integrodifferential equations. If, as in the atomic problems we mentioned, they can be reduced to equations in one variable ("radial equations"), then they are susceptible to solution by direct numerical integration [37], though often in practice with some further approximation, particularly for heavy atoms, to the integro part (the "exchange terms"), in order to make the solution easier.[15] Otherwise, such equations have simply not been solved. Indeed the situation here is precisely analogous to that which we discussed in Section 6 in the case of an infinite linear space and the response is similar: One confines attention to a finitely parameterized subset of Slater determinants. We will describe the standard procedure in the next section.

PROBLEMS

1. For two electrons, exhibit one antisymmetric function which is not a Slater determinant. Argue that this suffices to show that in general a sum of Slater determinants is not a Slater determinant.

2. The linear variation method was not our first example involving non-unique variation parameters. What was our first example? [Hint: See Section 4.]

3. Express the right hand side of (6) in terms of ϱ^{UHF}.

4. Be sure that you understand the details of the derivation of (8) and can prove (11).

5. For a two-electron system, use trial functions of the form $| u\!\uparrow, u\!\downarrow |$, where u is an arbitrary spatial function. However, do not require that u be normalized, so that a Lagrange multiplier is not needed. Show by

[15] For recent discussions and references to earlier literature, see Slater and Wood [38] and Folland [39].

detailed calculations that, as must be the case, the resultant equation for $u/(u, u)^{1/2}$ is identical to that which one derives by requiring from the outset that u be normalized.

6. $\hat{E}$ should be invariant to the choice of spin orbitals, canonical or otherwise. Show therefore that more generally the $\sum_i \varepsilon_i$ on the right hand side of (23) should be replaced by $\sum_i \varepsilon_{ii}$ and that this quantity is invariant to unitary transformations.

7. Show in detail that UHF for an isolated atom does not admit solutions of the form $|\ \hat{R}\uparrow, \hat{R}\downarrow, \hat{R}'\updownarrow\ |$.

8. Referring to (32), show that there is no loss in generality in assuming the ξ_i to be orthonormal.

9. Determine $\hat{f}_o$ and $\hat{f}_d{}'$ for the set of trial functions $|\ \xi_d\uparrow, \xi_d\downarrow, \xi_o\uparrow\ |$. Show that $|\ \hat{R}\uparrow, \hat{R}\downarrow, \hat{R}'\uparrow\ |$ is a solution for an isolated atom and derive the equations for $\hat{R}$ and $\hat{R}'$.

10. Generalize Problem 9, deriving the general formulas for $\hat{f}_o$ and $\hat{f}_d{}'$ using the trial functions (32).

9. The Unrestricted Self-Consistent Field Approximation

To get a finite problem, the standard procedure [30, 40] is to further restrict the φ_j by requiring that they can be expanded in finite sets of basis spin orbitals (which may also contain nonlinear parameters; however, we will *not* consider this possibility explicitly).[16] The expansion coefficients are then the A_i. Such further approximations can be made, and regularly are made, in all types of Hartree–Fock approximations, and usually go under the name of self-consistent field (SCF) approximations.

As we have introduced them, SCF approximations appear as approximations to HF approximations. However, one can, of course, regard them as approximations in their own right. In this section we will present some details of what we will call the unrestricted SCF approximation (USCF) in which, as in UHF, we look for optimal single Slater determinants but now with the further restriction that the φ_i will be expanded in the same finite set of basis spin orbitals u_σ; thus

$$\hat{\varphi}_j = \sum_{\sigma=1}^{M} u_\sigma \hat{b}_{\sigma j} \tag{1}$$

[16] For detailed equations, see Simonetti and Gianinetti [41], also Dehn [42] and Olive [43].

and

$$\delta\hat{\varphi}_j = \sum_{\sigma=1}^{M} u_\sigma \, \delta b_{\sigma j} \tag{2}$$

If we use (1) and (2) in (8-17) (the latter of course with $\delta\varphi_j$ replaced by $\delta\hat{\varphi}_j$), then we find

$$0 = \sum_{j=1}^{N} \sum_{\sigma=1}^{M} \sum_{\tau=1}^{M} \delta b_{\sigma j}^{*}[(u_\sigma, \hat{f}u_\tau)\hat{b}_{\tau j} - \sum_{i=1}^{N} (u_\sigma, u_\tau)\varepsilon_{ij}\hat{b}_{\tau i}] + \text{comp. conj.} \tag{3}$$

where now

$$(u_\sigma, \hat{f}u_\tau) = (u_\sigma, hu_\tau) + \sum_{l=1}^{N} \sum_{\varrho=1}^{M} \sum_{\delta=1}^{M} \hat{b}_{\varrho l}^{*}\hat{b}_{\delta l}[(u_\sigma u_\varrho, gu_\tau u_\delta) - (u_\sigma u_\varrho, gu_\delta u_\tau)] \tag{4}$$

To see the consequences of (3), it is helpful to introduce some matrix notation. Thus we introduce the $M \times M$ Hermitian matrix $\hat{\not{f}}$ whose elements are $(u_\tau, \hat{f}u_\sigma)$ and the $M \times M$ positive-definite Hermitian "overlap matrix" S whose elements are (u_τ, u_σ), and finally we introduce N M-element column vectors $B_{(j)}$, the elements of $B_{(j)}$ being b_{σ_j}. In terms of these quantities, then, (3) can be written

$$0 = \sum_{j=1}^{N} \delta B_{(j)}^{+} \cdot \left(\hat{\not{f}}\hat{B}_{(j)} - \sum_{i} S\hat{B}_{(i)}\varepsilon_{ij} \right) + \text{comp. conj.} \tag{5}$$

Also in this notation the orthonormality requirement

$$(\hat{\varphi}_i, \hat{\varphi}_j) = \delta_{ij} \tag{6}$$

becomes

$$\hat{B}_{(i)}^{+} \cdot S\hat{B}_{(j)} = \delta_{ij} \tag{7}$$

From this point on, the argument proceeds much as in Section 8. One is first led to the general equations

$$\hat{\not{f}}\hat{B}_{(j)} = \sum_{i} S\hat{B}_{(i)}\varepsilon_{ij} \tag{8}$$

and then to the canonical equations

$$\hat{\not{f}}\hat{B}_{(j)} = \varepsilon_j S\hat{B}_{(j)} \tag{9}$$

as a special case, the relationship between the general and the special cases being just as in Section 8, so we will not repeat the details.

Equations (9), together with (4), constitute a set of nonlinear algebraic equations for the $\hat{B}_j$ and ε_j. The usual solution procedures for such equations are of an iterative type, as is also the case with the numerical solution of (8-20) when it is possible. Most naively, one guesses N vectors, call them $\overset{1}{B}_j$, which one hopes are good approximations to the $\hat{B}_j$ in that they yield orthonormal $\hat{\varphi}_j$ of the expected shape, right number of nodes, symmetry properties, etc. From them, one computes $\overset{1}{\mathscr{F}}$ in the obvious way and solves the ordinary linear eigenvalue problem

$$\overset{1}{\mathscr{F}}\overset{2}{B} = \overset{2}{\varepsilon} S \overset{2}{B}$$

This problem has M orthonormal solutions. Hopefully, N of them resemble the $\overset{1}{B}_{(j)}$ from which we started. If so, one uses those N to compute an $\overset{2}{\mathscr{F}}$, etc. until, again hopefully, one has convergence (self-consistency) in that N of the $\overset{K}{B}$ differ negligibly from the N of the $\overset{K-1}{B}$ that were used to compute $\overset{K-1}{\mathscr{F}}$.

For details of practical procedures for doing this as well as a discussion of a way of handling nonlinear parameters, we refer the reader to an article by Roothaan and Bagus [44].[17] Also, there have been interesting studies of the totality (nature and number) of equation systems like (9) [51].

For molecules, especially for large molecules, the procedure we have outlined may not be practical at all, or it may only be practical using the largest of computers, because of the sheer number of integrals which must be calculated to evaluate $\hat{\mathscr{F}}$, and often, depending on the nature of u_σ, because of difficulty in calculating them. These difficulties, in addition to spawning extensive literature on the choice of the u_σ[18] and on integral evaluation,[19] have also led to the development of many methods which further approximate $\hat{\mathscr{F}}$ in some way or other. However, we will not attempt to review these methods here.[20]

As one final point, investigation of the nature of the stationary points of USCF and the like can, of course, be based on (4-21) with the A_i the $N \cdot M$ numbers $b_{\tau j}$. However, we will not enter any further into the extensive details except to indicate, at the end of the next section, a way of simplifying the analysis somewhat.

[17] See also Catana *et al.* [45]. For methods of solving SUSCF equations, see Roothaan [46], Berthier [47], Sleaman [48], Hunt *et al.* [49], and Albat and Gruen [50].

[18] For a recent review of SCF theories including a discussion of such matters, see Clementi [52].

[19] See, for example, Silverstone [53].

[20] Recent reviews: Jug [54], Brown and Roby [55], Pople and Beveridge [56], and Klopman and O'Leary [57].

PROBLEMS

1. Show that USCF is UHF for $\Pi H \Pi$, where Π is the projector onto the space of linear combinations of all the Slater determinants which can be formed from the u_σ.

2. "Derive" (9) from (8-20) in a manner analogous to that described following (5-17) for "deriving" that equation from the Schrödinger equation.

References

1. A. Wallis, D. L. S. McElwain, and H. O. Pritchard, *Int. J. Quantum Chem.* **3**, 711 (1969), and references therein.
2. G. S. Handler and H. W. Joy, *J. Chem. Phys.* **47**, 5074 (1967), and references therein.
3. J. Katriel and G. Adam, *Chem. Phys. Lett.* **10**, 533 (1971).
4. N. W. Winter, A. Laferrière, and V. McKoy, *Phys. Rev. A* **2**, 49 (1970).
5. C. G. Barraclough and J. R. Mooney, *J. Chem. Phys.* **54**, 35 (1971).
6. S. G. Mikhlin, "Variational Methods in Mathematical Physics." Pergamon, Oxford, 1964.
7. R. Courant, *Bull. Amer. Math. Soc.* **49**, 1 (1963).
8. B. A. Finlayson, "The Method of Weighted Residuals and Variational Principles." Academic Press, New York, 1972.
9. S. H. Gould, "Variational Methods for Eigenvalue Problems," 2nd ed. Univ. of Toronto Press, Toronto, 1966.
10. H. F. Schaefer III, "The Electronic Structure of Atoms and Molecules." Addison-Wesley, Reading, Massachusetts, 1972.
11. D. Pandres, Jr., and F. A. Matsen, *J. Chem. Phys.* **37**, 1008 (1962).
12. A. Messiah, "Quantum Mechanics," Vol. 1, Chapter 7. Wiley, New York, 1966.
13. E. A. Hylleraas and B. Undheim, *Z. Phys.* **65**, 759 (1930); J. K. L. MacDonald, *Phys. Rev.* **43**, 830 (1933); see also R. H. Young, *Int. J. Quantum Chem.* **6**, 596 (1972), and references therein.
14. H. S. Taylor, *Advan. Chem. Phys.* **18** (1970).
15. J. F. Perkins, *J. Chem. Phys.* **45**, 2156 (1966).
16. D. W. Davies, *J. Chem. Phys.* **33**, 781 (1960).
17. P.–O. Löwdin, *Phys. Rev.* **97**, 1474–1520 (1955); L. L. Foldy, *J. Math. Phys.* (*N.Y.*) **3**, 531 (1962).
18. J. F. Perkins, *J. Chem. Phys.* **42**, 3927 (1965); C. F. Melius and W. A. Goddard III, *ibid.* **56**, 3348 (1972).
19. P.–O. Löwdin, *Advan. Quantum Chem.* **5**, 185 (1970).
20. R. McWeeny, *Rev. Mod. Phys.* **32**, 335 (1960).
21. J. C. Slater, "Quantum Theory of Atomic Structure," Vol. 1, Section 12-5. McGraw Hill, New York, 1960.
22. W. England, L. S. Salmon, and K. Ruedenberg, *in* "Topics in Current Chemistry," (F. Boschke, ed.), Vol. 23. Springer-Verlag, Berlin and New York, 1971.
23. H. Weinstein, R. Pauncz, and M. Cohen, *Advan. At. Mol. Phys.* **7** (1971).

24. R. Manne and T. Åberg, *Chem. Phys. Lett.* **7**, 282 (1970).
25. D. A. Shirley, *Chem. Phys. Lett.* **16**, 220 (1972).
26. K. A. Brueckner, H. W. Melder, and J. D. Perez, *Phys. Rev. C* **6**, 773 (1972).
27. K. Siegbahn *et al.*, "ESCA Applied to Free Molecules," p. 67 et seq. North-Holland Publ., Amsterdam, 1969.
28. T. Koopmans, *Physica* (*Utrecht*) **1**, 105 (1933).
29. W. J. Hunt and W. A. Goddard III, *Chem. Phys. Lett.* **3**, 414 (1969); also E. R. Davidson, *J. Chem. Phys.* **57**, 1999 (1972), and references therein.
30. C. C. J. Roothaan, *Rev. Mod. Phys.* **23**, 69 (1951).
31. M. Reeken, *J. Math. Phys.* (*N.Y.*) **11**, 2505 (1970).
32. J. Paldus and J. Čižek, *Phys. Rev. A* **2**, 2268 (1970).
33. J. I. Musher, *Chem. Phys. Lett.* **7**, 397 (1970).
34. S. Lunell, *Phys. Rev. A* **7**, 1229 (1973), and references therein.
35. P.–O. Löwdin, *Rev. Mod. Phys.* **32**, 328 (1960).
36. R. Pauncz, "The Alternant Molecular Orbital Method." Saunders, Philadelphia, Pennsylvania, 1967.
37. D. R. Hartree, "The Calculation of Atomic Structures." Wiley, New York, 1957; C. Froese Fischer, *J. Comput. Phys.* **10**, 211 (1972), and references therein.
38. J. C. Slater and J. H. Wood, *Int. J. Quantum Chem. Symp.* **4**, 3 (1971).
39. N.–O. Folland, *Phys. Rev. A* **3**, 1535 (1971).
40. G. G. Hall, *Proc. Roy. Soc. Ser. A* **205**, 511 (1951).
41. M. Simonetti and E. Gianinetti, *in* "Molecular Orbitals in Chemistry, Physics, and Biology" (P.-O. Löwdin and B. Pullmann, eds.). Academic Press, New York, 1964.
42. J. T. Dehn, *J. Chem. Phys.* **37**, 2549 (1962).
43. J. P. Olive, *J. Chem. Phys.* **51**, 4340 (1969).
44. C. C. J. Roothaan and P. S. Bagus, *Methods Comput. Phys.* **2** (1963).
45. F. Catana, M. DiToro, E. Pace, and G. Schiffrer, *Nuovo Cimento* [10], *A* **11**, 733 (1972).
46. C. C. J. Roothaan, *Rev. Mod. Phys.* **32**, 1174 (1960).
47. G. Berthier, *in* "Molecular Orbitals in Chemistry, Physics, and Biology" (P.-O. Löwdin and B. Pullmann, eds.). Academic Press, New York, 1962.
48. D. H. Sleaman, *Theor. Chim. Acta* **11**, 135 (1968).
49. W. J. Hunt, W. A. Goddard III, and J. H. Dunning, Jr., *Chem. Phys. Lett.* **6**, 147 (1970).
50. R. Albat and N. Gruen, *Chem. Phys. Lett.* **18**, 572 (1973).
51. R. E. Stanton, *J. Chem. Phys.* **48**, 257 (1968); H. F. King and R. E. Stanton, *ibid.* **50**, 3789 (1969).
52. E. Clementi, *Chem. Rev.* **68**, 341 (1968).
53. H. J. Silverstone, *J. Chem. Phys.* **48**, 4098 (1968), and references therein.
54. K. Jug, *Theor. Chim. Acta* **14**, 91 (1969).
55. R. D. Brown and K. R. Roby, *Theor. Chim. Acta* **16**, 175 and 194 (1970).
56. J. A. Pople and D. L. Beveridge, "Approximate Molecular Orbital Theory." McGraw Hill, New York, 1970.
57. G. Klopman and B. J. O'Leary, *Fortschr. Chem. Forsch.* **15**, part 4.

Chapter III / The Generalized Brillouin Theorem

10. Derivation and Applications of the Theorem

In general, whatever sort of trial functions one uses, any $\hat{\psi}$ will almost certainly be only an approximation to an eigenfunction of H, and so the question naturally arises, how can we improve on that approximation? One approach is to enlarge the set of trial functions in some way. Another is to use Rayleigh–Schrödinger (RS) perturbation theory, and it is mainly with some aspects of this approach that we will be concerned in this section. In Section 36 we will return to the question again, and discuss it from a rather more general point of view.

In order to use RS perturbation theory, we must introduce a zeroth-order Hamiltonian $H^{(0)}$, which has $\hat{\psi}$ as an eigenfunction

$$H^{(0)}\hat{\psi} = \mathscr{E}^{(0)}\hat{\psi} \tag{1}$$

where $\mathscr{E}^{(0)}$, the eigenvalue, may or may not equal $\hat{E}$. Having chosen $H^{(0)}$, one now treats $H - H^{(0)}$ as a perturbation. The first-order correction to $\hat{\psi}$ is therefore given by the familiar formula (we will assume that $\mathscr{E}^{(0)}$ is either nondegenerate or effectively nondegenerate, for reasons of symmetry; also, for convenience, we will assume that $\hat{\psi}$ is normalized):

$$\Phi^{(1)} = -\sum_{\mu} \Theta_{\mu} \frac{(\Theta_{\mu}, (H - H^{(0)})\hat{\psi})}{\mathscr{E}_{\mu}^{(0)} - \mathscr{E}^{(0)}} \tag{2}$$

where the Θ_{μ} are the other orthonormal eigenfunctions of $H^{(0)}$ and the $\mathscr{E}_{\mu}^{(0)} \neq \mathscr{E}^{(0)}$ are the corresponding eigenvalues. Using (1) and the fact that the Θ_{μ} are orthogonal to $\hat{\psi}$, we can also write $\Phi^{(1)}$ as

$$\Phi^{(1)} = -\sum_{\mu} \Theta_{\mu} \frac{(\Theta_{\mu}, H\hat{\psi})}{\mathscr{E}_{\mu}^{(0)} - \mathscr{E}^{(0)}} \tag{3}$$

It should be kept in mind, however, that this perturbation approach involves a great deal of arbitrariness, since, using Dirac bra-ket notation, any operator of the form

$$H^{(0)} = \mathscr{E}^{(0)} \,|\, \hat{\varphi} \times \hat{\varphi} \,|\, + \sum_{\mu} \mathscr{E}_{\mu}^{(0)} \,|\, \Theta_{\mu} \times \Theta_{\mu} \,| \tag{4}$$

will satisfy (1) whatever we choose for the $\mathscr{E}_{\mu}^{(0)}$ and the Θ_{μ} so long as the latter are orthogonal to $\hat{\varphi}$. However, different choices of the $\mathscr{E}_{\mu}^{(0)}$ and the Θ_{μ} can make profound changes in $\Phi^{(1)}$ and hence in the second- and higher-order energy corrections. [Through first order, the energy is of course $\hat{E}$ whatever the choice of $H^{(0)}$ since

$$(\hat{\varphi}, H^{(0)}\hat{\varphi}) + (\hat{\varphi}, (H - H^{(0)})\hat{\varphi}) = (\hat{\varphi}, H\hat{\varphi}) = \hat{E}]$$

We will now show that as a consequence of the fact that $\hat{\varphi}$ is an optimal variational wave function, certain $(\Theta_{\mu}, H\hat{\varphi})$ which one would not otherwise expect to vanish—for example, they do not vanish for reasons of symmetry—may in fact vanish so that the corresponding Θ_{μ} will not occur in $\Phi^{(1)}$, though of course they may appear in $\Phi^{(2)}$ etc., and therefore will not affect the energy until at least fourth order. Namely suppose that

$$\eta \Theta_{\mu} \tag{5}$$

with η an arbitrary complex number, is among the variations of $\hat{\varphi}$ possible within the set of trial functions. Then it follows from (5-3) (which evidently applies to at least this variation) and the orthogonality of Θ_{μ} and $\hat{\varphi}$ that

$$(\Theta_{\mu}, H\hat{\varphi}) = 0 \tag{6}$$

which from (3) tells us that Θ_{μ} will not appear in $\Phi^{(1)}$.

In the context of UHF, where, as we will see, the Θ_{μ} satisfying (6) are one-electron excitations of $\hat{\varphi}$, this is known as Brillouin's theorem [1]. (By a one-electron excitation of a Slater determinant ψ we mean a determinant which differs from it by a single spin orbital orthogonal to all the spin orbitals in ψ.) Therefore we will call (6) the generalized Brillouin theorem.

More precisely, and quite apart from its application to perturbation theory, the generalized Brillouin theorem is the following: Let $\eta\Theta$, with η an arbitrary complex number and with Θ orthogonal to $\hat{\varphi}$, be a possible variation of $\hat{\varphi}$ within the set of trial functions. Then it follows from (5-3) that

$$(\Theta, H\hat{\varphi}) = 0 \tag{7}$$

Somewhat more generally, if we drop the requirement of orthogonality, then the theorem is the statement that if $\eta\Theta$, with η an arbitrary complex number, is a possible variation of $\hat{\psi}$ within the set of trial functions, then

$$(\Theta, (H - \hat{E})\hat{\psi}) = 0 \tag{8}$$

In many ways, then, the generalized Brillouin theorem is really not a new result but merely a restatement of the variation method, and has been so used by many authors.[1]

For the particular case of UHF, an obvious and natural choice for $H^{(0)}$ is[2]

$$H^{(0)} = F \equiv \sum_{s=1}^{N} \hat{f}(s) \tag{9}$$

with therefore

$$\mathscr{E}^{(0)} = \sum_{i=1}^{N} \varepsilon_i \tag{10}$$

Since F is a one-electron operator, this means that the Θ_μ are, or in the case of degeneracy can be chosen to be, single determinants involving 1-, 2-, . . ., N-electron excitations of $\hat{\psi}$. We will now show that the one-electron excitations do not appear in $\Phi^{(1)}$ (and therefore do not affect the energy until fourth order). The proof is as follows.

For $\hat{\psi}$, we have

$$\hat{\psi} = (1/\sqrt{N!}\,) \,|\, \hat{\varphi}_1, \ldots, \hat{\varphi}_N \,|$$

Therefore the most general $\delta\hat{\psi}$ is of the form

$$\delta\hat{\psi} = (1/\sqrt{N!}\,) \sum_{j=1}^{N} |\, \hat{\varphi}_1, \ldots, \delta\varphi_j, \ldots, \hat{\varphi}_N \,| \tag{11}$$

where the $\delta\varphi_j$ are arbitrary one-electron functions. If in particular we choose the $\delta\varphi_j$ to be orthogonal to all the $\hat{\varphi}_i$ but otherwise arbitrary, then (11) is an arbitrary sum of one-electron excitations of $\hat{\psi}$, which proves the point.

For SUHF, the Θ which satisfy the generalized Brillouin theorem are a general superposition of two types: one-electron excitations of the singly occupied orbitals without change in spin, and paired excitations of the doubly occupied orbitals, again without change in spin. Thus for three electrons and

$$\hat{\psi} = |\, \hat{\xi}_1\uparrow, \hat{\xi}_1\downarrow, \hat{\xi}_2\uparrow \,| \tag{12}$$

[1] For example, Nesbet [2], Le Febvre [3], Mayer [4], and Grein and Chang [5].

[2] Examples of other choices: Epstein [6], Manson [7], Silverstone and Yin [8], McWilliams and Huzinga [9], Pan and King [10], and Lunell [11].

the two types are

$$|\ \hat{\xi}_1\uparrow,\ \hat{\xi}_1\downarrow,\ \delta\xi_2\uparrow\ | \tag{13}$$

and

$$|\ \delta\xi_1\uparrow,\ \hat{\xi}_1\downarrow,\ \hat{\xi}_2\uparrow\ | + |\ \hat{\xi}_1\uparrow,\ \delta\xi_1\downarrow,\ \hat{\xi}_2\uparrow\ | \tag{14}$$

where for orthogonality to $\hat{\psi}$ we need

$$(\delta\xi_2,\ \hat{\xi}_2) = (\delta\xi_1,\ \hat{\xi}_1) = 0 \tag{15}$$

The second type of excitation may then be further classified according to whether or not $\delta\xi_1$ is orthogonal to $\hat{\xi}_2$.

Since, as we have discussed, RHF for closed shells and for certain open shells can be considered special cases of UHF and of SUHF, respectively, the preceding results apply to them as well. However, it is of some interest to derive them directly, thus giving a model for the discussion of other Hartree–Fock theories.[3] First let us consider closed shells, for example, the ground-state neon atom. Then, using standard spectroscopic notation,

$$\hat{\psi} = (1s)^2(2s)^2(2p)^6\ {}^1S$$

and it should be clear from (8-31) that (from varying $\hat{R}$)

$$\Theta = (1s)(ns)(2s)^2(2p)^6\ {}^1S$$

and (from varying $\hat{R}'$)

$$\Theta = (1s)^2(2s)(ns)(2p)^6\ {}^1S$$

and (from varying R'')

$$\Theta = (1s)^2(2s)^2(2p)^5(np)\ {}^1S$$

satisfy the condition of the generalized Brillouin theorem. (Note that in the language we have been using, these are sums of one-electron excitations.) Further, this is evidently all we learn from the variation method itself. However, we now note that any other (orthogonal) sums of one-electron excitations of $\hat{\psi}$ will automatically be orthogonal to $H\hat{\psi}$ for reasons of symmetry since the Θ just given are the only linear combinations of one-electron excitations of $\hat{\psi}$ which are ^1S. Thus whereas from the point of view of UHF the complete result that $(\Theta, H\hat{\psi}) = 0$ for Θ any one-electron excitation of $\hat{\psi}$ follows from the variation method alone, from the point of view of RHF it is a result partly of the variation method and partly of symmetry.

[3] See, for example, Bauche and Klapisch [12], Labarthe [13], also Levy and Berthier [14].

As an example of an open shell, consider the ground state of the lithium atom. Here

$$\hat{\varphi} = (1s)^2(2s)\ ^2S$$

and evidently from (8-30) the relevant Θ in RHF (= SUHF) are (from varying $\hat{R}'$)

$$\Theta = (1s)^2(ns)\ ^2S$$

and (from varying $\hat{R}$)

$$\Theta = (1s)(ns)(2s)\ ^2S$$

In particular, note that the second possibility includes $(1s)(2s)^2\ ^2S$. Other Θ which are in accord with the considerations we gave for SUHF, then, from this RHF point of view, yield $(\Theta, H\hat{\varphi}) = 0$, for reasons of symmetry.

In this discussion, we have spoken as though the Θ_μ and the Θ were all bound-state functions. However, this need not be the case. Indeed for the ground states of neutral systems and with $H^{(0)}$ given by (9), the Θ_μ are usually all in the continuum [15]. At first sight, this would seem to upset the entire development, since we assumed from the outset that our trial functions and hence presumably the $\delta\hat{\varphi}$ were normalizable. However, in fact this is not a difficulty. We will not attempt an elaborate proof, but a quick argument is the following. Any continuum wave function can be approximated by a normalizable one, in that one may multiply the continuum one by a function which is one essentially everywhere, but which goes to zero rapidly at infinity. However, clearly if $\hat{\varphi}$ is a bound-state wave function, then the value of an integral like

$$(\Theta, (H - \hat{E})\hat{\varphi})$$

is completely insensitive to whether Θ is a continuum function or a "normalizable continuum function" of the type which we have described.

Returning now to UHF, the fact that with the choice $H^{(0)} = F$, $\Phi^{(1)}$ contains no one-electron excitations of $\hat{\varphi}$ has an interesting consequence: There are no first-order corrections to the average value of any one-electron operator [16]. Thus if W is a one-electron operator, we have for its expectation value, correct through first order,

$$\frac{(\hat{\varphi} + \Phi^{(1)} + \cdots, W(\hat{\varphi} + \Phi^{(1)} + \cdots))}{(\hat{\varphi} + \Phi^{(1)} + \cdots, \hat{\varphi} + \Phi^{(1)} + \cdots)}$$
$$= (\hat{\varphi}, W\hat{\varphi}) + \{(\hat{\varphi}, [W - (\hat{\varphi}, W\hat{\varphi})]\Phi^{(1)}) + (\Phi^{(1)}, [W - (\hat{\varphi}, W\hat{\varphi})]\hat{\varphi}\} + \cdots \tag{16}$$

where we have again assumed that $\hat{\psi}$ is normalized. The quantity in the curly braces is evidently the first-order correction, and we will now show that it vanishes. The point is simply that with W a one-electron operator,

$$W = \sum_{s=1}^{N} w(s) \tag{17}$$

one has that

$$(\hat{\psi}, W\hat{\psi}) = \sum_{j=1}^{N} (\hat{\varphi}_j, w\hat{\varphi}_j) \tag{18}$$

Therefore

$$[W - (\hat{\psi}, W\hat{\psi})]\hat{\psi} = (1/\sqrt{N!}\,) \sum_{j=1}^{N} |\, \hat{\varphi}_1, \ldots, [w - (\hat{\varphi}_j, w\hat{\varphi}_j)]\hat{\varphi}_j, \ldots, \hat{\varphi}_N| \tag{19}$$

is a sum of one-electron excitations of $\hat{\psi}$, whence the second term in the braces in (16) vanishes. Also, if we write the first term as

$$([W^+ - (\hat{\psi}, W^+\hat{\psi})]\hat{\psi}, \Phi^{(1)})$$

where W^+ is the Hermitian conjugate of W, then a similar discussion shows that it vanishes as well.

If one defines the Hermitian one-body density matrix associated with any antisymmetric wave function ψ by

$$\varrho(1, 1') \equiv \int d2 \int d3 \cdots \int dN\, \psi(1, 2, \ldots, N)\, \psi^*(1', 2, \ldots, N) \tag{20}$$

then, as is well known, given ϱ, one can calculate the average value of any one-electron operator according to (we use Dirac bra-ket notation)

$$\int d1 \int d1' \langle 1' \mid W \mid 1\rangle \varrho(1, 1') \tag{21}$$

Thus we can infer from the vanishing of first-order corrections for any one-electron operator that $\hat{\varrho}^{\mathrm{UHF}}$, the explicit formula for which is given in (8-4) (with the addition of the caret), is accurate through first order. Alternatively, one can derive this last result from the fact[4] that ϱ itself is the expectation value of a one-electron operator.

The orthonormal eigenfunctions ω_τ and the eigenvalues n_τ of a one-body

[4] See, for example, McWeeny and Mizuno [17] and Yang [18].

density matrix (its natural spin orbitals and their occupation numbers) are of considerable physical interest.[5] These are defined by

$$\int d1'\, \varrho(1, 1')\omega_\tau(1') = n_\tau \omega_\tau(1) \tag{22}$$

so that one has

$$\varrho(1, 1') = \sum_\tau n_\tau \omega_\tau(1) \omega_\tau^*(1') \tag{23}$$

Comparison with (8-4) then shows that the $\hat{n}_\tau^{\rm UHF}$ are N one's and an infinite number of zeros, and from our theorem it follows that they are exact through first order. Also, from our theorem, the $\hat{\omega}_\tau^{\rm UHF}$ are exact through first order. However, because of the degeneracy of the $\hat{n}_\tau^{\rm UHF}$, the $\hat{\omega}_\tau^{\rm UHF}$ are not unique. Thus, comparing (23) with (8-4) and (8-5) we see that the $\hat{\varphi}_k$ and $\hat{\varphi}_k'$ are all $\hat{\omega}_\tau^{\rm UHF}$'s. Therefore to discover which $\hat{\omega}_\tau^{\rm UHF}$ are exact through first order, that is, in the language of degenerate perturbation theory, in order to find the correct zeroth-order functions, one must go on and examine at least the first-order effects of $(H - F)$ [20].

Because of the freedom in the choice of $H^{(0)}$, all of these results, though interesting, are rather more formal than physical unless supplemented by a more quantitative discussion of the order of magnitude of $(H - H^{(0)})$. Thus we could replace F by any other one-electron operator satisfying (1) and formally the results would be the same. Or, alternatively, we could certainly choose the Θ_μ and hence $H^{(0)}$ so that none of the Θ_μ was a pure one-electron excitation or sum thereof. In such a case, Brillouin's theorem would then have no especially interesting consequences for UHF. The one exception to this is of course the statement that $\hat{E}$ involves only a second-order error. As we have noted, and in accord with the variation principle, this is true for any choice of $H^{(0)}$.

However, for atoms, there is a more physical theorem concerning the accuracy of UHF (and of SUHF) which we will only indicate here, postponing detailed statement, proof, and discussion until Section 29. If one considers an isoelectronic sequence of atoms, then one can show that treating the Coulomb interaction between the electrons as a perturbation results in an expansion in powers of $\mathscr{Z}^{-1}$, where $\mathscr{Z}$ is the nuclear charge. Now, without the Coulomb interaction, H is a one-electron operator and therefore barring degeneracy problems, which would require that the correct zeroth-order function be some linear superposition of determinants, UHF and SUHF will be exact in lowest order. The theorem then states that under

[5] See Löwdin (Chapter II, ref. [17]). For recent reviews, see Davidson [19, 20].

these conditions, the UHF (SUHF) expectation value of any one-electron operator (spin-free one-electron operator) will be exact through terms of order $\mathscr{Z}^{-1}$ relative to the leading term.

From the point of view of the preceding theorems, the fact that $\hat{E}$ involves only a second-order error may appear anomalous since H contains the Coulomb interaction between the electrons, which is a two-electron operator. However, of course, there is no contradiction. Thus, as implied by our remark in the preceding paragraph, the average of the Coulomb interaction using zeroth-order wave functions is already of order $\mathscr{Z}^{-1}$ compared with the rest. Therefore the lack of accuracy in any further corrections to it is irrelevant to having $\hat{E}$ be accurate through terms of order $\mathscr{Z}^{-1}$ compared with the leading term.

Similarly, referring to our earlier formal theorem, if we would there explicitly introduce an order parameter ν whose physical value is one and write

$$H(\nu) = F + \nu(H - F)$$

then all the two-electron terms appear multiplied by ν. Therefore the fact that $\hat{E}$ is accurate through first order in ν is not in contradiction to having the average of $(H - F)$ accurate only through zero order in ν since in $\hat{E}$ that average appears multiplied by ν.

These results have obvious implications concerning the expected accuracy of UHF. Since none of the arguments applies to two-, three-, etc., electron operators, one expects that expectation values of one-electron operators are given more accurately by UHF than are expectation values of two-electron operators. Further, one expects this accuracy to be the same as that of $\hat{E}$. However, these arguments about the order of accuracy are a bit shaky. We have already noted the formal character of our first argument. Further, the $1/\mathscr{Z}$ theorem for atoms, when it applies, is a statement about the $\mathscr{Z} \to \infty$ limit and therefore not ovbiously of significance for neutral or near-neutral atoms. Nevertheless, as a general rule, UHF does quite well as regards one-electron properties,[6] but there is definite evidence that for some one-electron properties of molecules, for example, dipole moments, second- and higher-order corrections are not negligible and, in particular, that one-electron excitations of $\hat{\varphi}$, which, as we have seen, do not appear in $\Phi^{(1)}$ with the choice of $H^{(0)} = F$, but which can appear in $\Phi^{(2)}$, etc., can have an important effect [23].

[6] For an extensive compilation for atoms, see Fraga and Saxena [21]. For some molecular properties, see, for example, Geratt and Mills [22].

Finally, let us note an interesting and at first sight surprising corollary of the $\mathscr{Z}^{-1}$ theorem. We remarked in Section 8 that for lithiumlike atoms UHF does not yield a spin doublet ground state (except in the $\mathscr{Z} = \infty$ limit), whereas, by construction, SUHF (= RHF in this case) does. However, from the $\mathscr{Z}^{-1}$ theorem, it follows that at least for large $\mathscr{Z}$, UHF expectation values of spin-dependent one-electron operators will be more accurate than SUHF expectation values [24].

As another application of Brillouin's theorem, we will show that it can be used to simplify the much-studied $\delta_2\hat{E}$ for USCF, though we will not write out the detailed formulas. From Eq. (4-10), what we need to calculate is

$$\delta_2\hat{E}(\hat{\psi}, \hat{\psi}) = (\delta_2\hat{\psi}, (H-\hat{E})\hat{\psi}) + 2(\delta\hat{\psi}, (H-\hat{E})\, \delta\hat{\psi}) + (\hat{\psi}, (H-\hat{E})\, \delta_2\hat{\psi}) \quad (24)$$

where now

$$\delta\hat{\psi} = \sum_{i=1}^{N} | \hat{\varphi}_1, \ldots, \delta\hat{\varphi}_i, \ldots, \hat{\varphi}_N | \quad (25)$$

and

$$\delta_2\hat{\psi} = \sum_{\substack{i=1 \\ i>j}}^{N} \sum_{j=1}^{N} | \hat{\varphi}_1, \ldots, \delta\hat{\varphi}_i, \ldots, \delta\hat{\varphi}_j, \ldots, \hat{\varphi}_N | \quad (26)$$

The simplification comes by introducing a special basis for $\delta\hat{\varphi}_i$, much as we introduced a special basis for $\delta\hat{\psi}_k$ in discussing $\delta_2\hat{E}$ for the linear variation method. Namely let us introduce $M - N$ functions v_τ which are linear combinations of the u_τ and which are orthogonal to all the $\hat{\varphi}_i$. One common choice is the so-called virtual spin orbitals, which are defined as follows: Once the $\hat{B}_{(j)}$ have been determined, $\hat{\mathscr{f}}$ of (9-9) is a perfectly definite Hermitian matrix, so that the eigenvalue equation

$$(\hat{\mathscr{f}} - \varepsilon' S)B = 0 \quad (27)$$

has $M - N$ solutions other than the $\hat{B}_{(j)}$, and these yield the virtual spin orbitals.

In any case, given a set of v_τ, one can certainly write

$$\delta\hat{\varphi}_i = \sum_{j=1}^{N} \hat{\varphi}_j\, \delta b_{ji} + \sum_{\tau=1}^{M-N} v_\tau\, \delta b_{\tau i} \quad (28)$$

and what we are going to show is that we can ignore the δb_{ji} terms, thereby simplifying this situation by in effect reducing the number of variables from NM to $N(M - N)$. First we note that if we insert (28) into (25), then the δb_{ji} terms contribute only a multiple of $\hat{\psi}$ while the $\delta b_{\tau i}$ yield one-electron excitations of $\hat{\psi}$ within the space. However, for USCF, evidently the content of Brillouin's theorem is that $(H - \hat{E})\hat{\psi}$ is orthogonal to $\hat{\psi}$ and to all one-electron excitations of $\hat{\psi}$ within the space, and therefore the δb_{ji} make no contribution to $(\delta\hat{\psi}, (H - \hat{E})\, \delta\hat{\psi})$. Further, if we insert (28) into (26), then we see that the terms involving the δb_{ji} are either multiples of $\hat{\psi}$ or one-electron excitations and so, for the same reasons as before, their contribution to $(\delta_2\hat{\psi}, (H - \hat{E})\hat{\psi})$ and its complex conjugate vanish identically.

Leaving Hartree–Fock theories, let us consider instead cases in which the set of trial functions form a linear space. Then the possible $\delta\hat{\psi}_k$ can be any functions in the space, whence Brillouin's theorem tells us that $(H - \hat{E}_k)\hat{\psi}_k$ will be orthogonal to any function in the space, which of course is the content of (4-17) or of (6-7) and (6-5).

If we wish to improve the $\hat{\psi}_k$ by use of perturbation theory, a natural choice for $H^{(0)}$ is $\bar{H}$ or some linear combination of $\bar{H}$ and $(1 - \Pi)X(1 - \Pi)$ where X could be anything, for example, H [25]. We now note that with any of these choices, all the $\hat{\psi}_l$ (and we include here the continuum eigenfunctions of $\bar{H}$, if any, as well) will be eigenfunctions of $H^{(0)}$. Therefore we have the result that under these conditions, none of the $\hat{\psi}_l$ will appear in $\Phi^{(1)}$. Since the $\hat{\psi}_l$ span the linear space, another way of expressing this is to say that with this choice of $H^{(0)}$, $\Phi^{(1)}$ will be orthogonal to the space. Further, we then see from (16) that if $W\hat{\psi}_k$ and $W^{+}\hat{\psi}_k$ are in the space, then there will be no first-order corrections to the approximation to the expectation value of W provided by $\hat{\psi}_k$. In particular, therefore, in the case of the s limit in helium there will be no first-order correction to the expectation values of any purely radial operators.

As an alternative to using perturbation theory in an effort to improve $\hat{\psi}$, we mentioned at the outset the possibility of enlarging the set of trial functions. One common method of doing this is to do a further linear variation calculation with the basis set consisting of $\hat{\psi}$ and some other functions orthogonal to $\hat{\psi}$. The generalized Brillouin theorem then tells us that if any of these other functions satisfy the condition of the theorem, then they will not be directly coupled to $\hat{\psi}$ in the resulting secular equation [the analog of (5-17)] since $H\hat{\psi}$ will also be orthogonal to them. In particular, if *all* the functions satisfy the conditions of the theorem, then this procedure will lead to no improvement in $\hat{\psi}$ at all since $\hat{\psi}$ will still be an optimal trial function. In the context of the linear variation method, this last is just the statement that one must enlarge the original basis set to get a change, while within UHF, it is the statement that if the other functions consist of only one-electron excitations of $\hat{\psi}$, then there will be no change.

PROBLEMS

1. In connection with (1), given an $H^{(0)}$ with $\mathscr{E}^{(0)} \neq \hat{E}$, one can of course always add a constant to $H^{(0)}$ so as to make $\mathscr{E}^{(0)} = \hat{E}$. Show that such a change in $H^{(0)}$ leaves the perturbation analysis completely unchanged order by order. Show, however, that if $\mathscr{E}^{(0)} \neq \hat{E}$, then replacing $\mathscr{E}^{(0)}$ by $\hat{E}$ in (4) will produce order-by-order changes.

2. Given a $\hat{\psi}$, show that a possible $H^{(0)}$ is $\pi H \pi$, where π is the projector onto $\hat{\psi}$. What is $\mathscr{E}^{(0)}$? What are the $\mathscr{E}_\mu^{(0)}$? Show that this yields

$$\Phi^{(1)} = \frac{(H - \hat{E})}{\hat{E}} \hat{\psi}.$$

3. In Eqs. (9) and (10), we have taken it as obvious that, to put it quite generally, if a set of N spin orbitals φ_i satisfy $(h - \varepsilon_i)\varphi_i = 0$, then the Slater determinant Ψ formed from these orbitals will satisfy $(\sum_{s=1}^N h(s) - \sum_{i=1}^N \varepsilon_i)\Psi = 0$. However, if you do not find this obvious, please derive it.

4. Let $\psi = | \varphi_1, \varphi_2, \ldots, \varphi_N |$. Show that in general $| \varphi_1, \varphi_2, \ldots, \varphi_N' |$ is a linear combination of ψ and a one-electron excitation of ψ. Similarly, show that in general $| \varphi_1, \varphi_2, \ldots, \varphi_{N-1}', \varphi_N' |$ is a linear combination of ψ and one- and two-electron excitations of ψ.

References

1. L. Brillouin, *Actual. Sci. Ind.* No. **159** (1934).
2. R. K. Nesbet, *Rev. Mod. Phys.* **33**, 28 (1961).
3. R. LeFebvre, *in* "Modern Quantum Chemistry" (O. Sinanoğlu, ed.), Vol. 1. Academic Press, New York, 1965.
4. I. Mayer, *Chem. Phys. Lett.* **11**, 397 (1971), and references therein.
5. F. Grein and T. C. Chang, *Chem. Phys. Lett.* **12**, 44 (1971).
6. S. T. Epstein, *J. Chem. Phys.* **41**, 1045 (1964).
7. S. T. Manson, *Phys. Rev.* **145**, 35 (1966).
8. H. J. Silverstone and M.-L. Yin, *J. Chem. Phys.* **49**, 2026 (1968).
9. D. McWilliams and S. Huzinga, *J. Chem. Phys.* **55**, 2604 (1971).
10. K.-C. Pan and H. F. King, *J. Chem. Phys.* **56**, 4667 (1972).
11. S. Lunell, *Chem. Phys. Lett.* **15**, 27 (1972).
12. J. Bauche and M. Klapisch, *Proc. Phys. Soc. London* (*At. Mol. Phys.*) **5**, 29 (1972).
13. J. J. Labarthe, *Proc. Phys. Soc. London* (*At. Mol. Phys.*) **5**, L181 (1972).
14. B. Levy and G. Berthier, *Int. J. Quantum Chem.* **2**, 307 (1968).
15. H. P. Kelly, *Phys. Rev.* **131**, 684 (1963).

16. C. Møller and M. S. Plesset, *Phys. Rev.* **46**, 618 (1934); M. Cohen and A. Dalgarno *Proc. Phys. Soc. London* **77**, 748 (1961); G. G. Hall, *Phil. Mag.* **6**, 249 (1961).
17. R. McWeeny and Y. Mizuno, *Proc. Roy. Soc. Ser. A* **259**, 554 (1961).
18. C. N. Yang, *Rev. Mod. Phys.* **34**, 694 (1962).
19. E. R. Davidson, *Advan. Quantum Chem.* **6** (1972).
20. E. R. Davidson, *Rev. Mod. Phys.* **44**, 451 (1972).
21. S. Fraga and K. M. S. Saxena, *At. Data* **4**, 269 (1972), and references therein.
22. J. Geratt and I. M. Mills, *J. Chem. Phys.* **49**, 1719 (1968).
23. C. F. Bender and E. R. Davidson, *J. Chem. Phys.* **49**, 4222 (1968); *Phys. Rev.* **183**, 23 (1969); S. Green, *J. Chem. Phys.* **54**, 827 (1971).
24. M. Cohen and A. Dalgarno, *Proc. Roy. Soc. Ser. A* **275**, 492 (1963).
25. W. H. Adams, *J. Chem. Phys.* **45**, 3422 (1966).

Chapter IV / Special Theorems Satisfied by Optimal Trial Functions

11. Introduction

Atomic and molecular energy eigenfunctions satisfy various physically interesting and useful special theorems—hypervirial theorems, generalized Hellmann–Feynman theorems, etc. Also, they may be simultaneously eigenfunctions of other operators K which commute with the Hamiltonian H. In the next sections, we will discuss conditions under which one can be sure a priori that optimal trial functions will have analogous properties. (In Appendix C, we summarize the similar results for the time-dependent problems.) These theorems, when applicable, can then provide physical insight into the nature of the $\hat{\psi}$ and $\hat{E}$, and the degree to which they approximate the behavior of actual eigenfunctions and eigenvalues. Also, if within a given set of trial functions one cannot determine $\hat{\psi}$ and $\hat{E}$ exactly, the accuracy to which the applicable theorems are satisfied by the approximate $\hat{\psi}$ and $\hat{E}$ can give one an indication of how accurate the approximation is, for example, how accurately a USCF calculation approximates UHF.[1] [2] This last comment also raises the obviously very difficult question, which we will not attempt to discuss, that if the conditions we give are in some sense almost satisfied, then how closely can one expect the theorems to be satisfied?

One approach to ensuring that the $\hat{\psi}$ are eigenfunctions of K is of course simply to choose the trial functions so that each is already of the desired type. Analogously, in the last few years, there has developed a considerable literature in which, usually through the use of Lagrange multiplier techniques, the $\hat{\psi}$ are constrained to have various properties and to satisfy various theorems.[3] However, in what follows, we will be interested in more

[1] For use of tests of this type, see, for example, Matcha [1].

[2] For more detailed a posteriori comparisons, see, for example, Bagus *et al.* [2].

[3] See, for example, Loeb and Raisel [3].

general possibilities in which the theorems are, so to speak, satisfied naturally.

In all cases, we will give only sufficient conditions, and it seems that one can hardly do better than this in any useful way because any set of trial functions might contain, as one unique member, an eigenfunction which would then have all the desired properties. Nevertheless, it is empirically an excellent rule of thumb that if the sufficient conditions we give are not met, then unless some usually obvious symmetry or reality conditions intervene, it is very unlikely that the $\hat{\psi}$ will have the desired properties.

Incidentally, these conditions will always be stated in the form that the set of trial functions should be invariant to some transformation or other. However, it is clear, even before entering into the details, that if the trial functions have fixed overall scales, then it must be sufficient to have invariance only up to overall scale changes. That is, under the transformation, one member of the set goes into another, possibly multiplied by a constant. The point is that, as we have discussed in Section 4, the overall scale of trial functions is irrelevant to the final results. Indeed, "invariance up to overall scale changes," if true, could be replaced by "invariance" simply by conceptually enlarging the set of trial functions as described in Section 4 with no change in the final results. In any case, having made these points here, we will not explicitly mention them again. Rather, as implied at the outset, we will simply speak of invariance without further qualification.

In our discussions, we will assume implicitly that the reader is familiar with the relevant theorems for exact eigenfunctions. However, this is not strictly necessary since our sufficient conditions will always be met by Hilbert space (the set of trial functions for the variation principle) and hence our derivations apply to exact eigenfunctions as a special case. In this connection, it should be very much kept in mind that to infer results here and elsewhere for eigenfunctions and eigenvalues ψ' and E', *it is always sufficient to think that one uses as the set of trial functions the linear set* $\psi = A$, with A an *arbitrary* (or arbitrary to within certain symmetry requirements) function of all the variables since evidently in this case the variation method will, as desired, yield $\hat{\psi}$ $(= \hat{A}) = \psi'$ and $\hat{E} = E'$.

12. Reality

If, in the representation in which one is working, H is explicitly real, then if an eigenvalue is nondegenerate, the corresponding eigenfunction will automatically be essentially real, i.e., $\psi'^* = \beta\psi'$, where β is some constant, and hence can be made real by a suitable choice of phase. On the other hand, if an eigenvalue is degenerate, then the eigenfunctions are not automatically essentially real but one can find a set which is, an arbitrary degenerate eigenfunction then being some linear combination of the real ones. We will now show that if H is real, and if the set of trial functions is invariant to complex conjugation, then if $\hat{E}$ is nondegenerate, $\hat{\psi}$ will be essentially real.

We first note that since E is anyway real, then if H is real,

$$E = \frac{(\psi, H\psi)}{(\psi, \psi)} = \frac{(\psi, H\psi)^*}{(\psi, \psi)^*} = \frac{(\psi^*, H\psi^*)}{(\psi^*, \psi^*)} \tag{1}$$

that is, ψ and ψ^* yield the same energy. The other fact we need, and we will prove it in a moment, is that if the set of trial functions is invariant to complex conjugation, then the $\hat{\psi}^*$ are also the optimal trial functions. Therefore if $\hat{E}$ is nondegenerate, it follows that $\hat{\psi}$ and $\hat{\psi}^*$ must be proportional to one another, as we wanted to prove. UHF, SUHF, the linear variation method if the ϕ_k^* are also in the space, and SCF theories under similar conditions are examples involving sets of trial functions which are invariant to complex conjugation.

Returning to the proof that the $\hat{\psi}^*$ are optimal functions, since the set is invariant to complex conjugation, we must have for some A'

$$\psi(\hat{A})^* = \psi(A') \tag{2}$$

and also that

$$\psi(\hat{A} + \delta A)^* = \psi(A' + \delta' A) \tag{3}$$

for some $\delta' A$. Therefore from (1) and (3) it follows that as we change the variational parameters away from A' by an arbitrary amount $\delta' A$, E will run through the same sequence of values as when we change the values of the parameters away from $\hat{A}$ by the corresponding amounts δA, and hence, since $E(\hat{A})$ is a stationary point, $E(A')$ must also be, which proves the point.

If $\hat{E}$ is degenerate, there does not seem to be a simple general theorem about the reality of the $\hat{\psi}$ unless the set of trial functions forms a linear space. Then if the space is invariant to complex conjugation $\bar{H}$ will be real, whence the situation will be identical to that described at the outset of this chapter for exact eigenfunctions of H.

However, there is an obvious general converse theorem: If H is real and if $\hat{\psi}$ is not essentially real, then there surely is some degeneracy since $\hat{\psi}^*$ will yield the same energy. In such a case, and indeed whether or not $\hat{\psi}^*$ is in the set, it is interesting to point out that one can produce from $\hat{\psi}$ and $\hat{\psi}^*$

optimal real functions by doing a further linear variation calculation with $\hat{\psi}$ and $\hat{\psi}^*$ as the basis set, where we use the word optimal because, as we said, they are derived from a variation calculation. The point is simply that this linear space is obviously invariant to complex conjugation and therefore, from the discussion in the preceding paragraph, it follows that the resultant orthogonal combinations of $\hat{\psi}$ and $\hat{\psi}^*$ will be essentially real. Further, as a bonus, at least one of these combinations must have an energy less than (or at least not greater than) the $\hat{E}$ from which one started, since in doing the further calculation, we have in effect enlarged the set of trial functions.

A simpler way of producing real functions from $\hat{\psi}$ and $\hat{\psi}^*$ is of course just to take the real and imaginary parts of $\hat{\psi}$, that is, $\chi_1 = \frac{1}{2}(\hat{\psi} + \hat{\psi}^*)$ or $\chi_2 = \frac{1}{2}(i(\hat{\psi} - \hat{\psi}^*))$, and indeed it is also true that one of these functions will yield an energy which is less than (or at least not greater than) $\hat{E}$ [4]. To see this, we note that we can write $\hat{E}$ as follows:

$$\hat{E} = \frac{(\hat{\psi}, H\hat{\psi})}{(\hat{\psi}, \hat{\psi})} = \frac{(\chi_1 - i\chi_2, H(\chi_1 - i\chi_2))}{(\chi_1 - i\chi_2, \chi_1 - i\chi_2)} = \frac{(\chi_1, H\chi_1) + (\chi_2, H\chi_2)}{(\chi_1, \chi_1) + (\chi_2, \chi_2)} \tag{4}$$

where we have used the facts that since χ_1 and χ_2 are real

$$(\chi_1, \chi_2) = (\chi_2, \chi_1) \tag{5}$$

and that since in addition H is also real (and Hermitian)

$$(\chi_1, H\chi_2) = (\chi_2, H\chi_1) \tag{6}$$

Thus, as claimed, $\hat{E}$ is certainly not less than the smaller of $(\chi_1, H\chi_1)/(\chi_1, \chi_1)$ and $(\chi_2, H\chi_2)/(\chi_2, \chi_2)$. However, in general neither χ_1 nor χ_2 will be an optimal function in the sense of the preceding paragraph, since there is no reason to expect that χ_1 and χ_2 are orthogonal or that they diagonalize H. Indeed, physically they are really not well defined since if we simply replace $\hat{\psi}$ by the physically equivalent $e^{i\alpha}\hat{\psi}$, where α is a real constant, χ_1 and χ_2 change drastically.

PROBLEMS

1. Suppose more generally that $H^* = U^+HU$, where U is a unitary operator (see, for example, the discussion of time reversal invariance by Messiah [5]). If E' is nondegenerate, what then is the relation between ψ'^* and ψ'? If $\hat{E}$ is nondegenerate, under what conditions will a similar relation hold between $\hat{\psi}$ and $\hat{\psi}^*$?

2. To illustrate the results of this section, consider the following matrix example. Let

$$A = \begin{pmatrix} 1 & 0 & 0 \\ 0 & 2 & 0 \\ 0 & 0 & 3 \end{pmatrix}$$

and do a linear variation calculation using as trial basis vectors

$$X_{(1)} = \begin{pmatrix} 1 \\ i \\ 0 \end{pmatrix} \quad \text{and} \quad X_{(2)} = \begin{pmatrix} 1 \\ 0 \\ 1 \end{pmatrix}$$

Show that the space of trial vectors is not invariant to complex conjugation. Show that the resultant $\hat{\lambda}$ are nondegenerate but that the $\hat{X}$ are complex. Then do a similar calculation using

$$X_{(1)} = \begin{pmatrix} 1 \\ i \\ 0 \end{pmatrix} \quad \text{and} \quad X_{(2)} = \begin{pmatrix} 1 \\ -i \\ 0 \end{pmatrix}$$

which therefore yield a space which is invariant to complex conjugation.

3. Problem 9 of Section 13 also has to do with behavior under complex conjugation.

13. Unitary Invariance

Let U be a unitary operator. Then H^U defined by

$$H^U \equiv U^+ H U \tag{1}$$

has the same eigenvalues as H. That is, generalizing our usual notation,

$$E^{U\prime} = E' \tag{2}$$

Also the corresponding eigenfunctions are related by

$$U\psi^{U\prime} = \psi' \tag{3}$$

We will now show that if a set of trial functions is invariant to the transformation U, and if one uses that same set of trial functions for both H and H^U, then one will find that

$$\widehat{E^U} = \hat{E} \tag{4}$$

and that

$$U\widehat{\psi^U} = \hat{\psi} \tag{5}$$

The proof is very simple and follows somewhat the same pattern as that in the previous section. To find the $\hat{E}$, we look for the stationary points of

$$E = (\psi, H\psi)/(\psi, \psi) \tag{6}$$

as ψ ranges through the set. Similarly, to find the E^U, we look for the stationary points of

$$E^U = (\psi, H^U\psi)/(\psi, \psi) \tag{7}$$

which, since U is unitary, we can write as

$$E^U = (U\psi, HU\psi)/(U\psi, U\psi) \tag{8}$$

Now to say that the set of trial functions is invariant to U means that the set of functions $U\psi$ is the same as the set ψ. Therefore comparing (6) and (8), we see that in the searches, E and E^U go through the same set of values, with $U\psi$ giving the same result in the E search as ψ does in the E^U search. Hence in particular we have (4) and (5).

The invariance of energies to unitary transformations of the Hamiltonian as expressed by (2), and by (4) if applicable, becomes less abstract and therefore more interesting if the transformation U of the dynamical variables is equivalent to a transformation in which one leaves the dynamical variables alone and instead makes appropriate changes in some parameters Γ in the Hamiltonian; thus

$$U^+H(\Gamma)U = H(\Gamma') \tag{9}$$

where Γ' is some function of Γ,

$$\Gamma' = G(\Gamma) \tag{10}$$

In such a case, if the eigenvalues of $H(\Gamma)$ are denoted by $E'(\Gamma)$ then those of $H(\Gamma')$ are obviously $E'(\Gamma')$, that is,

$$E^{U\prime} = E'(\Gamma') \tag{11}$$

so that (2) becomes

$$E'(\Gamma') = E'(\Gamma) \tag{12}$$

which tells us that the functional form of E' is invariant to the transformation (10) (actually this conclusion, though a possible one, is not a necessary one; however, we leave the discussion of some of the subtleties to the problems).

We will now derive analogous results in the variational case, under the condition that the set of trial functions is invariant to U and to the transformation (10). Allowing for the possibility that the individual trial functions may depend on the parameters Γ, we write (6) and (7) in more detail as

$$E = (\psi(\Gamma), H(\Gamma)\psi(\Gamma))/(\psi(\Gamma), \psi(\Gamma)) \tag{13}$$

and

$$E^U = (\psi(\Gamma), H(\Gamma')\psi(\Gamma))/(\psi(\Gamma), \psi(\Gamma))$$

If the set is invariant to the transformation (10), then it follows that we may equally well write the last formula as

$$E^U = (\psi(\Gamma'), H(\Gamma')\psi(\Gamma'))/(\psi(\Gamma'), \psi(\Gamma')) \tag{14}$$

since the right hand sides go through the same set of values as one ranges through the set of trial functions. Therefore comparing (14) and (13), it follows that if we write

$$\hat{E} = \hat{E}(\Gamma) \tag{15}$$

then we will have

$$\widehat{E^U} = \hat{E}(\Gamma') \tag{16}$$

Having derived (16), it follows from (3) that if the set is also invariant to U, then, in complete analogy to (11),

$$\hat{E}(\Gamma') = \hat{E}(\Gamma) \tag{17}$$

We will now discuss some particular unitary transformations within the framework of our standard example of the clamped nuclei Hamiltonian (1-1). For example, in the absence of external fields, the unitary transformations of rigid translations, rotations, and inversions of the electrons are equivalent to leaving the electrons alone and inverting, rotating, and translating the nuclei the other way. Therefore (11), and (17) if applicable, tell us what is physically obvious, namely that the energy can depend on the nuclear coordinates $\mathbf{R}_a$ only in a way which is invariant to such inversion, rotation, and translation. That is, the energy can be a function only of the "bond lengths"

$$R_{(ab)} = |\mathbf{R}_a - \mathbf{R}_b| \tag{18}$$

and the cosines of the "bond angles"

$$\cos\theta_{(ab,cd)} = \frac{(\mathbf{R}_a - \mathbf{R}_b)\cdot(\mathbf{R}_c - \mathbf{R}_d)}{|\mathbf{R}_a - \mathbf{R}_b|\,|\mathbf{R}_c - \mathbf{R}_d|} \tag{19}$$

In particular, these considerations apply to both UHF and SUHF. First of all, the set of trial functions is obviously the same whatever the Hamiltonian and whatever the parameter values. Further, since all these transformations act on each electron separately, the U's have the form

$$U = \exp\left[i \sum_{s=1}^{N} \omega(s)\right] = \prod_{s=1}^{N} \exp(i\omega(s)) \tag{20}$$

Thus for translations

$$U = \exp\left(i \sum_{s=1}^{N} \mathbf{d} \cdot \mathbf{p}_s\right)$$

where $\mathbf{d}$ is the displacement in question, while for rotations about the coordinate origin

$$U = \exp\left[i \sum_{s=1}^{N} \mathbf{r}_s \times \mathbf{p}_s \cdot \mathbf{n}\right]$$

where $\mathbf{n}$ is a vector in the direction of the axis of rotation. Therefore it follows that

$$U \mid \varphi_1, \varphi_2, \ldots, \varphi_N \mid = \mid e^{i\omega}\varphi_1, e^{i\omega}\varphi_2, \ldots, e^{i\omega}\varphi_N \mid$$

That is, U sends determinants into determinants and so the set of all Slater determinants (UHF) is invariant to all such U's. Further, SUHF will also be invariant if ω is spin independent, which is the case in the examples which we have discussed.

Another example of (9) is provided by gauge transformations in the presence (or indeed also in the absence) of a magnetic field. Namely the unitary transformation

$$U = \exp\left[(i/c) \sum_{s=1}^{N} \lambda(\mathbf{r}_s)\right] \tag{21}$$

with λ a real function, is equivalent to the gauge transformation

$$\mathscr{A}(\mathbf{r}) \rightarrow \mathscr{A}(\mathbf{r}) + \nabla\lambda(\mathbf{r})$$

so that the vector potential $\mathscr{A}$ is the Γ of this example. Thus in this case, (11), and (17) if applicable, states that energies are gauge invariant. In particular, since (21) is again of the form (20) with a spinless ω, we also have the result that the $\hat{E}$ of UHF and SUHF are invariant to all gauge transformations.

We have been talking about quite general gauge transformations of quite general magnetic fields. However, discussions in the literature are usually confined to a limited class of gauge transformations. Namely the $\mathscr{A}$ which is usually used to represent a uniform magnetic field $\mathscr{B}^0$ (which may be only part of the total field) is

$$\mathscr{A} = \tfrac{1}{2}\mathscr{B}^0 \times \mathbf{r} \tag{22}$$

and the class of λ's considered is

$$\lambda = \tfrac{1}{2}\mathscr{B}^0 \times \mathbf{d}\cdot\mathbf{r} \tag{23}$$

where $\mathbf{d}$ is some vector which is independent of $\mathbf{r}$. The corresponding transformation of $\mathscr{A}$ is then

$$\mathscr{A} \to \tfrac{1}{2}\mathscr{B}^0 \times (\mathbf{r} + \mathbf{d})$$

that is, a change in "gauge origin."

We have based our discussion on the use of a single set of trial functions throughout. However, one may well argue that that is in general an unnatural requirement in that since H and H^U are different Hamiltonians, it is more natural to use different sets of trial functions for each. After all, according to (3), the exact eigenfunctions are different, therefore why not the set of trial functions? Further, as we will see, one may in this way easily ensure the invariance (17).

To illustrate the situation, we will consider the completely trivial example of an isolated one-electron, one-dimensional "atom" with Γ the coordinate X of the nucleus. Then since a translation of the electron is equivalent to a translation of the nucleus the other way, (11) says that E' is independent of X. Then the point is that it is very natural to let the trial functions depend on X according to

$$\psi(x - X, A) \tag{24}$$

where x is the electron coordinate and where the A are some variational parameters. Further, it is obvious that with such functions, and whether or not the set is invariant, either to changes in X or to the action of U, each E will be independent of X and therefore the $\hat{E}$ will certainly be independent of X in accord with (17).

The key point about (24), of course, is the fact that the combination $x - X$ is invariant to simultaneous translation of x and X and hence is a natural combination to use. More generally, if one uses functions $\psi(\Gamma)$

which depend on Γ in such a way that the *set* of $\psi(\Gamma)$ and the *set* of $\psi(\Gamma')$ are related by

$$\psi(\Gamma') = U^{+}\psi(\Gamma) \tag{25}$$

that is, in just the way exact eigenfunctions are related, then whether or not the set of $\psi(\Gamma')$ is the same as the set of $\psi(\Gamma)$, and whether or not either set individually is invariant to the transformation U, still (14) will be satisfied.

To see this, we first note that in general

$$E(\Gamma) = \frac{(\psi(\Gamma), H(\Gamma)\psi(\Gamma))}{(\psi(\Gamma), \psi(\Gamma))} = \frac{(U^{+}\psi(\Gamma), H(\Gamma')U^{+}\psi(\Gamma))}{(U^{+}\psi(\Gamma), U^{+}\psi(\Gamma))} \tag{26}$$

Therefore, if (25) holds we can write

$$E(\Gamma) = \frac{(\psi(\Gamma'), H(\Gamma')\psi(\Gamma'))}{(\psi(\Gamma')\psi(\Gamma'))} \tag{27}$$

in that the right hand side goes through the same set of values as the right hand side of (26), whence (14) follows.

However, in assessing the significance of this invariance, the following considerations should be kept in mind. First, we point out that in a general way, (25) is really only a rule for relating different sets of trial functions in order to ensure invariance. (One has invariance by constraint.) Thus, returning to our simple example, it says that if, with the nuclear coordinate equal to, say, X_0 Bohr units, one used the set of trial functions

$$\phi(x, A), \tag{28}$$

then when the nuclear coordinate is equal to X Bohr units, one should use the set

$$\phi(x + X_0 - X, A) \tag{29}$$

However, this can be done for any set ϕ, adequate or not, and hence the resulting invariance, in and of itself, lends no special credence to the calculation at any one value of the nuclear coordinate. Rather, one's expectations must be based on the detailed nature of the set ϕ.

In contrast, for *the* set of trial functions to be invariant to a transformation U is nontrivial, and provides some evidence concerning the mathematical and physical completeness of the set. In this connection, it is interesting to consider the following question: Given a set $\phi(x, A)$, to what value of the nuclear coordinate is it best adapted? One way to answer this question is to let the variation method decide by enlarging the set of trial functions at arbitrary X by letting X_0 in (29) be a variational parameter. We now note that if one does this, then the enlarged set of trial functions in fact meets both of our invariance criteria. Namely the set is invariant separately to translations of X and to translations of x since either is equivalent to a change in X_0 and hence sends one trial function into another.

In Section 16, we will show that if (at a given value of Γ) the set of trial functions is invariant to continuous families of unitary transformations, for example, to continuous families of translations, or rotations, or gauge transformations, then the $\hat{\psi}$ have other physically desirable properties in that they satisfy certain hypervirial theorems. In contrast, if the set is not invariant, then probably (as usual the conditions are only sufficient) the hypervirial theorems will not be satisfied. Now, if one uses sets with the property (25), and if the set $\psi(\Gamma')$ is different from the set of $\psi(\Gamma)$, so that one has invariance by constraint, then obviously neither set can be invariant to U (incidentally note that invariance to U also implies invariance to U^+, to U^2, etc.) since U just sends one set into the other. Therefore when one has invariance by constraint to a continuous family of unitary transformations, one does not expect the corresponding hypervirial theorem to be satisfied, and, conversely, if it is not satisfied and one nevertheless has invariance, then it certainly is invariance by constraint.

The use of different sets of trial functions for H and H^U is most often implicit rather than explicit. An explicit example is the use of "gauge-invariant atomic orbitals"[4] when a uniform magnetic field is present. More descriptively, these should perhaps be called gauge-variant atomic orbitals since they change under the gauge transformation (23) in just such a way that (25) is satisfied. However, the associated hypervirial theorems (see Section 16) are not in general satisfied [7]. The outstanding example of implicit use involves the geometric properties of $\hat{E}$, especially for diatomic molecules. For example, for an isolated molecule, having chosen a coordinate system and having calculated $\hat{E}$ for some configuration of the nuclei, one will, simply and quite naturally, implicitly assume that $\hat{E}$ will be the same for all other configurations which differ from the original one by a rotation, or translation (or inversion) of the nuclei. However, the point is that if pressed to produce, using the same coordinate system, the same $\hat{E}$ for these other configurations, one would certainly be required to use different sets of trial functions for the different configurations if, as is

[4] Ditchfield [6] gives an extensive bibliography.

often the case, the associated hypervirial theorems are not satisfied in the original calculation. Indeed, this use of different sets is quite explicit when in USCF-type calculations the basis spin orbitals are tied to the nuclear frame[5] as in our trivial atomic example.

PROBLEMS

1. Verify that the U's given following (20) do produce translation and rotation, respectively. Show that the U of (21) does produce a gauge transformation.

2. Show that the transformation from coordinates space to momentum space is a unitary transformation. Show that it sends a Slater determinant into a Slater determinant. Show that it sends a spherical harmonic into a spherical harmonic. Draw conclusions. (For Hartree–Fock theories in momentum space, see Epstein and Lipscomb [9].)

3. This and the next four problems have reference to the comment following Eq. (12). A close reading will show that what we actually proved in the text was that under the given conditions, the *set* of $\hat{E}(\Gamma')$ will be the same as the set of $\hat{E}(\Gamma)$. From this, we concluded that individually $\hat{E}(\Gamma') = \hat{E}(\Gamma)$. However this conclusion, though a possible one, is obviously not a necessary one since given any two sets, one can derive many rules for linking the elements. Thus show that we implicitly, and very naturally, assumed that the formula $\hat{E}(\Gamma)$ should be a continuous function of Γ.

4. The requirement of continuity is evidently sufficient to imply $\hat{E}(\Gamma) = \hat{E}(\Gamma')$ only if there is no crossing of the levels as the parameters vary between Γ and Γ'. However, if there is crossing, then there is no algebraically unique way to link the levels on one side of the crossing with those on the other, and although an "invariant linkage" is algebraically possible, it may not be correct. Thus consider an atom in a uniform electric field $\mathscr{E}$. With the coordinate origin at the nucleus, show that the set of levels should be invariant to rotation and inversion of $\mathscr{E}$. Show that, through first order in $\mathscr{E}$, an invariant formula must therefore have the form

$$E' = E'^{(0)} + a\,|\,\mathscr{E}\,| + \cdots$$

[5] See, for example, Csizmadia *et al.* [8].

where $E'^{(0)}$ and a are independent of $\mathscr{E}$ but may vary from state to state. However, writing

$$\mathscr{E} = \varepsilon \mathbf{k}$$

where $\mathbf{k}$ is a fixed unit vector, we have with this formula that $\partial E'/\partial \varepsilon$ is discontinuous at $\varepsilon = 0$, which seems quite unphysical and therefore is presumably not correct. Show that another possible (noninvariant) formula is

$$E' = E'^{(0)} + b\varepsilon + \cdots$$

provided that if $b \neq 0$, $E'^{(0)}$ is degenerate and the nonzero b values come in pairs of equal magnitudes but opposite signs.

5. Illustrate the general results of Problem 4 by discussing in detail the first-order Stark effect for the $n = 2$ state of hydrogen atom, showing that both formulas are possible algebraically. However, show that the noninvariant formula is correct in that it yields the same connection across $\mathscr{E} = 0$ as is made by inverting the electron coordinates. That is, with ε positive, show that $U\psi'_{E'^{(0)}+b\varepsilon+\cdots} = \psi'_{E'^{(0)}-b\varepsilon+\cdots}$, where U is the inversion operator, or alternatively that $U\psi'_{E'^{(0)}+a|\mathscr{E}|+\cdots} = \psi_{E'^{(0)}-a|\mathscr{E}|+\cdots}$.

6. If we assume that the invariant formulas cannot be correct, then a corollary of the considerations of Problem 4 is evidently the well-known result that an atom in a nondegenerate state cannot exhibit a first-order Stark effect. Show that the basic assumptions and facts from which we have derived the result, though at first sight they may seem somewhat different, are in fact identical to those of the usual derivation (first-order perturbation theory plus the fact that for $\mathscr{E} = 0$ a nondegenerate state must have a definite parity under inversion through the nucleus).

7. The perturbation in the previous three problems was a rather well-behaved one. However, if the perturbation is very singular, then the situation can change again. Thus consider the one-dimensional diatomic delta-function molecule [10]. Thus

$$V = -\mathscr{X}_1\,\delta(x - X_1) - \mathscr{X}_2\,\delta(x - X_2)$$

Argue that the energy should depend on X_1 and X_2 only through $|X_1 - X_2|$ and that therefore, since the ground state is nondegenerate for all finite $|X_1 - X_2|$, its energy, when expanded in powers of $|X_1 - X_2|$, should contain only even powers. Then find the exact solution of this problem and show that for the ground state, E' can formally be expanded in a series in $| X_1 - X_2 |$ which involves *all* powers of $| X_1 - X_2 |$. What was the flaw in the earlier argument? [Hint: Change variable from x to $y = x - X_1$ and try to expand V in powers of $X_1 - X_2$.]

8. To put the results of Problem 7 in perhaps even stronger focus, consider the homonuclear case $\mathscr{Z}_1 = \mathscr{Z}_2 = \mathscr{Z}$ and put the coordinate origin midway between the nuclei. Then

$$V = -\mathscr{Z}[\delta(x - \tfrac{1}{2}(X_1 - X_2)) + \delta(x + \tfrac{1}{2}(X_1 + X_2))]$$

which is an even function of $X_1 - X_2$. "Therefore" a perturbation expansion should involve only even powers of $X_1 - X_2$. Is this true? If not, what is the flaw in the argument? [Hint: try expanding V in powers of $X_1 - X_2$.] For related results for H_2^+, where also $\ln| R_1 - R_2 |$ appears, see Byers Brown and Steiner [11].

9. Generalize the discussion in the text to cases in which $H^U = H(\Gamma')$ + const. Show that the Hamiltonian (1-1) for a nonneutral molecule in a uniform electric field has this property under translations of the electrons. Show further that inverting or rotating the electrons is equivalent to leaving the electrons alone and inverting or rotating the nuclei and $\mathscr{E}$. From your results, show, for a diatomic molecule in a uniform electric field, that if the set of trial functions has the requisite invariance properties, and in the absence of any degeneracy which is broken by the field, that, through second order in the field

$$\hat{E} = \hat{E}^{(0)} - \gamma \frac{\mathbf{R} \cdot \mathscr{E}}{|\mathbf{R}|} - q\mathbf{R}_1 \cdot \mathscr{E} + \alpha \mathscr{E} \cdot \mathscr{E} + \beta \frac{\mathscr{E} \cdot \mathbf{R} \, \mathscr{E} \cdot \mathbf{R}}{|\mathbf{R}|^2} + \cdots$$

where $\mathbf{R}$ is the internuclear separation, q is the net charge on the molecule and where $E^{(0)}$, γ, and β are numbers which depend only on $|\mathbf{R}|$.

10. Consider the Hamiltonian (1-1), now in the presence of a uniform magnetic field, and use the gauge

$$\mathscr{A} = \tfrac{1}{2}(\mathscr{B} \times \mathbf{r})$$

Show that rotating the electrons is equivalent to rotating the nuclei and $\mathscr{B}$ the other way. Show that inverting the electrons is equivalent to inverting the nuclei, leaving $\mathscr{B}$ alone ($\mathscr{B}$ is a pseudovector). Show that to in effect translate the nuclei, we must both translate the electrons *and* carry out a suitable gauge transformation (see, for example, McWeeny [12]).

11. Continuing on, show that $H^*(\mathscr{B}) = H(-\mathscr{B})$. Show in general that if $H^*(\Gamma) = H(\Gamma')$ and if the set of trial functions is invariant to complex conjugation and to the transformation $\Gamma \to \Gamma'$, then the set of $\hat{E}$ will be invariant to $\Gamma \to \Gamma'$.

14. Symmetry

If H commutes with a Hermitian or unitary operator K and if an eigenvalue of H is nondegenerate, then the corresponding eigenfunction will automatically be an eigenfunction of K. If the eigenvalue is degenerate, then, although the eigenfunctions are not automatically also eigenfunctions of K, one can find a set of simultaneous eigenfunctions of H and K, an arbitrary degenerate eigenfunction then being some linear combination of them.

If the set of trial functions forms a linear space which is invariant to K, then, as we have already discussed in Section 6, the $\hat{\psi}_k$ have the same properties as eigenfunctions. On the other hand, in nonlinear cases, and if there is degeneracy, bizarre things can happen, as we will shortly see. First, however, we will consider the nondegenerate case, and prove in general that if the set of trial functions is invariant to a unitary transformation U and if U commutes with H, then if $\hat{E}$ is nondegenerate, $\hat{\psi}$ will automatically be an eigenfunction of U.

If, therefore, K is unitary as well as Hermitian, it then follows that if the set of trial functions is invariant to K then, analogously to the situation for exact eigenfunctions, a nondegenerate $\hat{\psi}$ will automatically be an eigenfunction of K. If K is only Hermitian but not unitary, for example, a linear or angular momentum operator, then what is required is that the set be invariant to all

$$U(\alpha) \equiv e^{i\alpha K}$$

where α is an arbitrary real number, since if $\hat{\psi}$ is an eigenfunction of the $U(\alpha)$ for all α, it is then obviously also an eigenfunction of K.

We first note that since H commutes with U and since $U^+U = 1$, we can write

$$\hat{E} = \frac{(\hat{\psi}, H\hat{\psi})}{(\hat{\psi}, \hat{\psi})} = \frac{(\hat{\psi}, U^+HU\hat{\psi})}{(\hat{\psi}, U^+U\hat{\psi})} = \frac{(U\hat{\psi}, HU\hat{\psi})}{(U\hat{\psi}, U\hat{\psi})} \tag{1}$$

that is, $\hat{\psi}$ and $U\hat{\psi}$ yield the same energy. Also, we need the fact that $U\hat{\psi}$ is an optimal trial function, the proof being essentially the same as that of the analogous theorem in Section 12. From these facts, we can then conclude that if $\hat{E}$ is nondegenerate, then $\hat{\psi}$ and $U\hat{\psi}$ must be proportional to one another, which is what we wanted to prove.

We now turn to the degenerate case. For an exact eigenvalue, degeneracy can usually be immediately traced to the existence of K's which commute with H but not with each other. Further, it is usually obvious how to remove such degeneracies in favor of one subset of commuting K's or another by applying suitable external fields as perturbations. Therefore one might expect, since one could imagine doing variational calculations with a given set of trial functions in various external fields and then letting the fields tend to zero, that among the degenerate $\hat{\psi}$ there will always be eigenfunctions of any mutually commuting set of K's which commute with H and whose U's leave the set of trial functions invariant.

Unhappily, as we will see by example, the situation is not really so simple and indeed this author knows of no general theorems. Nevertheless, the converse sort of theorem does seem to be true, namely that if the set is not invariant, then one does not expect to find eigenfunctions. Thus consider UHF for an isolated atom with the Hamiltonian (1-1). Then L^2 (square of the angular momentum about the nucleus) and S^2 (square of the total spin) commute with H. However, since they are two-electron operators, the set of Slater determinants is not invariant to the corresponding U's. Therefore one does not expect to find eigenfunctions of L^2 and S^2 and, as we discussed in Section 8, this is in general accord with the facts. Indeed, we can even rationalize the apparent exceptions. Thus, for example, we saw that UHF admits closed shell solutions, and these are eigenfunctions of L^2 and S^2 with eigenvalues zero. However, this can be regarded as a consequence of the fact that these $\hat{\psi}$ are nondegenerate. Namely all components of $\mathbf{L}$ and $\mathbf{S}$ commute with H and, being one-electron operators, their corresponding U's leave the set of Slater determinants invariant. Therefore any nondegenerate $\hat{\psi}$ must be a simultaneous eigenfunction of all $\mathbf{L}$ and $\mathbf{S}$, that is must be ^{1}S. Similarly, in a general way one is not surprised to find orbital S states or spin singlets since these can also be characterized as simultaneous eigenfunctions of the one-electron operators $\mathbf{L}$ and $\mathbf{S}$, respectively. Similarly, an eigenfunction of a component of $\mathbf{S}$ with eigenvalue $\frac{1}{2}N$, where N is the number of electrons, is automatically an eigenfunction of S^2 with eigenvalue $\frac{1}{2}N(\frac{1}{2}N + 1)$.

That the unexpected can happen is revealed, however, by the simple example[6] of a one-dimensional, homonuclear, diatomic, delta-function

[6] See also Katriel [13].

molecule,

$$H = \tfrac{1}{2}p^2 - \mathscr{X}[\delta(x - R) + \delta(x + R)]$$

with the set of trial functions being

$$\psi(x, A) = \exp[-(x - A)^2] \tag{2}$$

where the number A is the variational parameter. The only obvious symmetry of H is its invariance to the transformation

$$x \to -x \tag{3}$$

which we will call I. Therefore since the set of trial functions is invariant to I $[I\psi(x, A) = \psi(x, -A)]$, we would expect the $\hat{E}$ to be nondegenerate, and the $\hat{\psi}$ to be eigenfunctions of I (as is the case with the exact eigenvalues and eigenfunctions except in the limit $R = \infty$, where there is a two-fold degeneracy). Moreover, from (2) it is clear that only if A equals zero can we have an eigenfunction of I and therefore we would expect that $\hat{A} = 0$ would be the unique solution. However, in fact this is not the case. Thus it is easy to show that

$$E(A) = \alpha - \beta\{\exp[-(R - A)^2] + \exp[-(R + A)^2]\} \tag{4}$$

where α, the average kinetic energy, and $\beta = \mathscr{X}/(\psi, \psi)$ are positive numbers independent of both A and R. Then one finds that

$$\partial\hat{E}/\partial\hat{A} = 0$$

yields as the equation for $\hat{A}$

$$\hat{A}/R = \tanh 2\hat{A}R \tag{5}$$

Now clearly one solution of (5) is always the expected $\hat{A} = 0$. However, the point is, as one readily sees by graphing the two sides of (5), that if R is large enough, there are two more solutions which are negatives of one another and which lead to a doubly degenerate $\hat{E}$ different from $E(0) = \alpha - 2\beta\exp(-R^2)$. Indeed for R large, since $\tanh(\pm\infty) = \pm 1$, these solutions are approximately $\pm R$ and hence lead to a doubly degenerate $\hat{E} \simeq \alpha - \beta$, which is less than $E(0) \simeq \alpha$.

Given that this degeneracy exists, the reason that the variational calculation cannot resolve it is of course that we have not allowed linear combinations of the $\psi(x, A)$, since if we did so, we would obviously break the degeneracy and produce one symmetric function and one antisymmetric one.

More generally, in analogy to our discussion in Section 12, and regardless of whether or not the functions $U\hat{\psi}$, $U^2\hat{\psi}$, ..., $U^+\hat{\psi}$, ..., all of which yield the same energy if U commutes with H, are in the set, one can in any case produce an optimal set of functions which are eigenfunctions of U by doing a linear variation calculation in the space spanned by these functions. That this will produce eigenfunctions of U is then guaranteed by the fact that this linear space is obviously invariant to the action of U, and therefore the Π appropriate to it will commute with U. Indeed it is easy to see what the result of this calculation will be. Namely we can expand $\hat{\psi}$ in normalized eigenfunctions ζ_ω of U belonging to different eigenvalues ω:

$$\hat{\psi} = \sum_{\omega} C_\omega \zeta_\omega \tag{6}$$

where the sum may be infinite, though hopefully it is only finite. Then since $U^n\hat{\psi}$ and $U^{+n}\hat{\psi}$ are simply linear combinations of these same functions with different coefficients, it is clear that the functions ζ_ω involved in the sum span the linear space formed from $\hat{\psi}$, $U\hat{\psi}$, ... Further, since from symmetry

$$(\zeta_\omega, \zeta_{\omega'}) = \delta_{\omega,\omega'}, \qquad (\zeta_\omega, H\zeta_{\omega'}) = 0, \qquad \omega \neq \omega' \tag{7}$$

it follows from the "converse theorem" of Section 6 that the ζ_ω will be the $\hat{\psi}_k$ which would result from the linear variation calculation. Thus instead of doing the linear variation calculation, we can "simply" project out of $\hat{\psi}$ the various symmetry components which it contains [14–17]. This procedure, and approximations to it, have been extensively applied to UHF functions [18],[7] particularly to produce functions of a definite total spin S^2 (this is a case in which $U\hat{\psi}$, $U^2\hat{\psi}$, ... do not belong to the original set, since, as noted earlier, S^2 is not a one-electron operator). Of course, in general one can do even better if, as mentioned in Section 8, one were to project before carrying out the original variation calculations [14–17].

PROBLEM

1. Choose a value for $\mathscr{X}$ and plot $E(A)$ of Eq. (4) as a function of A for various R values in order to examine the nature of the stationary points.

[7] A recent reference is Claxton and McWilliams [19].

15. Generalized Hellmann–Feynman Theorems

Suppose that H contains a real parameter σ. Then by differentiating

$$(\hat{\psi}, (H - \hat{E})\hat{\psi}) = 0$$

with respect to σ, we find

$$\left(\frac{\partial \hat{\psi}}{\partial \sigma}, (H - \hat{E})\hat{\psi}\right) + \left(\hat{\psi}, (H - \hat{E})\frac{\partial \hat{\psi}}{\partial \sigma}\right) + \left(\hat{\psi}, \left(\frac{\partial H}{\partial \sigma} - \frac{\partial \hat{E}}{\partial \sigma}\right)\hat{\psi}\right) = 0 \quad (1)$$

where in carrying out the differentiation, we have of course kept the integration variables fixed. In this connection, it should be especially noted that if one changes variables, then in general $\partial H/\partial\sigma$ will change if the change of variable is σ dependent.[8] Also, we have assumed that the volume element in the integration does not depend on σ. This is not necessarily true. However, for many cases of interest in which the volume element does depend on σ, the dependence is only multiplicative and hence cancels out of (1), so we will not pursue the additional complication here [21].

We will now show that if the set of trial functions is invariant to changes in the value of σ, that is, if the set is independent of σ, then the sum of the first two terms in Eq. (1) will be separately equal to zero. Therefore under this condition we will have

$$\left(\hat{\psi}, \left(\frac{\partial H}{\partial \sigma} - \frac{\partial \hat{E}}{\partial \sigma}\right)\hat{\psi}\right) = 0$$

or

$$\frac{\partial \hat{E}}{\partial \sigma} = \left(\hat{\psi}, \frac{\partial H}{\partial \sigma}\hat{\psi}\right) \Big/ (\hat{\psi}, \hat{\psi}) \quad (2)$$

which is the variational version of the generalized Hellmann–Feynman theorem.[9] We will often write (2) as

$$\partial \hat{E}/\partial \sigma = \langle \partial H/\partial \sigma \rangle \quad (3)$$

where from now on angular brackets around an operator will mean its average value calculated using $\hat{\psi}$.

To be quite general, suppose that each member of the set may depend on σ. Then since the $\hat{A}$ will also usually depend on σ, we have

$$\hat{\psi} = \psi(\hat{A}(\sigma), \sigma)$$

[8] See, for example, Epstein [20].

[9] For some discussion and references to early history, see Epstein [22] and Musher [23].

If now we change σ, we change $\hat{\psi}$, both because we change $\hat{A}$ and because we change the explicit σ dependence. However, if the set of trial functions is invariant to changes in σ, it must be that a change in the explicit σ dependence is equivalent to a further change in the $\hat{A}$ since it must send $\hat{\psi}$ into another member of the set. Thus in such a case,

$$\psi(\hat{A}(\sigma + \delta\sigma), \sigma + \delta\sigma) = \psi(\hat{A}(\sigma) + \bar{A}, \sigma)$$

for some $\bar{A}$ which in general depends on $\hat{A}$, σ, and $\delta\sigma$, its dependence on the latter presumably having the form

$$\bar{A} = (\delta\sigma)\bar{A}^{(1)} + (\delta\sigma)^2\bar{A}^{(2)} + \cdots$$

Therefore we have

$$\frac{\partial\hat{\psi}}{\partial\sigma} = \lim_{\delta\sigma\to 0} \frac{\psi(\hat{A}(\sigma + \delta\sigma), \sigma + \delta\sigma) - \psi(\hat{A}(\sigma), \sigma)}{\partial\sigma} = \frac{\partial\hat{\psi}(\sigma)}{\partial\hat{A}(\sigma)}\bar{A}^{(1)} \tag{4}$$

Now, as we mentioned at the end of the discussion of notation in Section 4, the δA form a real linear space. Therefore since $\bar{A}$ is evidently a possible δA, so is $\bar{A}/\delta\sigma$ and therefore so is

$$\lim_{\delta\sigma\to 0} (\bar{A}/\delta\sigma) = \bar{A}^{(1)}$$

However, it then follows from (4) that $\partial\hat{\psi}/\partial\sigma$ is a possible $\delta\hat{\psi}$ and thus must satisfy

$$\left(\frac{\partial\hat{\psi}}{\partial\sigma}, (H - \hat{E})\hat{\psi}\right) + \left(\hat{\psi}, (H - \hat{E})\frac{\partial\hat{\psi}}{\partial\sigma}\right) = 0 \tag{5}$$

which is the announced result.

The theorem which we have just proven, that when the set of trial functions is invariant to changes in σ, the generalized Hellmann–Feynman theorem for σ will be satisfied, is in its essentials due to Hurley [24]. We would emphasize its simplicity and its generality since this does not seem to be widely enough appreciated, possibly because he did not originally state it for a general σ. In any case, there are in the literature, subsequent to Hurley's work, many very detailed derivations of special Hellmann–Feynman theorems (i.e., special choices of σ) for particular variational approximations; derivations which are quite unnecessary since the results are immediate consequences of Hurley's theorem. Also, there have been "proofs" that certain variation methods do not satisfy (2), whereas Hurley's theorem shows immediately that they do.

Obvious examples of sets of trial functions which are invariant to changes in interesting σ's are (i) the trial functions of most Hartree–Fock (UHF, SUHF, RHF, multiconfigurational Hartree–Fock approximations, etc.) type approximations since there are usually no a priori requirements as to how the spin orbitals should depend on possible σ's like nuclear charge, nuclear configuration, strength of external fields, etc.; (ii) the trial functions of SCF-type approximations if the basis spin orbitals are independent of σ, and this will usually be the case for most σ's except nuclear coordinates (and magnetic fields if one uses the gauge-invariant atomic orbitals mentioned in Section 13); (iii) linear spaces in which the basis set is independent of σ.

Certain σ's are of particular interest. We will meet *the* Hellmann–Feynman theorem (σ a nuclear coordinate) in Section 18. Here we will consider some examples in which σ is an external field and where H is given by (1-1).

First suppose that there is a uniform electric field $\mathscr{E}$ present. Then from Eq. (1-1) with $\Phi(\mathbf{r}) = -\mathscr{E} \cdot \mathbf{r}$

$$\partial H / \partial \mathscr{E} = -\mathscr{D} \tag{6}$$

where $\mathscr{D}$ is the dipole moment operator

$$\mathscr{D} = -\sum_{s=1}^{N} \mathbf{r}_s + \sum_{a} \mathscr{Z}_a \mathbf{R}_a \tag{7}$$

Thus if one uses a set of trial functions which is independent of $\mathscr{E}$, then one is assured of the familiar and very physical result that the negative derivative of the energy with respect to the field is the average electric dipole moment.

If E' is an exact eigenvalue, then the permanent electric dipole moment of the system can be defined as

$$-(\partial E' / \partial \mathscr{E})_{\mathscr{E}=0} \tag{8}$$

Therefore it is natural to take as an approximation to this moment the quantity (8) but with E' replaced by $\hat{E}$. However, for exact eigenfunctions an equivalent definition is of course that it is the term in the average of $\mathscr{D}$ which is independent of $\mathscr{E}$. Hence it would be equally natural to use $\langle \mathscr{D} \rangle_{\mathscr{E}=0}$, that is, $\langle \mathscr{D} \rangle$ evaluated with $\mathscr{E} = 0$, as an approximation. Now in general these two approximations will be different, with no obvious a priori way of choosing between them. However, and this is the point we would make, if one uses an $\mathscr{E}$-independent set of trial functions, then, as we have just shown, they will agree.

Similarly, the electric polarizability is defined by

$$-(\partial^2 E'/\partial \mathscr{E}^2)_{\mathscr{E}=0} \tag{9}$$

which suggests that one approximate it by replacing E' by $\hat{E}$. However, we can also write (9) as

$$\left(\frac{\partial}{\partial \mathscr{E}}\left(\frac{\partial E'}{\partial \mathscr{E}}\right)\right)_{\mathscr{E}=0} \tag{10}$$

which suggests approximating it instead by $(\partial\langle\mathscr{D}\rangle/\partial \mathscr{E})_{\mathscr{E}=0}$. Again, with an $\mathscr{E}$-independent set of trial functions, these two approximations will agree, but probably not otherwise.

If there is a uniform magnetic field $\mathscr{B}^0$ present, then, using the symmetric form of the vector potential

$$\mathscr{A}^0 = \tfrac{1}{2}\mathscr{B}^0 \times \mathbf{r} \tag{11}$$

one finds from Eq. (1-1) that

$$\partial H/\partial \mathscr{B}^0 = -\mathbf{M} \tag{12}$$

where $\mathbf{M}$ is the magnetic dipole moment operator in the presence of the total (if, as in the next paragraph, $\mathscr{B}^0$ is only part of $\mathscr{B}$) magnetic field. The discussion now proceeds as in our first example, simply replacing $\mathscr{E}$ by $\mathscr{B}^0$, electric by magnetic, and electric polarizability by magnetic susceptibility.

As our final example, let us suppose that there is a point magnetic dipole $\mathbf{k}$ present in addition to the uniform magnetic field. Then, with the usual gauge for the magnetic dipole,

$$\mathscr{A} = \tfrac{1}{2}\mathscr{B}^0 \times \mathbf{r} + (\mathbf{k} \times \mathbf{r}/r^3) \tag{13}$$

and one finds that

$$\partial H/\partial \mathbf{k} = -\mathbf{b} \tag{14}$$

where $\mathbf{b}$ is the operator corresponding to the magnetic field produced at the dipole by the electrons. Now a quantity of interest in this situation is the magnetic shielding tensor, which is defined by

$$\left(\frac{\partial^2 E'}{\partial \mathbf{k}\, \partial \mathscr{B}^0}\right)_{\mathbf{k}=0,\mathscr{B}^0=0} \tag{15}$$

If we rewrite (15) as

$$\left(\frac{\partial}{\partial \mathscr{B}^0}\left(\frac{\partial E'}{\partial \mathbf{k}}\right)_{\mathbf{k}=0}\right)_{\mathscr{B}^0=0} \tag{16}$$

or as

$$\left(\frac{\partial}{\partial \mathbf{k}}\left(\frac{\partial E'}{\partial \mathscr{B}^0}\right)_{\mathscr{B}^0=0}\right)_{\mathbf{k}=0} \tag{17}$$

then each of the three forms suggests a potentially different way of approximating the shielding tensor [25]. Thus one might simply use (15) but with E' replaced by $\hat{E}$. Alternatively, based on (16), one might calculate

$$\left(\frac{\partial\langle\mathbf{b}\rangle_{\mathbf{k}=0}}{\partial \mathscr{B}^0}\right)_{\mathscr{B}^0=0} \tag{18}$$

Finally, based on (17), one might calculate

$$\left(\frac{\partial\langle\mathbf{M}\rangle_{\mathscr{B}^0=0}}{\partial \mathbf{k}}\right)_{\mathbf{k}=0} \tag{19}$$

If the set of trial functions is independent of both $\mathscr{B}^0$ and **k**, then all these approximations will agree.[10] If the set is independent of **k** ($\mathscr{B}^0$) but not of $\mathscr{B}^0$ (**k**),[11] then only the first and second (third) are guaranteed to agree. Finally, if the set is dependent on both **k** and $\mathscr{B}^0$,[12] then all three approximations will probably be different.

PROBLEMS

1. Consider a linear harmonic oscillator of spring constant K. What is the generalized Hellmann–Feynman theorem for $\sigma = K$ in the usual Cartesian coordinates x? Now change coordinates to $y = K^{1/4}x$. What is the theorem for $\sigma = K$ in these coordinates? By combining these two theorems, derive the virial theorem.

2. Show that, in general, whatever the nature of the trial functions,

$$\frac{\partial \hat{E}}{\partial \sigma} = \left\langle \frac{\partial H}{\partial \sigma} \right\rangle + \left\{ \left(\frac{\partial' \hat{\psi}}{\partial \sigma}, (H - \hat{E})\hat{\psi} \right) + \left(\hat{\psi}, (H - \hat{E}) \frac{\partial' \hat{\psi}}{\partial \sigma} \right) \right\} \Big/ (\hat{\psi}, \hat{\psi})$$

where the prime means that one need differentiate only the explicit σ dependence of $\hat{\psi}$.

3. In the linear variation method, since the $\hat{\psi}_k$ are eigenfunctions of $\bar{H}$, we certainly have

$$\partial \hat{E}_k / \partial \sigma = \langle \partial \bar{H} / \partial \sigma \rangle$$

[10] For example, Stevens *et al.* [26].
[11] For example, Amos and Roberts [27].
[12] For example, Pople [28] and McWeeny [29].

Show that, as must be the case, this is equivalent to (3) if the linear space is independent of σ. [Hint: What then is $\partial \Pi / \partial \sigma$?]

4. Verify (12) and (14). That is, show that **M** and **b** are in fact what they are claimed to be in the text.

5. Referring to Problem 1 of Section 5, derive the consequences of the assumption that $\hat{\phi}$ and $\hat{\psi}$ are derived from sets of trial functions which are independent of σ.

16. Hypervirial Theorems: General

Let $\mathscr{G}$ be a Hermitian operator, and suppose that the set of trial functions is invariant to the continuous family of unitary transformations

$$U(\alpha) = e^{i\alpha\mathscr{G}} \tag{1}$$

parametrized by the real number α. This then implies that

$$e^{i\alpha}\psi(\hat{A}) = \psi(\hat{A} + \bar{A})$$

for some $\bar{A} = \alpha\bar{A}^{(1)} + \alpha^2\bar{A}^{(2)} + \cdots$, where the $\bar{A}^{(n)}$ in general will depend on $\hat{A}$. Differentiating with respect to α, we have

$$i\mathscr{G}\, e^{i\alpha\mathscr{G}}\psi(\hat{A}) = \frac{\partial\psi(\hat{A} + \bar{A})}{\partial(\hat{A} + \bar{A})} \frac{\partial \bar{A}}{\partial\alpha}$$

which, putting $\alpha = 0$, becomes

$$i\mathscr{G}\hat{\psi} = \frac{\partial\psi(\hat{A})}{\partial\hat{A}} \bar{A}^{(1)}$$

Now by an argument analogous to that given after Eq. (15-4), it follows that $\bar{A}^{(1)}$ is a possible δA. Therefore we see that under these conditions

$$i\mathscr{G}\hat{\psi} \tag{2}$$

must be a possible variation of $\hat{\psi}$ within the set and hence must satisfy

$$(i\mathscr{G}\hat{\psi}, (H - \hat{E})\hat{\psi}) + (\hat{\psi}, (H - \hat{E})i\mathscr{G}\hat{\psi}) = 0 \tag{3}$$

or

$$(\hat{\psi}, [H, \mathscr{G}]\hat{\psi}) = 0 \tag{4}$$

where $[H, \mathscr{G}]$ denotes the commutator of H and $\mathscr{G}$

$$[H, \mathscr{G}] \equiv H\mathscr{G} - \mathscr{G}H \tag{5}$$

Therefore if we divide by $(\hat{\psi}, \hat{\psi})$ and write (4) as

$$\langle [H, \mathscr{G}] \rangle = 0 \tag{6}$$

we have that under the given conditions, $\hat{\psi}$ will satisfy Eq. (6), the so-called [30] "hypervirial theorem" for $\mathscr{G}$.

In particular we have shown in Section 13 that UHF (SUHF) is invariant to all unitary transformations with one-electron $\mathscr{G}$'s (spin-free one-electron $\mathscr{G}$'s) and hence we have the result that the hypervirial theorems for all such $\mathscr{G}$'s will be satisfied by the $\hat{\psi}$ of UHF (SUHF). Indeed one can characterize the $\hat{\psi}$ of UHF as those single Slater determinants which satisfy (4) for all one-electron $\mathscr{G}$'s since then (2) is the most general $\delta\hat{\psi}$ in UHF. On the other hand, since UHF and SUHF are not invariant when $\mathscr{G}$ is a two-, three-, ..., electron operator, we do not expect that (4) will be satisfied for such $\mathscr{G}$'s, this barring symmetry considerations or special relationships between $\mathscr{G}$ and H. [Thus, for example, (4) is obviously satisfied by any $\mathscr{G}$ which commutes with H, and hence in particular for $\mathscr{G} = H$.] However, one can usually arrange to satisfy such theorems by imposing them as further constraints on the set of trial functions.[13]

Though we said that U should be unitary, formally the argument proceeds as well with α pure imaginary, except that (3) will now yield

$$(\hat{\psi}, (H\mathscr{G} + \mathscr{G}H)\hat{\psi}) = 2\hat{E}(\hat{\psi}, \hat{\psi}) \tag{7}$$

which, assuming that (4) is also satisfied, we can combine with the latter, to yield

$$(\hat{\psi}, H\mathscr{G}\hat{\psi}) = \hat{E}(\hat{\psi}, \mathscr{G}\hat{\psi}) = (\hat{\psi}, \mathscr{G}H\hat{\psi}) \tag{8}$$

Further, since formally UHF (SUHF) *is* invariant to all such transformations in which $\mathscr{G}$ is a one-electron operator (spin-free one-electron operator), we would conclude that (8) should be satisfied in UHF (SUHF) for all such $\mathscr{G}$'s. However, since nonunitary transformations are often not pleasant to deal with (they may transform a normalizable trial function into a nonnormalizable one), it should be noted that all we really need is that $i\mathscr{G}\hat{\psi}$ and $\mathscr{G}\hat{\psi}$ be possible variations of $\hat{\psi}$ and obviously this is the case in UHF (SUHF) so long as $\mathscr{G}$ is a one-electron (spin-free one-electron) operator and so the announced results do follow.

Turning to linear spaces, (2) and any multiple, real or complex, thereof will certainly be a possible variation of $\hat{\psi}$ if the space is invariant under the action of $\mathscr{G}$. Thus, for example, for the s limit in helium, (4), (7),

[13] See, for example, Bjorna [31].

and (8) will all be satisfied for any purely radial $\mathscr{G}$. However, finite linear spaces of trial functions are rarely invariant under the action of physically interesting $\mathscr{G}$'s. One readily sees that for a finite linear space to be invariant means that there is a finite set of eigenfunctions of $\mathscr{G}$ which form a basis for the space. Then the point is that most interesting $\mathscr{G}$'s have only non-normalizable eigenfunctions, with, however, orbital angular momentum being an obvious exception, while trial functions must be normalizable. As a final point concerning linear spaces, we may note that if the space is invariant to $\mathscr{G}$, then it is also invariant to $\mathscr{G}^2$, $\mathscr{G}^3$, ... and therefore to the transformation (1).

The family of transformations (1) is somewhat special in that the exponent is linear in the transformation parameter α, though many interesting transformations are of this type; for example, rotations, translations, and the special gauge transformations (13-23). Nevertheless, more generally, one might have a set of trial functions invariant to a family of transformations

$$e^{iG(\alpha)}$$

where G is a Hermitian operator and where α is some real, continuous numerical parameter which for convenience we will assume has been normalized so that

$$G(0) = 1$$

Then following the same pattern as before, one concludes that under these conditions the hypervirial theorem

$$\left\langle \left[H, \left(\frac{\partial G}{\partial \alpha} \right)_{\alpha=0} \right] \right\rangle = 0 \tag{9}$$

will be satisfied.

In particular, if G is a gauge transformation $(i/c) \sum_{s=1}^{N} \lambda(\mathbf{r}_s)$, then one finds after some integration by parts that (9), with H given by (1-1), becomes

$$\int d\mathbf{r}\, \boldsymbol{\nabla} \cdot \mathbf{j} \left(\frac{\partial \lambda}{\partial \alpha} \right)_{\alpha=0} = 0 \tag{10}$$

where $\mathbf{j}$ is the single-particle current (including the effect of the magnetic field). Thus if, as in UHF or SUHF the set is invariant to all gauge transformations, then (10) will be true for all $(\partial\lambda/\partial\alpha)_{\alpha=0}$, and since that latter could be a perfectly arbitrary function of $\mathbf{r}$, it follows that

$$\boldsymbol{\nabla} \cdot \mathbf{j} = 0 \tag{11}$$

that is, current will be conserved.

In the next few sections, we will consider other special $\mathscr{G}$'s of physical interest, though always in the absence of magnetic fields. The Hamiltonian is again taken to be the clamped nuclei Hamiltonian (1-1) but now with $\mathscr{A} \equiv 0$. In Appendix D, we will then briefly summarize the generalization of these results to the situation in which a magnetic field is present.

PROBLEMS

1. Using $(H - E')\psi' = 0$ and the hermiticity of H, show directly that eigenfunctions satisfy the hypervirial theorem for an arbitrary $\mathscr{G}$. (But see Problem 1 of Section 17.)

2. Suppose that $\hat{\psi}$ is derived from a set of trial functions which is invariant to the transformations (1). Consider the set of trial functions $e^{iA\mathscr{G}}\hat{\psi}$, where A is an arbitrary real number. Argue that $\hat{A}$ must be zero. From this, rederive (4). Also, give an analogous derivation of (15-2) by considering the set of trial functions generated from $\psi(\hat{A}(\sigma), \sigma)$ by replacing σ by a variational parameter.

3. Referring to Problem 5 of Section 15, what if the sets of trial functions are invariant to (1)?

4. We certainly have that in the linear variation method

$$(\hat{\psi}_k, \bar{H}\mathscr{G}\hat{\psi}_k) = \hat{E}_k(\hat{\psi}_k, \mathscr{G}\hat{\psi}_k) = (\hat{\psi}_k, \mathscr{G}\bar{H}\hat{\psi}_k)$$

for any $\mathscr{G}$. Show that this is, as it must be, equivalent to (8) if the space is invariant to the action of $\mathscr{G}$.

5. Verify (10). In particular, show that $\mathbf{j}$ is what it is claimed to be in the text.

6. Derive the result analogous to (11) if $\alpha\mathscr{G} = \alpha\Lambda(r_1, \ldots, r_N)$ where Λ is an arbitrary symmetric function of all electronic coordinates.

7. For UHF, derive (11) directly from the equation for the $\hat{\varphi}_i$. Show in the process that $\mathbf{j}$ is the sum of individual spin–orbital currents, and that although $\mathbf{j}$ is conserved, the individual currents in general are not.

8. Consider a single particle in one dimension in a potential V. Let $\mathscr{G} = \frac{1}{2}(pf + fp)$, where f is a function of x alone. Show that the corresponding hypervirial theorem is

$$\int_{-\infty}^{\infty}\left[-\hat{\psi}^*\hat{\psi}\frac{dV}{dx}+\frac{1}{4}\frac{\partial}{\partial x}\left(\frac{\partial^2\hat{\psi}^*}{\partial x^2}\hat{\psi}-2\frac{\partial\hat{\psi}^*}{\partial x}\frac{\partial\hat{\psi}}{\partial x}+\hat{\psi}^*\frac{\partial^2\hat{\psi}}{\partial x^2}\right)\right]f\,dx=0 \quad (*)$$

Assume that $\hat{\psi}$ is real and show that if this theorem is satisfied for all f, then $\hat{\psi}$ is an eigenfunction [32]. Now consider many electrons with

$$\mathscr{G}=\tfrac{1}{2}\sum_{s=1}^{N}[\mathbf{p}_s\cdot f(\mathbf{r}_s)+f(\mathbf{r}_s)\cdot\mathbf{p}_s]$$

and write the corresponding hypervirial theorem in a form analogous to (*).

17. Momentum Theorems

Let **D** be the negative of the electronic part of the electric dipole moment operator, that is,

$$\mathbf{D}\equiv\sum_{s=1}^{N}\mathbf{r}_s \quad (1)$$

Then in the absence of magnetic fields

$$i[H,\mathbf{D}]=\sum_{s}(\partial H/\partial\mathbf{p}_s)=\mathbf{P} \quad (2)$$

where **P** is the total electron–momentum operator

$$\mathbf{P}=\sum_{s=1}^{N}\mathbf{p}_s \quad (3)$$

Thus if the hypervirial theorem for D_k is satisfied, then the average of P_k will vanish. We will call this a momentum theorem. In particular since **D** is a spinless one-electron operator, UHF and SUHF will satisfy all momentum theorems.

One way to produce a set of trial functions which is invariant to the transformations generated by **D** is to explicitly introduce as variational parameters three real numbers **A** according to [33]

$$\psi(\mathbf{A})=[\exp(i\mathbf{A}\cdot\mathbf{D})]\Theta \quad (4)$$

where the functions Θ are independent of **A**. That this works then follows from the observation that

$$[\exp(i\boldsymbol{\alpha}\cdot\mathbf{D})]\psi(\mathbf{A})=\{\exp[i(\boldsymbol{\alpha}+\mathbf{A})\cdot\mathbf{D}]\}\Theta=\psi(\mathbf{A}+\boldsymbol{\alpha}) \quad (5)$$

so that, as desired, the set is invariant to the unitary transformations produced by **D**.

However, having made these remarks, it is very important to point out that often the momentum theorems will be satisfied simply for reasons of reality, or of symmetry of one kind or another. For example, since **P** is a pure imaginary Hermitian operator, its average will automatically vanish if $\hat{\psi}$ is real:

$$(\hat{\psi}, \mathbf{P}\hat{\psi}) = (\hat{\psi}, \mathbf{P}\hat{\psi})^* = (\hat{\psi}^*, \mathbf{P}^*\hat{\psi}^*) = -\ (\hat{\psi}, \mathbf{P}\hat{\psi}) \tag{6}$$

and therefore

$$(\hat{\psi}, \mathbf{P}\hat{\psi}) = 0 \tag{7}$$

More generally, if H is real, as it is in the absence of magnetic fields, then by the same argument the hypervirial theorems for all real Hermitian $\mathscr{G}$'s will automatically be satisfied if $\hat{\psi}$ is real since each $i[H, \mathscr{G}]$ will then be a pure imaginary Hermitian operator. For example, the theorems for all $\mathscr{G}$'s which, like **D**, are real functions of coordinates will automatically be satisfied [30, 34].

Rotation and inversion symmetries can also result in momentum theorems. Thus for an isolated (i.e., no external fields) atom, if $\hat{\psi}$ has a definite parity under inversion through the nucleus, then since **P** changes sign under inversion, $(\hat{\psi}, \mathbf{P}\hat{\psi})$ will vanish. Or consider an isolated diatomic or a diatomic in an electric field that is symmetric around the internuclear axis, or, which really is the same thing, an atom in an electric field which is symmetric about an axis through the nucleus. If $\hat{\psi}$ has a definite component of angular momentum along the internuclear axis, then it will have a definite parity under rotations of 180° about the internuclear axis, and since the components of **P** perpendicular to the axis change sign under such a rotation, averages of these components will automatically vanish. Also, they will evidently vanish if $\hat{\psi}$, instead of having a definite component of angular momentum along the axis, has a definite parity for reflection in two orthogonal planes containing the internuclear axis since, if we call this axis the z axis, P_x changes sign on reflection in the yz plane and P_y changes sign on reflection in the xz plane. For a homonuclear diatomic, $\hat{\psi}$ will usually have definite parity with respect to reflection in a plane perpendicular to the internuclear axis and through the midpoint, whence the average of P_z will vanish automatically. However, for a general diatomic, the vanishing of the average of P_z will usually not be automatic except for reasons of reality.

PROBLEMS

1. Consider a particle in a box with periodic boundary conditions (Born–von Karman boundary conditions). Show that there are energy eigenfunctions which are also eigenfunctions of **p** and which therefore do not satisfy the momentum theorems. Explain. [Hint: Recall Problem 1 of Section 16.]

2. With H and $\hat{\psi}$ both real, we showed in the text that the hypervirial theorem for any pure imaginary Hermitian operator would be automatically satisfied. Derive the analogous result under the conditions of Problem 1, Section 12 [33].

18. Force Theorems

Continuing, we have

$$i[H, \mathbf{P}] = -\sum_{s} (\partial H/\partial \mathbf{r}_s) = \mathbf{F} \tag{1}$$

where **F** is the operator for the total force on the electrons. That is,

$$\mathbf{F} = \sum_{s=1}^{N} \sum_{a} \mathscr{Z}_a \frac{(\mathbf{R}_a - \mathbf{r}_s)}{|\mathbf{R}_a - \mathbf{r}_s|^3} - \sum_{s=1}^{N} \mathscr{E}(\mathbf{r}_s) \equiv \mathbf{F}_{\text{eN}} + \mathbf{F}_{\text{e}\mathscr{E}} \tag{2}$$

the double sum yielding the force on the electrons due to the nuclei and the single sum the force due to a possible electric field, the electron–electron forces cancelling. Therefore if the hypervirial theorem for P_k is satisfied, then the average of F_k will vanish. This we will call a force theorem. In particular since **P** is a spinless one-electron operator, UHF and SUHF satisfy all force theorems.

More generally, since $e^{i\alpha P_k}$ produces a rigid translation of the electrons in the direction k,[13a] it follows that if the set of trial function is invariant to rigid translation in a particular direction, then the corresponding force theorem will be satisfied. Thus one way to ensure that the force theorems will be satisfied is to explicitly introduce real variational parameters **A** attached in an additive way to each electron coordinate:

$$\psi(\mathbf{r}, \mathbf{A}) = \Theta(\mathbf{r}_1 + \mathbf{A}, \mathbf{r}_2 + \mathbf{A}, \ldots, \mathbf{r}_N + \mathbf{A}) = [\exp(i\mathbf{A}\cdot\mathbf{P})]\psi(\mathbf{r}, 0) \tag{3}$$

since then

$$\psi(\mathbf{r} + \mathbf{d}, \mathbf{A}) = \psi(\mathbf{r}, \mathbf{A} + \mathbf{d}) \tag{4}$$

and therefore the set is invariant to translation.

[13a] This was mentioned in Section 13.

However, just as with **P**, this elaborate machinery may be unnecessary in that force theorems can often be satisfied simply by reason of symmetry. Thus consider again an isolated atom. Since **F** is odd under inversion through the nucleus, $(\hat{\psi}, \mathbf{F}\hat{\psi})$ will vanish if $\hat{\psi}$ has a definite parity under inversion through the nucleus. Also, for an isolated diatomic molecule or a diatomic in an electric field that is symmetric around the internuclear axis (or an atom in an external electric field that is axially symmetric), force theorems perpendicular to the internuclear axis will be satisfied under the same symmetry conditions as the corresponding momentum theorems of the preceding section. However, along the axis, symmetry is usually of no help except in the case of a homonuclear diatomic molecule. Hence to satisfy the force theorem along the axis usually requires the use of a set of trial functions which is explicitly invariant to translation along the axis.

As an interesting application of the force theorem, consider the average total force on all the nuclei. Since the operator for this total force is

$$\mathscr{F} = \sum_a \sum_{s=1}^N \mathscr{Z}_a \frac{(\mathbf{r}_s - \mathbf{R}_a)}{|\mathbf{r}_s - \mathbf{R}_a|^3} + \sum_a \mathscr{Z}_a \mathscr{E}(\mathbf{R}_a) \equiv \mathbf{F}_{\mathrm{Ne}} + \mathbf{F}_{\mathrm{N}\mathscr{E}} \tag{5}$$

and since

$$\mathbf{F}_{\mathrm{Ne}} = -\mathbf{F}_{\mathrm{eN}} \tag{6}$$

we have that

$$\langle \mathscr{F} \rangle = -\langle \mathbf{F}_{\mathrm{eN}} \rangle + \langle \mathbf{F}_{\mathrm{N}\mathscr{E}} \rangle \tag{7}$$

However, if the force theorems are satisfied, then

$$0 = \langle \mathbf{F}_{\mathrm{eN}} \rangle + \langle \mathbf{F}_{\mathrm{e}\mathscr{E}} \rangle \tag{8}$$

which combined with (7) yields

$$\langle \mathscr{F} \rangle = \langle \mathbf{F}_{\mathrm{N}\mathscr{E}} \rangle + \langle \mathbf{F}_{\mathrm{e}\mathscr{E}} \rangle \tag{9}$$

or in words

average total force on the nuclei
= force on the nuclei due to the external field
+ average force on the electrons due to the external field (10)

The special cases of a uniform electric field and of no electric field lead to especially simple and interesting results. For example, consider first an atom in a uniform electric field. Then the force on the nucleus is $\mathscr{Z}\mathscr{E}$, while the average force on the electrons due to the field is $-N\mathscr{E}$, so that (10) yields

$$\text{average force on nucleus} = (\mathscr{Z} - N)\mathscr{E} = \mathscr{Z}\left(1 - \frac{N}{\mathscr{Z}}\right)\mathscr{E} \qquad (11)$$

Thus when the force theorems are satisfied the "dipole shielding factor" will be correctly [35] $N/\mathscr{Z}$. Similarly, for molecules we will have, correctly and exactly [36],

$$\text{average total force on the nuclei} = \left(\sum_a \mathscr{Z}_a - N\right)\mathscr{E} \equiv q\mathscr{E} \qquad (12)$$

where q is the net charge on the molecule.

If there is no external field, and if the force theorems are satisfied, then we see that the average force on the nucleus in an atom will be zero, and similarly that the total force on all the nuclei in a molecule will be zero. In particular, for a diatomic molecule, this means that the force on one nucleus should be equal and opposite to that on the other, a property which is often used to test approximate wave functions ψ. More generally, in the presence of a uniform electric field, the wave functions are usually given in the form of an expansion in powers of the electric field

$$\psi = \sum_n \lambda^n \psi^{(n)} \qquad (13)$$

where $\lambda^n \psi^{(n)}$ is of nth order in the field, whence the wave function tests provided by (8) are

$$(\psi^{(0)}, \mathbf{F}_{\mathrm{eN}}\psi^{(0)}) \overset{?}{=} 0$$

$$\lambda(\psi^{(0)}, \mathbf{F}_{\mathrm{eN}}\psi^{(1)}) + \lambda(\psi^{(1)}, \mathbf{F}_{\mathrm{eN}}\psi^{(0)}) + (\psi^{(0)}, \mathbf{F}_{\mathrm{e}\mathscr{E}}\psi^{(0)}) \overset{?}{=} 0$$

$$\lambda^2(\psi^{(0)}, \mathbf{F}_{\mathrm{eN}}\psi^{(2)}) + \lambda^2(\psi^{(1)}, \mathbf{F}_{\mathrm{eN}}\psi^{(1)}) + \lambda^2(\psi^{(2)}, \mathbf{F}_{\mathrm{eN}}\psi^{(0)}) + \lambda(\psi^{(0)}, \mathbf{F}_{\mathrm{e}\mathscr{E}}\psi^{(1)}) + \lambda(\psi^{(1)}, \mathbf{F}_{\mathrm{e}\mathscr{E}}\psi^{(0)}) \overset{?}{=} 0$$

etc., since (8) should be true for any value of $\mathscr{E}$, and hence contributions of each order in $\mathscr{E}$ must vanish separately.

PROBLEMS

1. Referring to Problem 9 of Section 13, under certain conditions $\gamma(\mathbf{R}/R) + q\mathbf{R}_1$ will equal $\langle \mathscr{D} \rangle_{\mathscr{E}=0}$, the permanent electric dipole moment of the molecule. What are these conditions? Are they true in UHF? In RHF? In the linear variation method?

2. Show that we can write

$$\mathscr{D} = -\sum_{s=1}^{N} [\mathbf{r}_s - \alpha\mathbf{R}_1 - (1-\alpha)\mathbf{R}_2] + (\mathscr{Z}_1 - N_\alpha)\mathbf{R}_1 + [\mathscr{Z}_2 - N(1-\alpha)]\mathbf{R}_2$$

the sum then being the electronic part of the dipole moment referred to the point $\alpha\mathbf{R}_1 + (1-\alpha)\mathbf{R}_2$. Under certain conditions, the average of the sum for $\mathscr{E} = 0$ will be of the form $\gamma'\mathbf{R}/|\mathbf{R}|$ where γ' is a function only of $|\mathbf{R}|$ and α. What are these conditions? Are they met in UHF? In RHF? In the linear variation method?

3. Referring again to Problem 9 of Section 13 and also to Problem 1 here, under certain conditions that part of the force on nucleus a which is linear in $\mathscr{E}$ will be given by $(\partial\langle\mathscr{D}\rangle_{\mathscr{E}=0}/\partial\mathbf{R}_a)\cdot\mathscr{E}$. What are these conditions? Are they met in UHF? In RHF? In the linear variation method?

4. Assuming that the conditions in Problems 2 and 3 are met, express this force in terms of γ' and $d\gamma'/d|\mathbf{R}|$. In particular, by choosing $\alpha = \mathscr{Z}_1/(\mathscr{Z}_1 + \mathscr{Z}_2)$ so that the electronic dipole moment is referred to the center of nuclear charge, derive the results given by Sambe [36].

19. Torque Theorems

Let $\mathbf{L}$ be the operator for the total electronic orbital angular momentum,

$$\mathbf{L} = \sum_{s=1}^{N} \mathbf{r}_s \times \mathbf{p}_s \tag{1}$$

Then one readily finds that $i[H, L_k]$ is the corresponding component of the operator for the net torque on the electrons, this net torque being provided by the nuclei and whatever external electron fields may be present, the electron–electron contribution cancelling. Thus if the hypervirial theorem for L_k is satisfied, then the average of the corresponding component of the net torque will vanish. This we will call a torque theorem.

In particular since the components of $\mathbf{L}$ are spinless one-electron operators, UHF and SUHF satisfy all torque theorems. Further, since L_k generates a rigid rotation of the electrons about the kth coordinate axis,[13a] it follows that in general if the set of trial functions is invariant to rigid rotations about a particular axis, then the corresponding torque theorem will be satisfied.

The angular momentum and torque which we have been talking about are calculated about the origin of the coordinate system. If, however, we calculate the angular momentum and torque about another point (or if we use a different coordinate origin), that is, if we replace the $\mathbf{r}_s$ by $\mathbf{r}_s + \mathbf{d}$ then the angular momentum operator changes by

$$\mathbf{d} \times \mathbf{P} \tag{2}$$

and hence the torque operator changes by

$$\mathbf{d} \times \mathbf{F}$$

Therefore, only if all three momentum theorems are satisfied is the average angular momentum independent of origin, and only if all three force theorems are satisfied is the average torque independent of origin. Of course, $\mathbf{D}$ of the preceding section is also origin dependent, changing by $N\mathbf{d}$ with a change in origin.

Turning to symmetry considerations, for an isolated atom, the operator for the torque about the nucleus vanishes identically ($\mathbf{L}$ about the nucleus is conserved) and therefore the torque theorems about the nucleus are trivially satisfied for any $\hat{\psi}$. If the atom is in a uniform electric field, and if again we use the nucleus as the coordinate origin, then the operator for the torque about the nucleus is simply

$$-\mathbf{D} \times \mathscr{E} \tag{3}$$

Therefore the torque theorem in the direction of $\mathscr{E}$ is again trivially satisfied ($\mathbf{L} \cdot \mathscr{E}$ is conserved), while it will be satisfied in directions perpendicular to $\mathscr{E}$ (denote them by the subscript $\perp$) if $\langle \mathbf{D}_\perp \rangle = 0$, which, since $\mathbf{D}$ has the same transformation properties as $\mathbf{P}$, will automatically be the case if $\hat{\psi}$ is in eigenfunction of $\mathbf{L} \cdot \mathscr{E}$ or if it has a definite parity under reflection in two orthogonal planes containing $\mathscr{E}$ and the nucleus.

Similar remarks apply if $\mathscr{E}$, instead of being completely uniform, is simply symmetric about an axis through the nucleus, and thus also apply to a diatomic, either isolated or in an axially symmetric electric field, if the torques are computed about one of the nuclei. Namely in a general electric field (in the diatomic, this would include the field due to the other nucleus) the operator for the torque about the nucleus is

$$\tau = -\sum_{s=1}^{N} \mathbf{r}_s \times \mathscr{E}(\mathbf{r}_s) \tag{4}$$

One then readily finds that if $\mathscr{E}$ is symmetric about some axis through the nucleus, call it the z axis, so that

$$\mathscr{E} = -\boldsymbol{\nabla}\Phi(z, (x^2 + y^2)^{1/2}) \tag{5}$$

then $\tau_z \equiv 0$ ($\mathbf{L} \cdot \mathscr{E}$ is conserved), while τ_x and τ_y change signs under a 180° rotation about the z axis. Also, one finds that τ_x changes sign on reflection in the xz plane, while τ_y changes sign on reflection in the yz plane, and these observations, of course, suffice to prove the point.

For a molecule, the total electronic torque operator about a point $\mathbf{d}$ in a uniform electric field is

$$\boldsymbol{\tau}(d) = \sum_{s=1}^{N} (\mathbf{r}_s - \mathbf{d}) \times \mathbf{F}_{s\mathrm{N}} - (\mathbf{D} - N\mathbf{d}) \times \mathscr{E} \equiv \boldsymbol{\tau}_{\mathrm{eN}}(d) + \boldsymbol{\tau}_{\mathrm{e}\mathscr{E}}(d) \tag{6}$$

where $\mathbf{F}_{s\mathrm{N}}$ is the operator for the force on the sth electron due to the nuclei, that is,

$$\mathbf{F}_{s\mathrm{N}} \equiv \sum_{a} \mathscr{Z}_a \frac{(\mathbf{R}_a - \mathbf{r}_s)}{|\mathbf{R}_a - \mathbf{r}_s|^3} \tag{7}$$

Thus we can also write

$$\boldsymbol{\tau}(d) = \sum_{s=1}^{N} \sum_{a} \mathscr{Z}_a \frac{\mathbf{r}_s \times \mathbf{R}_a}{|\mathbf{R}_a - \mathbf{r}_s|^3} - \mathbf{d} \times \sum_{s} \mathbf{F}_{s\mathrm{N}} - (\mathbf{D} - N\mathbf{d}) \times \mathscr{E} \tag{8}$$

On the other hand, the operator for the net torque about the point $\mathbf{d}$ on all the nuclei is

$$\boldsymbol{\tau}_{\mathrm{N}}(d) = \sum_{a} (\mathbf{R}_a - \mathbf{d}) \times \sum_{s=1}^{N} \mathscr{Z}_a \frac{(\mathbf{r}_s - \mathbf{R}_a)}{|\mathbf{R}_a - \mathbf{r}_s|^3} + \sum_{a} \mathscr{Z}_a (\mathbf{R}_a - \mathbf{d}) \times \mathscr{E} \tag{9}$$

or

$$\boldsymbol{\tau}_{\mathrm{N}}(d) = \sum_{a} \sum_{s=1}^{N} \mathscr{Z}_a \frac{\mathbf{R}_a \times \mathbf{r}_s}{|\mathbf{R}_a - \mathbf{r}_s|^3} + \mathbf{d} \times \sum_{s} \mathbf{F}_{s\mathrm{N}} + \sum_{a} \mathscr{Z}_a (\mathbf{R}_a - \mathbf{d}) \times \mathscr{E} \tag{10}$$

Therefore, from (8), we have

$$\boldsymbol{\tau}_{\mathrm{N}}(d) = -\boldsymbol{\tau}(d) - (\mathbf{D} - N\mathbf{d}) \times \mathscr{E} + \sum_{a} \mathscr{Z}_a (\mathbf{R}_a - \mathbf{d}) \times \mathscr{E} \tag{11}$$

or

$$\boldsymbol{\tau}_{\mathrm{N}}(d) = -\boldsymbol{\tau}(d) + \mathscr{D}(d) \times \mathscr{E} \tag{12}$$

where $\mathscr{D}(d)$ is the total dipole moment operator of the molecule with respect to the point $\mathbf{d}$:

$$\mathscr{D}(d) \equiv \sum_{a} \mathscr{Z}_a (\mathbf{R}_a - \mathbf{d}) - \sum_{s} (\mathbf{r}_s - \mathbf{d}) = \sum_{a} \mathscr{Z}_a \mathbf{R}_a - \sum_{s} \mathbf{r}_s + q\mathbf{d} \tag{13}$$

and where again q is the net charge of the molecule. Therefore, we see from (12) that if the torque theorems are satisfied about the point **d**, that is, if

$$\langle \tau(d) \rangle = 0 \tag{14}$$

then it will follow that the net torque on the nuclei about the point **d** will be given by [36, 37]

$$\langle \tau_N(d) \rangle = \langle \mathscr{D}(d) \rangle \times \mathscr{E} \tag{15}$$

As we have shown, for a diatomic in an axial electric field, the torque theorems about nucleus 1, say, will usually be satisfied automatically. Also, in this case, $\langle \mathscr{D}(\mathbf{R}_1) \rangle$ will also be axial, so that (15) says that under these conditions the average torque on nucleus 2 about nucleus 1 will automatically vanish. However, if the electric field is not axial, or more generally in a polyatomic even in zero field, the torque theorems will usually not be satisfied automatically about any point and (14), or equivalently (15), can be used to test wave functions. In particular, if the latter are given in the form (18-13), then from (6), the torque theorem yields the tests

$$\begin{aligned} (\psi^{(0)}, \tau_{eN}(d)\psi^{(0)}) &\overset{?}{=} 0 \\ \lambda(\psi^{(0)}, \tau_{eN}(d)\psi^{(1)}) + \lambda(\psi^{(1)}, \tau_{eN}(d)\psi^{(0)}) + (\psi^{(0)}, \tau_{e\mathscr{E}}(d)\psi^{(0)}) &\overset{?}{=} 0 \\ \lambda^2(\psi^{(0)}, \tau_{eN}(d)\psi^{(2)}) + \lambda^2(\psi^{(1)}, \tau_{eN}(d)\psi^{(1)}) + \lambda^2(\psi^{(2)}, \tau_{eN}(d)\psi^{(0)}) & \\ + \lambda(\psi^{(0)}, \tau_{e\mathscr{E}}(d)\psi^{(1)}) + \lambda(\psi^{(1)}, \tau_{e\mathscr{E}}(d)\psi^{(0)}) &\overset{?}{=} 0 \end{aligned} \tag{16}$$

etc.

20. Virial Theorems

Consider the Hermitian "virial operator"

$$\mathscr{V} \equiv \tfrac{1}{2} \sum_{s=1}^{N} (\mathbf{r}_s \cdot \mathbf{p}_s + \mathbf{p}_s \cdot \mathbf{r}_s) \tag{1}$$

or

$$\mathscr{V} = \sum_{s=1}^{N} \mathbf{r}_s \cdot \mathbf{p}_s - 3i(N/2) \tag{2}$$

Before discussing the content of the hypervirial theorem for $\mathscr{V}$, let us first consider the unitary transformation produced by $\mathscr{V}$. The key observation is that

$$e^{i\alpha\mathscr{V}} \psi(\mathbf{r}_1, \ldots, \mathbf{r}_N) = \tau^{3N/2} \psi(\tau\mathbf{r}_1, \ldots, \tau\mathbf{r}_N) \tag{3}$$

where $\tau = e^{\alpha}$. The proof follows from the observation that $X \equiv e^{i\alpha\mathscr{V}}\psi$ can also be defined by

$$\frac{1}{i}\frac{dX}{d\alpha} = \mathscr{V}X, \qquad X = \psi \qquad \text{when} \quad \alpha = 0 \tag{4}$$

It is then easy to verify that the right hand side of (3) satisfies (4); however, we leave the details as a problem.

Thus we see from (3) that $\mathscr{V}$ produces a positive scaling of the electronic coordinates, and therefore we have the result that if the set of trial functions is invariant to such scaling, then the hypervirial theorem for $\mathscr{V}$ will be satisfied. In particular, since $\mathscr{V}$ is a spinless one-electron operator, UHF and SUHF will satisfy the theorem. Note, however, that like **L**, the operator $\mathscr{V}$ is not invariant to change of coordinate origin. Namely if one makes such a change, then $\mathscr{V}$ acquires an additional term $\mathbf{d}\cdot\mathbf{P}$. Therefore the hypervirial theorem for $\mathscr{V}$ will be coordinate origin independent if and only if the force theorems are satisfied.

One common way to ensure that a set of trial functions will be invariant to positive scaling is to explicitly include, as a numerical variational parameter, a coordinate scaling parameter which we will denote by ζ. That is, one uses trial functions of the form

$$\psi = \zeta^{3N/2}\Theta(\zeta\mathbf{r}_1, \zeta\mathbf{r}_2, \ldots, \zeta\mathbf{r}_N) \tag{5}$$

with ζ an arbitrary real number. (The factor of $\zeta^{3N/2}$, since it is only a constant multiplier, is of course optional.) That such a set is invariant to positive scaling then follows from the observation that replacing the $\mathbf{r}_s$ by $\tau\mathbf{r}_s$ in ψ and multiplying by $\tau^{3N/2}$ is equivalent to leaving the $\mathbf{r}_s$ alone and replacing ζ by $\tau\zeta$, and therefore produces another member of the set.

The hypervirial theorem for $\mathscr{V}$ is essentially the virial theorem, which is so often invoked in discussions of chemical binding,[14] force constants,[15] etc., and which is often used to test approximate wave functions. To show this, we first note that we can calculate $i[H, \mathscr{V}]$ quite generally as

$$i[H, \mathscr{V}] = \sum_{s=1}^{N} \left(\frac{\partial H}{\partial \mathbf{p}_s}\cdot\mathbf{p}_s - \mathbf{r}_s\cdot\frac{\partial H}{\partial \mathbf{r}_s} \right) \tag{6}$$

Now let us specialize to the clamped nuclei Hamiltonian (1-1) of an

[14] See, for example, Ruedenberg [38].

[15] See, for example, Clinton [39]. General survey: Schutte [40].

isolated molecule ($\mathscr{A} = \Phi = 0$). Then, schematically,

$$H = T + V \tag{7}$$

where T, the kinetic energy operator, is a homogeneous function of degree two in the $\mathbf{p}_s$, while V, the potential energy operator (including the nuclear repulsions), is a homogeneous function of degree -1 in the $\mathbf{r}_s$ and the $\mathbf{R}_a$. Thus from Euler's theorem on homogeneous functions, we have

$$\sum_{s=1}^{N} \frac{\partial H}{\partial \mathbf{p}_s} \cdot \mathbf{p}_s = 2T \tag{8}$$

and

$$\sum_{s=1}^{N} \frac{\partial H}{\partial \mathbf{r}_s} \cdot \mathbf{r}_s = -V - \sum_a \mathbf{R}_a \cdot \frac{\partial H}{\partial \mathbf{R}_a} \tag{9}$$

Further, $-\partial H/\partial \mathbf{R}_a$ is the operator representing the force on nucleus a, thus

$$-\frac{\partial H}{\partial \mathbf{R}_a} \equiv \mathbf{F}_a = \sum_{s=1}^{N} \mathscr{Z}_a \frac{(\mathbf{r}_s - \mathbf{R}_a)}{|\mathbf{r}_s - \mathbf{R}_a|^3} + \sum_{b \neq a} \mathscr{Z}_a \mathscr{Z}_b \frac{(\mathbf{R}_a - \mathbf{R}_b)}{|\mathbf{R}_a - \mathbf{R}_b|^3} \tag{10}$$

Combining (6), and (8)–(10), we therefore have that the hypervirial theorem for $\mathscr{V}$ is the statement that

$$2\langle T \rangle + \langle V \rangle - \sum_a \mathbf{R}_a \cdot \langle \mathbf{F}_a \rangle = 0 \tag{11}$$

Before proceeding to a detailed discussion of Eq. (11), we will briefly mention a few generalizations of it. First of all, $\mathscr{V}$ can be thought of as the trace of the "tensor virial operator"

$$\mathscr{V}_{jk} = \sum_{s=1}^{N} r_{sj} p_{sk} - \tfrac{1}{2} iN \, \delta_{jk} \tag{12}$$

We have shown following (5) that by use of $\zeta \mathbf{r}_s$, thereby scaling all components of the $\mathbf{r}_s$ equally, we can satisfy the hypervirial theorem for $\mathscr{V}$. Similarly one can show that by using $\zeta \cdot \mathbf{r}_s$, that is, "vector scaling," one can separately satisfy the hypervirial theorems for the diagonal components of $\mathscr{V}_{ij}$ [41], and finally one can show that by using tensor scaling, that is, $\sum_{k=1}^{3} \zeta_{jk} r_{sk}$, one can guarantee the theorems for all components [42]. However, as always, symmetry alone may be enough to guarantee some of these theorems (or to guarantee some, given others).

As a second generalization, let us introduce a uniform electric field. This means from (1-1) that one adds a term $-\mathscr{D} \cdot \mathscr{E}$ to H, where $\mathscr{D}$ is the dipole moment operator (15-6). Thus since this term is homogeneous of degree one in the $\mathbf{r}_s$ and the $\mathbf{R}_a$, it follows that (11) will be replaced by

$$2\langle T\rangle + \langle V\rangle + \langle\mathscr{D}\rangle \cdot \mathscr{E} - \sum_a \mathbf{R}_a \cdot \langle\mathbf{F}_a\rangle = 0 \tag{13}$$

where V is the potential energy operator exclusive of the $-\mathscr{D}\cdot\mathscr{E}$ term and $\mathbf{F}_a$ is given by the right hand side of (10) plus the additional term $\mathscr{Z}_a\mathscr{E}$.

We now return to Eq. (11). To reduce it to more familiar form, suppose further that the generalized Hellmann–Feynman theorems in $\mathbf{r}_s$ coordinates for σ equal to the components of the $\mathbf{R}_a$,

$$\partial\hat{E}/\partial\mathbf{R}_a = -\langle\mathbf{F}_a\rangle \tag{14}$$

are also satisfied. Then (11) becomes the more familiar

$$2\langle T\rangle + \langle V\rangle + \sum_a \mathbf{R}_a\cdot(\partial\hat{E}/\partial\mathbf{R}_a) = 0 \tag{15}$$

Equations (14) are the variational version of *the* Hellmann–Feynman theorem [43]. From our point of view, they state, when applicable, that one can calculate the average force on a nucleus by calculating the negative derivative of the energy with respect to the Cartesian coordinates of the nucleus. However, in order to avoid confusion, we should point out that usually the theorem is read the other way around. Namely because of the use which is made of $\hat{E}$ in approximating *the* Born–Oppenheimer approximation,[16] and contrary to the language which we have been using, $-\partial\hat{E}/\partial\mathbf{R}_a$ rather than $\langle\mathbf{F}_a\rangle$ is considered a priori to be the average force on nucleus a. Thus (14), when applicable, is usually read to say that this force can also be calculated as the average of the operator $\mathbf{F}_a$. Therefore $\langle\mathbf{F}_a\rangle$ is often called the Hellmann–Feynman force, and analogously one might call $-\partial\hat{E}/\partial\mathbf{R}_a$ the Born–Oppenheimer force. Moreover, the theorem often appears in other forms, forms which are all equivalent if various other theorems are satisfied. Thus consider a diatomic molecule. Then if the force theorems are satisfied, we can replace the right hand side of (14) by

$$-(\alpha\langle\mathbf{F}_a\rangle + (\alpha - 1)\langle\mathbf{F}_b\rangle)$$

where α is an arbitrary real number, and indeed various choices have been used in the literature.

[16] For a concise summary of various versions of the Born–Oppenheimer approximation, see Hirschfelder and Meath [44].

Equation (15) can be further simplified if we make the assumption that the set of trial functions is invariant to rotation, translation, and inversion of the electrons. Then, since presumably the set is quite independent of the $\mathbf{R}_a$ in order to satisfy (14), it follows from the discussion in Section 13 that $\hat{E}$ can depend only on the $R_{(ab)}$ and $\cos \Theta_{(ab,cd)}$ of (13-18) and (13-19). Now these quantities are in general not independent, so there is no unique formula for $\hat{E}$ in terms of them. Nevertheless, having chosen a formula, then since the $R_{(ab)}$ are homogeneous functions of degree one in the $\mathbf{R}_a$ while the $\cos \Theta_{(ab,cd)}$ are homogeneous of degree zero, one readily finds that under the given conditions, (15) can be written as [45]

$$2\langle T\rangle + \langle V\rangle + \sum_{(ab)} R_{(ab)}(\partial \hat{E}/\partial R_{(ab)}) \tag{16}$$

For a diatomic molecule, this becomes the familiar

$$2\langle T\rangle + \langle V\rangle + R(\partial \hat{E}/\partial R) \tag{17}$$

where R is the internuclear separation, and for an isolated atom ($R = 0$), it is the equally familiar

$$2\langle T\rangle + \langle V\rangle = 0 \tag{18}$$

Equation (16), or its further specializations (17) and (18), is what is usually called *the* virial theorem.

Thus if one uses a set of trial functions which is invariant to positive scaling, inversion, translation, and rotation of the electronic coordinates, and is invariant to changes in the nuclear coordinates, then (11), (14), and (16) will be satisfied. In addition, it follows from results in earlier sections that these conditions will also guarantee that the average net force on the electrons is zero, that the average net torque on the electrons is zero, and that the average net force and average net torque on all the nuclei is zero. Also, the average force on each nucleus in a diatomic molecule will be only along the internuclear axis since if $\hat{E}$ is a function only of $R = |\mathbf{R}_2 - \mathbf{R}_1| = |\mathbf{R}|$, then

$$-\frac{\partial \hat{E}}{\partial \mathbf{R}_1} = \frac{\partial \hat{E}}{\partial \mathbf{R}_2} = \frac{\mathbf{R}}{R}\,\frac{\partial \hat{E}}{\partial R} \tag{19}$$

As we have seen, UHF and SUHF satisfy all these conditions. For other sorts of trial functions, one often explicitly introduces variation parameters to do the job. Since there are various ways of doing this, it will help to avoid notational confusion if we consider a simple example—a one-electron diatomic molecule in a simple LCAO type of approximation involving two atomic orbitals. Generalizations should be obvious. A first choice (Hund–Mulliken) for the set of trial functions might be the functions

$$\exp(-\mathscr{Z}_1 \mid \mathbf{r} - \mathbf{R}_1 \mid) + c \exp(-\mathscr{Z}_2 \mid \mathbf{r} - \mathbf{R}_2 \mid) \tag{20}$$

where c is a variational parameter. However, this set has none of the properties we want. It is not invariant to scaling, it depends explicitly on $\mathbf{R}_1$ and $\mathbf{R}_2$, and it is not invariant to either rotation or translation of the electron's coordinates. To take care of all these deficiencies, while still keeping to the same sort of functions, it is then quite natural to introduce two real, positive variational parameters ζ and τ and two real, vector variational parameters $\boldsymbol{\mu}$ and $\boldsymbol{\mu}'$ and use the set of trial functions

$$\exp(-\mid \zeta\mathbf{r} - \boldsymbol{\mu} \mid) + c \exp(-\tau \mid \zeta\mathbf{r} - \boldsymbol{\mu}' \mid) \tag{21}$$

This set is then obviously invariant to changes in $\mathbf{R}_1$ and $\mathbf{R}_2$ since each member of the set is separately invariant. Also, it is obviously invariant to positive scaling. Further, the set is invariant to translations since replacing $\mathbf{r}$ by $\mathbf{r} + \mathbf{d}$ is equivalent to replacing $\boldsymbol{\mu}$ and $\boldsymbol{\mu}'$ by $\boldsymbol{\mu} - \zeta\mathbf{d}$ and $\boldsymbol{\mu}' - \zeta\mathbf{d}$, respectively, and also it is invariant to inversion since $\mathbf{r} \to -\mathbf{r}$ is equivalent to $\boldsymbol{\mu} \to -\boldsymbol{\mu}$, $\boldsymbol{\mu}' \to -\boldsymbol{\mu}'$. Finally, the set is invariant to rotations since replacing $\mathbf{r}$ by $J\mathbf{r}$, where J is a rotation dyadic, is equivalent to replacing $\boldsymbol{\mu}$ and $\boldsymbol{\mu}'$ by $J^{-1}\boldsymbol{\gamma}$ and $J^{-1}\boldsymbol{\mu}$, as one sees from

$$\mid J\zeta\mathbf{r} - \boldsymbol{\gamma} \mid = \mid J(\zeta\mathbf{r} - J^{-1}\boldsymbol{\gamma}) \mid = \mid (\zeta\mathbf{r} - J^{-1}\boldsymbol{\gamma}) \mid$$

the last equality following from the fact that the length of a vector is invariant to rotation.

The atomic orbitals in (20) are centered on the nuclei and are invariant to rotation about the internuclear axis. For reasons of symmetry, one might expect the latter also to be true of the optimal orbitals derived from (21), that is, that the points $\hat{\boldsymbol{\mu}}$ and $\hat{\boldsymbol{\mu}}'$ will be on the internuclear axis, and probably in practice one would put them there from the outset. However, there is no reason to expect that $\hat{\boldsymbol{\mu}}/\zeta$ will equal $\mathbf{R}_1$, or that $\hat{\boldsymbol{\mu}}'/\zeta$ will equal $\mathbf{R}_2$. Therefore, as a price one pays for axial translational invariance, and hence for the axial force theorem (the other force theorems being satisfied by symmetry) with such a simple set of trial functions, the atomic orbitals may have their cusps off the nuclei. Following Hurley [24], sets like (21) are often called "floating wave functions." Eigenfunctions for this problem, of course, have cusps at the nuclei.

In the previous discussion, we have used a coordinate system which is quite independent of the nuclear framework. However, for atoms and diatomic molecules, for example, one has special symmetries which are usually taken advantage of in any variational calculation. Thus for an atom, one would certainly use the nucleus as the coordinate origin, in which case (15) immediately becomes the virial theorem

$$2\langle T\rangle + \langle V\rangle = 0 \tag{22}$$

Conversely, therefore, if one chooses the nucleus as the coordinate origin, then the virial theorem will be guaranteed simply by having the set of trial functions be invariant to positive scaling of electronic coordinates. In particular, then, it follows that most restricted Hartree–Fock methods for atoms will satisfy the virial theorem since positive scaling affects only the radial coordinates and not the angles, and in RHF, though angular dependence is fixed, the radial functions are quite flexible. Therefore sets of RHF trial functions for atoms will be invariant to positive scaling.

Since it illustrates a technique which is often used in the literature, we will now show in a different way than before how the use of trial functions like (5) for an isolated atom with coordinate origin at the nucleus leads to the virial theorem [46]. By changing variables in the integrals from the $\mathbf{r}_s$ to the $\zeta\mathbf{r}_s$ and using the homogeneity properties of T and V, it is easy to show that

$$E(\zeta) = \zeta^2\tilde{T}(1) + \zeta\tilde{V}(1) \tag{23}$$

where $\tilde{T}(1)$ is the average kinetic energy when $\zeta = 1$, etc. Requiring that

$$\partial E(\hat{\zeta})/\partial\hat{\zeta} = 0$$

then yields

$$2\hat{\zeta}\tilde{T}(1) + \tilde{V}(1) = 0 \tag{24}$$

and this, when multiplied by $\hat{\zeta}$, becomes the virial theorem

$$2\langle T\rangle + \langle V\rangle = 0 \tag{25}$$

which proves the point. Note also that if $\psi(\mathbf{r}_1, \ldots, \mathbf{r}_N)$ was derived from some other variational calculation, then to scale it so as to satisfy the virial theorem would require no extra work since $\tilde{T}(1)$ and $\tilde{V}(1)$ would already have been calculated [47]. Scaling of molecular wave functions requires somewhat more work but also is relatively straightforward [48].

Further, one can adapt the preceding argument to yield another general proof that invariance to scaling implies the virial theorem. The point is that if the set is invariant to scaling, then with α a positive number, $\psi(\alpha\mathbf{r}, \hat{A})$

must be a member of the set, and hence a change of α away from one must correspond to some variation of $\hat{A}$. If we denote the energy as calculated with this wave function by $E(\alpha, \hat{A})$, with therefore $\hat{E} = E(1, \hat{A})$, then from (23) we have

$$E(\alpha, \hat{A}) = \alpha^2 \hat{T} + \alpha \hat{V} \tag{26}$$

Further, from what we have said, it must be that

$$\left(\frac{\partial E(\alpha, \hat{A})}{\partial \alpha} \right)_{\alpha=1} = 0 \tag{27}$$

which of course yields (25).

Turning now to diatomic molecules, here one would almost certainly use the internuclear axis as one of the coordinate axes, the x' axis say, and in addition the coordinate origin is usually placed at some fixed fractional distance along the internuclear axis (at one of the nuclei, at the mid-point, at the nuclear center of mass, etc.). Then in such a coordinate system, the nuclear coordinates are

$$\mathbf{R}_1' = \left(-\frac{1}{\alpha + 1} R, 0, 0 \right) \tag{28}$$

$$\mathbf{R}_2' = \left(\frac{\alpha}{\alpha + 1} R, 0, 0 \right) \tag{29}$$

where R is the nonzero component of the nuclear separation $\mathbf{R}_2' - \mathbf{R}_1'$ which for simplicity of discussion we will assume is positive, and where the numerical value of the number α depends on the exact choice of origin.

In such coordinates V is a homogeneous function of degree -1 in the $\mathbf{r}_s'$ and R, while T is a homogeneous function of degree two in the $\mathbf{p}_s'$. Therefore if the set of trial functions is invariant to positive scaling of the $\mathbf{r}_s'$, it will follow that

$$2\langle T \rangle + \langle V \rangle + R\langle \partial H / \partial R \rangle = 0 \tag{30}$$

where, as a short calculation shows,

$$\frac{\partial H}{\partial R} = \frac{1}{\alpha + 1} F_{1x'} - \frac{\alpha}{\alpha + 1} F_{2x'} \tag{31}$$

If, further, the generalized Hellmann–Feynman theorem for $\sigma = R$ in these coordinates,

$$\partial \hat{E} / \partial R = \langle \partial H / \partial R \rangle \tag{32}$$

is satisfied, then we will have the virial theorem

$$2\langle T \rangle + \langle V \rangle + R(\partial \hat{E}/\partial R) = 0 \tag{33}$$

However, it should be kept in mind that these invariances to positive scaling and to changes in R will not in general ensure that the axial force theorem is satisfied. Thus in particular one expects that without further variational flexibility,

$$\hat{F}_{1x'} + \hat{F}_{2x'} \neq 0$$

On the other hand, the Born–Oppenheimer forces on the nuclei are equal and opposite essentially by definition since, as discussed in Section 13, whatever the set of trial functions, having chosen α and having made calculations at various R values, one will automatically assume that the results yield $\hat{E}(|\mathbf{R}_2 - \mathbf{R}_1|)$ independent of coordinate system, so that

$$\frac{\partial \hat{E}}{\partial \mathbf{R}_1} + \frac{\partial \hat{E}}{\partial \mathbf{R}_2} \equiv 0$$

Along these same lines, it is of interest to observe that there is a simple prescription which leads directly to the virial theorem with no additional assumptions concerning generalized Hellmann–Feynman theorems. The procedure [49] is to use trial functions of the form

$$\zeta^{3N/2} \psi(\zeta \mathbf{r}_1', \ldots, \zeta \mathbf{r}_N', \zeta R) \tag{34}$$

the factor of $\zeta^{3N/2}$ being optional. To see that this works, we first note that

$$\zeta \frac{\partial \psi}{\partial \zeta} = \sum_s \mathbf{r}_s' \cdot \frac{\partial \psi}{\partial \mathbf{r}_s'} + \frac{3N}{2} \psi + R \frac{\partial \psi}{\partial R} = i \mathscr{V}' \psi + R \frac{\partial \psi}{\partial R} \tag{35}$$

Therefore from

$$\left(\frac{\partial \hat{\psi}}{\partial \zeta}, (H - \hat{E}) \hat{\psi} \right) + \left(\hat{\psi}, (H - \hat{E}) \frac{\partial \hat{\psi}}{\partial \zeta} \right) = 0 \tag{36}$$

it follows that [recall Eq. (30)]

$$2\langle T \rangle + \langle V \rangle + R \left\langle \frac{\partial H}{\partial R} \right\rangle$$
$$+ R \left[\left(\frac{\partial \hat{\psi}}{\partial R}, (H - \hat{E}) \hat{\psi} \right) + \left(\hat{\psi}, (H - \hat{E}) \frac{\partial \hat{\psi}}{\partial R} \right) \right] \Big/ (\hat{\psi}, \hat{\psi}) = 0 \tag{37}$$

[In (35), $\partial \psi/\partial R$ meant $(\partial \psi/\partial R)_\zeta$. However, in (37), $\partial \hat{\psi}/\partial R$ can and does mean the total derivative $(\partial\hat{\psi}/\partial R)_\zeta + (\partial\hat{\psi}/\partial\hat{\xi})(\partial\hat{\xi}/\partial R)$ since, from (36), the extra term makes no contribution.]

We now observe that

$$\left(\frac{\partial\hat{\psi}}{\partial R}, (H-\hat{E})\hat{\psi}\right) = \frac{\partial}{\partial R}(\hat{\psi}, (H-\hat{E})\hat{\psi}) - \left(\hat{\psi}, \frac{\partial}{\partial R}(H-\hat{E})\hat{\psi}\right)$$

which, since

$$(\hat{\psi}, (H-\hat{E})\hat{\psi}) \equiv 0$$

we can write as

$$\begin{aligned}\left(\frac{\partial\hat{\psi}}{\partial R}, (H-\hat{E})\hat{\psi}\right) &= -\left(\hat{\psi}, \frac{\partial}{\partial R}(H-\hat{E})\hat{\psi}\right)\\ &= -\left(\hat{\psi}, \frac{\partial H}{\partial R}\hat{\psi}\right) + \frac{\partial\hat{E}}{\partial R}(\hat{\psi}, \hat{\psi}) - \left(\hat{\psi}, (H-\hat{E})\frac{\partial\hat{\psi}}{\partial R}\right)\end{aligned} \tag{38}$$

and this when combined with (37) yields (33) again.

However, unless the individual ψ's do not depend on R at all, or involve R only in conjunction with an additional parameter, such a set is in general not invariant to positive scaling, and is not invariant to changes in R, since neither separately is equivalent to a change in ζ alone. Therefore in the absence of further variational flexibility, in general the Hellmann–Feynman theorem for $\sigma = R$ in these coordinates will not be satisfied.

An interesting specialization of (34) is provided by trial functions of the form

$$\psi(\mathbf{r}_1'/R, \mathbf{r}_2'/R, \ldots, \zeta R) \tag{39}$$

Such a set then has the additional property that for fixed values of

$$\boldsymbol{\rho}_s \equiv \mathbf{r}_s'/R$$

it is independent of R. Therefore with such a set the generalized Hellmann–Feynman theorem for $\sigma = R$ in $\boldsymbol{\rho}_s$ coordinates will be satisfied, and hence also in any coordinates derived from the $\boldsymbol{\rho}_s$ by an R-independent transformation, for example, the often used orthogonal confocal elliptic coordinates (with the nuclei at the foci). We will now show that, not surprisingly, the generalized Hellmann–Feynman theorem for $\sigma = R$ in such coordinates is precisely the virial theorem [50].

In $\boldsymbol{\rho}_s$ coordinates, H takes the form

$$H = \frac{t}{R^2} + \frac{v}{R} \tag{40}$$

where t and v are independent of R. Therefore the generalized Hellmann–Feynman theorem for $\sigma = R$ in these coordinates yields (the factor of R^3 in the volume element cancels out)

$$\frac{\partial \hat{E}}{\partial R} = -\frac{2}{R^3}\langle t\rangle - \frac{1}{R^2}\langle v\rangle$$

and, multiplying through by R, we have the virial theorem. Thus, to repeat, to satisfy the virial theorem, it is sufficient to use a set of trial functions which in $\boldsymbol{\rho}_s$-type coordinates is independent of R.

As a final exercise in scaling, we will simply state the following, leaving the detailed proof as a problem. Consider a general molecule in general Cartesian coordinates, and refer the nuclear configuration to a fixed similar configuration according to

$$\mathbf{R}_a = S\mathbf{R}_a{}^0, \qquad \text{same } S \text{ for all } a, \qquad S > 0 \tag{41}$$

It is then easy to show that use of trial functions of the form

$$\psi = \psi(\zeta\mathbf{r}_1, \zeta\mathbf{r}_2, \ldots, \zeta\mathbf{r}_N, \zeta\mathbf{R}_1, \zeta\mathbf{R}_2, \ldots)$$

will yield directly

$$2\langle T\rangle + \langle V\rangle + S(\partial\hat{E}/\partial S) = 0 \tag{42}$$

which is also one consequence of (15) [51], since if we use (41), then obviously

$$S\frac{\partial \hat{E}}{\partial S} = S\sum \frac{\partial \hat{E}}{\partial R_{ab}} R_{ab}^0 = \sum \frac{\partial \hat{E}}{\partial R_{ab}} R_{ab}$$

PROBLEMS

1. A simple approximation to the ground state of helium is provided by use of trial functions of the form $\psi = e^{-A(r_1+r_2)}$ times a spin function, where the real number A is a variational parameter (effective nuclear charge). Show that the virial theorem will be satisfied by $\hat{\psi}$. Referring to Section 10 show that with the obvious choice of $H^{(0)}$, one will automatically have $\mathscr{E}^{(0)} = \hat{E}$. (For a generalization to many-electron atoms and to molecules, see Hansen [52], G. Gliemann [53].)

2. Referring to the discussion of finite linear spaces in Section 16, what are the eigenfunctions and eigenvalues of $\mathbf{r}\cdot\mathbf{p} + \mathbf{p}\cdot\mathbf{r}$?

3. Formally, one can write (5) as $\psi = e^{i \ln \zeta \mathscr{V}} \Theta$. Suppose that Θ is a given function which one wants to scale [recall the discussion following (25)]. Show that $\partial^2 \hat{E}/\partial \hat{\zeta}^2$ can be written as the average of a certain double commutator. Show that for an atom, this average is always positive. What conclusions do you draw?

4. Generalizing the virial operator in another way (recall also Problem 8 of Section 16), consider for simplicity a single particle in one dimension and

$$\mathscr{V}_f \equiv \tfrac{1}{2}(pf + fp), \qquad f = f(x)$$

Show that

$$e^{iA\mathscr{V}_f}\psi(x) = f^{1/2}(\sigma) f^{-1/2}(x)\psi(\sigma)$$

where $\sigma(x)$ is related to x by

$$\partial\sigma/\partial A = f(\sigma), \qquad \sigma = x \qquad \text{when} \quad A = 0$$

[33]. Thus $\mathscr{V}_f$ generates a "point transformation" $x \to \sigma(x)$. (See also, for example, Eger and Gross [54], and Trickey *et al.* [55].)

5. Rederive the result that $\hat{\psi}$ derived from trial functions of the form (34) satisfy the virial theorem by following a procedure analogous to that of Eq. (23) et seq. [49].

21. Orthogonality and Related Theorems

Eigenfunctions belonging to different eigenvalues of H are automatically orthogonal. As we will see in a moment, it is easy enough to give sufficient conditions such that $\hat{\psi}_a$ and $\hat{\psi}_b$ will be automatically orthogonal if $\hat{E}_a \neq \hat{E}_b$. However, it must be stated that so far, the only known way of realizing these conditions a priori is to draw $\hat{\psi}_a$ and $\hat{\psi}_b$ from a common linear space, and for this case, we have already discussed the orthogonality properties of the $\hat{\psi}_k$ in Section 11. Also, there are various other theorems of a similar type for which the same situation prevails. Conversely, if the set of trial functions involves nonlinear parameters, as in UHF and SUHF and CI with nonlinear parameters, then, barring symmetry considerations, we do not expect the theorems to be satisfied, and this is in agreement with experience,[17] although, for example, orthogonality, though not exact, may be nearly realized in some SCF calculations.[18]

[17] See Chong and Benston [56]. They also describe procedures for satisfying such theorems by constraint. Also Epstein *et al.* [57].

[18] For example Bagus [58]. See also Cohen and Dalgarno [59].

To cover all the theorems at once, let $\hat{\psi}_a$ be an optimal trial function for a Hamiltonian H_a, and let $\hat{\psi}_b$ be an optimal trial function for a Hamiltonian H_b. Then suppose that among the variations of $\hat{\psi}_a$ which were possible in the set from which it was drawn were $i\mathscr{G}\hat{\psi}_b$ and $\mathscr{G}\hat{\psi}_b$, where $\mathscr{G}$ is a Hermitian operator. Then from (5-3) applied to $H = H_a$, we find, with obvious changes in notation,

$$(\hat{\psi}_b, \mathscr{G}H_a\hat{\psi}_a) = \hat{E}_a(\hat{\psi}_b, \mathscr{G}\hat{\psi}_a) \tag{1}$$

Similarly, if $i\mathscr{G}\hat{\psi}_a$ and $\mathscr{G}\hat{\psi}_a$ were possible variations of $\hat{\psi}_b$, then from the complex conjugate of (5-3) applied to $H = H_b$, we find

$$(\hat{\psi}_b, H_b\mathscr{G}\hat{\psi}_a) = \hat{E}_b(\hat{\psi}_b, \mathscr{G}\hat{\psi}_a) \tag{2}$$

Subtracting these, we have as our basic result

$$(\hat{\psi}_b(\mathscr{G}H_a - H_b\mathscr{G})\hat{\psi}_a) = (\hat{E}_a - \hat{E}_b)(\hat{\psi}_b, \mathscr{G}\hat{\psi}_a) \tag{3}$$

Various special cases are now of interest, as follows.

(i) $\mathscr{G} = 1$; $H_a = H_b = H$. Then Eq. (3) tells us that if $\hat{E}_a \neq \hat{E}_b$, then $\hat{\psi}_a$ will be orthogonal to $\hat{\psi}_b$. However, as we noted at the outset of the section, the one way we know to implement this sufficient condition in an a priori manner is to draw $\hat{\psi}_a$ and $\hat{\psi}_b$ from a common linear space, and conversely if they are not drawn from a common linear space, one does not expect orthogonality. Thus, for example, in general the $\hat{\psi}$ of UHF are not one-electron excitations of one another, and hence not possible variations of one another, and indeed are usually not orthogonal. (However, there is a $\mathscr{Z}^{-1}$ theorem for orthogonality—see Section 29.)

(ii) $\mathscr{G} \neq 1$; $H_a = H_b = H$. For $\hat{\psi}_a = \hat{\psi}_b$, (3) is of course the hypervirial theorem for $\mathscr{G}$, which we have already discussed in detail. For $\hat{\psi}_a \neq \hat{\psi}_b$, Eq. (3) is the sort of relation which is useful in discussing transition probabilities. The one way we know to implement the sufficient condition in an a priori manner is to draw $\hat{\psi}_a$ and $\hat{\psi}_b$ from a common linear space which is invariant to the action of $\mathscr{G}$, in which case we can derive the theorem more directly, as follows:

$$(\hat{\psi}_b, (\mathscr{G}\bar{H} - \bar{H}\mathscr{G})\hat{\psi}_a) = (\hat{E}_a - \hat{E}_b)(\hat{\psi}_b, \mathscr{G}\hat{\psi}_a)$$

But $\Pi\hat{\psi}_a = \hat{\psi}_a$, $\Pi\hat{\psi}_b = \hat{\psi}_b$, $\Pi\mathscr{G}\hat{\psi}_a = \mathscr{G}\hat{\psi}_a$, and $\Pi\mathscr{G}\hat{\psi}_b = \mathscr{G}\hat{\psi}_b$. Therefore we can replace $\bar{H}$ by H and we have the theorem. However, physically useful finite linear spaces are rarely invariant to physically interesting $\mathscr{G}$'s (recall the analogous comment in Section 16).

(iii) $\mathscr{G} = 1$; $H_a \neq H_b$. Equation (3) is now the variational version of the so-called [60] integral Hellmann–Feynman theorem. Again, the one known way to implement the sufficient condition in an a priori manner is to draw $\hat{\psi}_a$ and $\hat{\psi}_b$ from a common linear space [57]. However, in such a case, we can derive the theorem directly, as follows:

$$(\hat{\psi}_b, (\bar{H}_a - \bar{H}_b)\hat{\psi}_a) = (\hat{E}_a - \hat{E}_b)(\hat{\psi}_a, \hat{\psi}_b)$$

But $\Pi\hat{\psi}_a = \hat{\psi}_a$ and $\Pi\hat{\psi}_b = \hat{\psi}_b$ (note that under our assumption, the same Π is involved in $\bar{H}_a$ and $\bar{H}_b$), so that we may replace $\bar{H}_a$ and $\bar{H}_b$ by H_a and H_b, thereby deriving the desired theorem.

In practice this common linear space is sometimes formed from optimal trial functions derived from separate calculations, using different sets of trial functions for H_a and H_b, respectively. If one envisages doing this for a continuous sequence of H_a's (a, for example, might describe the nuclear configuration), then one should note that the resultant linear space will be a continuous one, and hence the homogeneous algebraic equations (6-1) are replaced by a homogeneous integral equation [61].

(iv) $\mathscr{G} \neq 1$; $H_a \neq H_b$. We leave it to the reader to name and discuss this case.

PROBLEM

1. In this chapter we have several times, without making any special remarks about it, considered more than one hypervirial theorem at a time. The present problem deals with some general results concerning simultaneous hypervirial theorems. Consider trial functions of the form $(\exp iA_1\mathscr{G}_1)(\exp iA_2\mathscr{G}_2)\Phi$ where $\mathscr{G}_1$ and $\mathscr{G}_2$ are hermitian operators, and where Φ does not depend explicitly on the numerical variational parameters A_1 and A_2. Show that $\partial\hat{E}/\partial\hat{A}_1 = 0$ and $\partial\hat{E}/\partial\hat{A}_2 = 0$ imply that the hypervirial theorems for both $\mathscr{G}_1$ and $\mathscr{G}_2$ are satisfied if $\mathscr{G}_1$ and $\mathscr{G}_2$ commute. Cite examples from the text which illustrate this theorem. Show that if they do not commute, then the theorems for $\mathscr{G}_1$ and $\mathscr{G}_2' = (\exp i\hat{A}_1\mathscr{G}_1)\mathscr{G}_2[\exp(-i\hat{A}_1\mathscr{G}_1)]$ will still be satisfied [33]. Show that if $\mathscr{G}_1$ and $\mathscr{G}_2$ form a Lie algebra, that is if $[\mathscr{G}_1, \mathscr{G}_2]$ is a linear combination of $\mathscr{G}_1$ and $\mathscr{G}_2$, then again the theorems for $\mathscr{G}_1$ and $\mathscr{G}_2$ will be satisfied [62]. [Hint: Show that then $\mathscr{G}_2'$ is a linear combination of $\mathscr{G}_1$ and $\mathscr{G}_2$.] Generalize these results to more than two $\mathscr{G}$'s. Show that the example discussed in [42] involves a Lie algebra and the sort of trial functions which we have been discussing. (Also see Problem 5, Appendix D.)

References

1. R. L. Matcha, *J. Chem. Phys.* **47**, 4595 (1967).
2. P. S. Bagus, B. Liu, and H. Schaefer III, *Phys. Rev. A* **2**, 555 (1970).
3. R. S. Loeb, and Y. Raisel, *J. Chem. Phys.* **52**, 4995 (1970), and references therein.
4. E. Brändas, *J. Mol. Spectrosc.* **27**, 236 (1968); C. A. Coulson and R. J. White, *Mol. Phys.* **18**, 577 (1970).
5. A. Messiah, "Quantum Mechanics," Vol. II. Wiley, New York, 1966.
6. R. Ditchfield, *J. Chem. Phys.* **56**, 5688 (1972).
7. S. T. Epstein, *J. Chem. Phys.* **58**, 1592 (1973).
8. I. G. Csizmadia, M. C. Harrison, J. W. Moskowitz, and B. T. Sutcliffe, *Theor. Chim. Acta* **6**, 191 (1966).
9. I. R. Epstein and W. N. Lipscomb, *J. Chem. Phys.* **53**, 4418 (1970), and references therein.
10. A. A. Frost, *J. Chem. Phys.* **25**, 1150 (1956).
11. W. Byers Brown and E. Steiner, *J. Chem. Phys.* **44**, 3934 (1966), and references therein.
12. R. McWeeny, *Chem. Phys. Lett.* **9**, 341 (1971).
13. J. Katriel, *Int. J. Quantum Chem.* **6**, 541 (1972), and references therein.
14. S. Lunell, *Phys. Rev. A* **7**, 1229 (1973), and references therein.
15. P.-O. Löwdin, *Rev. Mod. Phys.* **32**, 328 (1960).
16. R. Pauncz, "The Alternant Molecular Orbital Method." Saunders, Philadelphia, Pennsylvania, 1967.
17. N. MacDonald, *Advan. Phys.* **19**, 79 (1970).
18. T. Amos and G. G. Hall, *Proc. Roy. Soc. Ser. A* **263**, 483 (1962).
19. T. A. Claxton and D. McWilliams, *Theor. Chim. Acta* **16**, 346 (1970).
20. S. T. Epstein, *J. Chem. Phys.* **42**, 3813 (1965), and references therein.
21. S. T. Epstein, *J. Chem. Phys.* **46**, 571 (1967).
22. S. T. Epstein, *Amer. J. Phys.* **22**, 613 (1954).
23. J. I. Musher, *Amer. J. Phys.* **34**, 267 (1966).
24. A. C. Hurley, *Proc. Roy. Soc. Ser. A* **226**, 179 (1954).
25. N. F. Ramsey, *Phys. Rev.* **78**, 699 (1950); **83**, 540 (1951); **86**, 243 (1952).
26. R. M. Stevens, R. M. Pitzer, and W. N. Lipscomb, *J. Chem. Phys.* **38**, 550 (1963)
27. T. Amos and H. G. Ff. Roberts, *Mol. Phys.* **20**, 1073, 1081, 1089 (1971).
28. J. A. Pople, *Mol. Phys.* **1**, 175 (1958).
29. R. McWeeny, *Mol. Phys.* **1**, 175 (1958).
30. J. O. Hirschfelder, *J. Chem. Phys.* **33**, 1462 (1960).
31. N. Bjorna, *Mol. Phys.* **24**, 1 (1972), and references therein.
32. C. A. Coulson, *Quart. J. Math. Oxford Ser.* **16**, 279 (1965).
33. S. T. Epstein and J. O. Hirschfelder, *Phys. Rev.* **123**, 1495 (1961).
34. E. Brändas, *J. Mol. Spectrosc.* **27**, 236 (1968); J. Katriel and G. Adam, *Proc. Phys. Soc. London* (*At. Mol. Phys.*) **5**, 1 (1972).
35. R. Sternheimer, *Phys. Rev.* **96**, 951 (1954).
36. H. Sambe, *J. Chem. Phys.* **58**, 4779 (1973).
37. R. H. Young, *Mol. Phys.* **16**, 509 (1969).
38. K. Ruedenberg, *Rev. Mod. Phys.* **34**, 326 (1962).
39. W. Clinton, *J. Chem. Phys.* **33**, 1603 (1960).
40. C. J. H. Schutte, *Struct. Bonding* (*Berlin*) **9**, 213 (1971).

41. W. J. Meath and J. O. Hirschfelder, *J. Chem. Phys.* **39**, 1135 (1963).
42. D. Pandres, Jr., *Phys. Rev.* **131**, 886 (1963).
43. H. Hellmann, "Einfuhrung in die Quantenchemie," p. 285. Denticke, Leipzig, 1937; R. P. Feynman, *Phys. Rev.* **56**, 340 (1934).
44. J. O. Hirschfelder and W. J. Meath, *in* "Intermolecular Forces" (J. O. Hirschfelder, ed.). Academic Press, New York, 1967.
45. B. Nelander, *J. Chem. Phys.* **51**, 469 (1969).
46. E. A. Hylleraas, *Z. Phys.* **54**, 347 (1929); V. Fock, *ibid.* **63**, 855 (1930).
47. P.-O. Löwdin, *J. Mol. Spectrosc.* **3**, 46 (1959) and references therein; also W. A. Bingel, *Ann. Phys.* (*Leipzig*) **18**, 6 (1966); R. T. Brown, *J. Chem. Phys.* **48**, 4698 (1968).
48. J. O. Hirschfelder, and J. F. Kincaid, *Phys. Rev.* **52**, 658 (1937); A. D. McLean, *J. Chem. Phys.* **40**, 2774 (1964), and references therein.
49. C. A. Coulson and R. P. Bell, *Trans. Faraday Soc.* **41**, 141 (1945); P.-O. Löwdin, *J. Mol. Spectrosc.* **3**, 46 (1959), and references therein.
50. C. A. Coulson and A. C. Hurley, *J. Chem. Phys.* **37**, 448 (1962); P. Phillipson, *ibid.* **39**, 3010 (1963).
51. A. C. Hurley, *Proc. Roy. Soc. Ser. A* **226**, 170 (1954); *J. Chem. Phys.* **37**, 449 (1962).
52. K. H. Hansen, *Theor. Chim. Acta* **6**, 87 (1966).
53. G. Gliemann, *Theor. Chim. Acta* **11**, 75 (1968), and references therein.
54. M. Eger and E. P. Gross, *Ann. Phys.* (*Leipzig*) **24**, 63 (1963).
55. S. B. Trickery, N. M. Witriol, and G. L. Morley, *Phys. Rev. A* **7**, 1663 (1973), and references therein.
56. D. P. Chong and M. L. Benston, *J. Chem. Phys.* **49**, 1302 (1968), and references therein.
57. S. T. Epstein, A. C. Hurley, R. E. Wyatt, and R. G. Parr, *J. Chem. Phys.* **47**, 1275 (1967).
58. P. S. Bagus, *Phys. Rev. A* **139**, 619 (1965).
59. M. Cohen and A. Dalgarno, *Rev. Mod. Phys.* **35**, 506 (1963).
60. R. G. Parr, *J. Chem. Phys.* **40**, 3726 (1964).
61. A. C. Hurley, *Int. J. Quantum Chem.* **1**, S677 (1967); C. M. Farmer, *ibid.* **3**, 1027 (1969).
62. J. Katriel and E. Domany, *Chem. Phys. Lett.* **21**, 448 (1973).

Chapter V / Perturbation Theory and the Variation Method: General Theory

22. The Variation Principle and Perturbation Theory

There are many connections between the variation method and perturbation theory. In the succeeding sections, we will be concerned in the main with using the variation method to approximate solutions of the perturbation equations, and before that, with the closely related problem of the perturbation analysis of optimal trial functions and energies within the variation method. However, the variation principle can play other roles. For example, it can often offer insight into exact perturbation theory results. Thus let ν be the perturbation parameter. Then if $(\psi - \psi')$ is of order ν^{n+1}, where ψ' is an eigenfunction (not necessarily of the lowest state), it is an immediate consequence of (2-7) that $E - E'$ is of order ν^{2n+2}, and we have the well-known result that to calculate the energy through order ν^{2n+1}, it is sufficient to know the wave function only through order ν^{n} [1]. On the other hand, to derive this same result strictly within perturbation theory itself requires clever and intricate manipulations of the perturbation equations.[1] In Section 27, we will present a generalization of this theorem.

In this section, we wish to discuss yet another kind of connection; one which, so to speak, goes from perturbation theory to a variation principle rather than the other way round. The point is simply the following. The mark of a variation principle is that it leads to errors of second order in the quantity to be calculated (we have been concentrating on the energy). This suggests that if, using perturbation theory, we can give a formula for calculating a quantity correct through first order, that is, so that the error is of second order, then from this formula we should be able to infer a variation principle for the quantity in question. (This also suggests that

[1] See, for example, Dalgarno and Stewart [2] and Dupont-Bourdelet *et al.* [3].

starting from the results of higher-order perturbation theory, one could infer "supervariation principles.")[2]

Rather than discussing in general how to carry out this program [9], we will give one illustration. Thus consider the energy itself. Then given a zero-order wave function $\psi^{(0)}$, a zero-order energy $E^{(0)}$, and a zero-order Hamiltonian $H^{(0)}$ such that

$$(H^{(0)} - E^{(0)})\psi^{(0)} = 0 \tag{1}$$

one has the familiar result that, through first order, the energy eigenvalue for a Hamiltonian H is given by

$$E^{(0)} + \frac{(\psi^{(0)}, (H - H^{(0)})\psi^{(0)})}{(\psi^{(0)}, \psi^{(0)})} \tag{2}$$

However, using (1), the result (2) can be written as

$$(\psi^{(0)}, H\psi^{(0)})/(\psi^{(0)}, \psi^{(0)}) \tag{3}$$

In this last form, all explicit reference to $H^{(0)}$ has disappeared. Further, it is a formula which is correct through first order whatever $\psi^{(0)}$ may be. Hence we can infer that

$$(\psi, H\psi)/(\psi, \psi)$$

will yield a variational approximation to the energy; and we are back where we started from at the beginning of Section 2.

23. Perturbation Analysis of the Variation Method: Introduction

Let us suppose that the Hamiltonian takes the form

$$H = H^{(0)} + \nu H^{(1)} + \nu^2 H^{(2)} + \cdots \tag{1}$$

where the $H^{(n)}$ are independent of the real number ν, and that we are interested in exhibiting the $\hat{\psi}$ and the $\hat{E}$ and other expectation values as power series in ν,

$$\hat{\psi} = \hat{\psi}^{(0)} + \nu\hat{\psi}^{(1)} + \nu^2\hat{\psi}^{(2)} + \cdots \tag{2}$$

$$\hat{E} = \hat{E}^{(0)} + \nu\hat{E}^{(1)} + \nu^2\hat{E}^{(2)} + \cdots \tag{3}$$

[2] See, for example, Kikuta [4], Biedenharn and Blatt [5], Kostin and Brooks [6]. Hirschfelder [7], and Turinsky [8].

Of course such expansions may not exist, but assuming that they do, one way to derive them is to first carry out the complete calculation and then expand the results. The other approach, and the one we will explore in this and the next few sections, is to try to produce the successive terms in the series by a direct iterative method.

Before getting involved in the detailed formalism, we should first say a few words about the various possibilities for the $H^{(n)}$ that one encounters in applications. First of all, $H^{(0)}$ will usually be either the "accurate" Hamiltonian for an isolated system or some approximation to it. In saying this, we qualified the word accurate by using quotation marks because it is a relative term. Thus for an isolated atom or molecule, under many circumstances, the clamped nuclei Hamiltonian (1-1) in the absence of external fields is the standard of accuracy. Corrections are then adequately taken care of by *first*-order perturbation theory (and the adiabatic approximation for nuclear motion in the case of molecules). However, in other circumstances, this may well not be the case, and one may wish to include spin, relativistic, and/or nuclear motion effects in $H^{(0)}$.

If $H^{(0)}$ is an accurate Hamiltonian, then the perturbation will usually represent the effects of external fields and/or, in molecular applications, the effects of changing the nuclear configuration. On the other hand, if $H^{(0)}$ is an approximate Hamiltonian, then the perturbation may also include corrections for the difference between $H^{(0)}$ and the "exact" Hamiltonian of the isolated system. Thus for complete generality we should really replace each $\nu^n H^{(n)}$ by a sum of terms. However, since one can generalize the formulas at any time by the replacements

$$\nu H^{(1)} \rightarrow \nu_1 H^{(100\ldots)} + \nu_2 H^{(010\ldots)} + \cdots \tag{4}$$

$$\nu^2 H^{(2)} \rightarrow \nu_1{}^2 H^{(200\ldots)} + \nu_1\nu_2 H^{(110\ldots)} + \cdots$$
$$+ \nu_2{}^2 H^{(020\ldots)} + \nu_2\nu_3 H^{(0110\ldots)} + \cdots \quad \text{etc.} \tag{5}$$

we will forego this further notational complication for the present.

In general there are two main (though not always distinct) reasons for wanting to make the expansion of $\hat{E}$ and $\hat{\psi}$ at all. On the one hand, with $H^{(0)}$ an accurate Hamiltonian, if the perturbations are due to external fields or to changes in other parameters describing the system, then the individual expansion coefficients in $\hat{E}$ will often have a direct physical significance as approximations to permanent moments, polarizabilities, susceptibilities, force constants, etc. Now in the past, one has usually argued that it is probably better to calculate such quantities directly rather than, to take polarizability as an example, by calculating $\hat{E}$ for a series of weak fields and then trying to extract the polarizability by differencing or fitting the results to a polynomial in the field strength. For some reason, though, this latter approach was standard in calculating force constants. Recently, however, the counter argument has been made that it is more efficient to repeat the same calculation several times for different values of ν than to do somewhat different, separate calculations for the individual expansion coefficients, and thus more and more "curve fitting" is being done.[3] At the same time, though, in the calculation of force constants, there has been somewhat of a trend in the other direction,[4] with perturbation theory being used more and more as an alternative to the more traditional curve fitting.

On the other hand, with $H^{(0)}$ an approximate Hamiltonian, perturbation theory (perhaps augmented by techniques like the use of the Padé approximants) [19], is also used as a numerical technique for approximating the solution of the variational equations, especially in the linear variation method.[5] A situation in which these two aspects merge is in the use of expansions in $\mathscr{Z}^{-1}$ for atoms. Thus $\mathscr{Z}^{-1}$ is a real physical parameter (albeit not continuously variable) and the coefficients in the series for $\hat{E}$ and other expectation values can be compared with data on isoelectronic series. However, the $\mathscr{Z}^{-1}$ expansion also offers a rapidly converging technique for solving the variation equations of the linear variation method[6] and of Hartree–Fock approximations[7] for particular values of $\mathscr{Z}$. (See also Section 29.)

Returning to the first use, it should be kept in mind that these approximate permanent moments, polarizabilities, etc., will in general not have special variational relationships to their exact counterparts. Thus we know in general that

$$\hat{E} = E' + O(\psi' - \psi)^2$$

from which it follows that $\hat{E}$ will be a variational approximation to E' also for $\nu = 0$. However, unless $(\psi' - \hat{\psi})$ vanishes to some order as $\nu \to 0$, we can draw no conclusions about the relationship between the derivatives

[3] For example: Watson and Freeman [10], Cohen and Roothaan [11], McLean and Yoshimine [12], Grasso *et al.* [13], Pople *et al.* [14], Billingsley and Krauss [15].

[4] See Bishop and Macias [16], also Schutte [17] and Yde *et al.* [18].

[5] For example, Diner *et al.* [20], Gershgorn and Shavitt [21], Roos [22], and Huron *et al.* [23].

[6] Many papers by Midtdal and collaborators, for example, Aashamar *et al.* [24]. Also many papers by Dalgarno, Drake, and collaborators, for example, Doyle *et al.* [25].

[7] Many papers by Dalgarno, Cohen, and collaborators and by Coulson, Sharma, and collaborators. Some recent references are by Cohen [26] and Sharma [27].

of $\hat{E}$ at $\nu = 0$ and those of E'. Thus, in particular, for a ground state, increasing the flexibility of the trial functions, though it will certainly improve $\hat{E}^{(0)}$, need not, at least based on these simple considerations alone, improve $\hat{E}^{(1)}$, $\hat{E}^{(2)}$, etc., by bringing them into closer agreement with $E'^{(1)}$, $E'^{(2)}$, etc. This is not to say, of course, that empirically one may not find that this is systematically the case in certain calculations.

In what follows, our language will almost always be appropriate only to situations which are nondegenerate or effectively nondegenerate in zero order. Thus we will talk as though, aside from possible normalization ambiguities, that given $\hat{E}^{(0)}$, $\hat{\psi}^{(0)}$ is unique; and as though the first-order equations (24-15) and (24-17) to be given, then completely determine the first-order quantities, etc. Nevertheless, these equations are valid in any case. In our discussion of multiple perturbations, we will presumably, however, be quite explicitly dealing only with essentially nondegenerate situations, since we will assume that $\hat{E}$ and $\hat{\psi}$ can be expanded in a multiple power series in the ν_a. (Where there is degeneracy and, say, two ν_a, then already in exact Rayleigh–Schrödinger perturbation theory of ψ' the correct zeroth-order solutions for $\nu_1 \equiv 0$, $\nu_2 \to 0$ and for $\nu_1 \to 0$, $\nu_2 \equiv 0$ may be different,[8] so that simple power series may not exist.)

24. Perturbation Analysis of the Variation Method: General Formalism

Let us now return to our problem, the expansion of $\hat{E}$ and $\hat{\psi}$ in powers of ν.[9] In general, $\hat{\psi}$ will depend on ν, first because the optimal variational parameters $\hat{A}$ depend on ν, and in addition there may be some further explicit ν dependence; thus

$$\hat{\psi} = \psi(\hat{A}(\nu), \nu) \tag{1}$$

We will now assume that $\hat{\psi}$ can be expanded in a power series in ν. A sufficient condition for this (beyond the obvious requirement that the explicit ν dependence be of an analytic type) is that the $\hat{A}$ can also be so expanded, thus

$$\hat{A} = \hat{A}^{(0)} + \nu\hat{A}^{(1)} + \nu^2\hat{A}^{(2)} + \cdots \tag{2}$$

and we will assume this as well. However, although this is the most common situation, it does not cover all the cases which one meets in practice, as we will mention briefly in Section 26.

[8] See, for example, Hirschfelder *et al.* [28, Section IV E].

[9] For related discussions, see Silverman and van Leuvan [29], and Harriman [30].

Using (2) in (1), we can now expand $\hat{\psi}$ in a power series in ν as

$$\hat{\psi} = \hat{\psi}^{(0)} + \nu\hat{\psi}^{(1)} + \nu^2\hat{\psi}^{(2)} + \cdots \tag{3}$$

where

$$\hat{\psi}^{(0)} \equiv \psi(\hat{A}^{(0)}, 0) \tag{4}$$

$$\hat{\psi}^{(1)} \equiv \frac{\partial\hat{\psi}^{(0)}}{\partial\hat{A}^{(0)}}\hat{A}^{(1)} + \left(\frac{\partial\hat{\psi}}{\partial\nu}\right)_0 \tag{5}$$

$$\hat{\psi}^{(2)} \equiv \frac{\partial\hat{\psi}^{(0)}}{\partial\hat{A}^{(0)}}\hat{A}^{(2)} + \frac{1}{2}\frac{\partial^2\hat{\psi}^{(0)}}{\partial\hat{A}^{(0)}\,\partial\hat{A}^{(0)}}\hat{A}^{(1)}\hat{A}^{(1)} + \left(\frac{\partial^2\hat{\psi}}{\partial\nu\,\partial\hat{A}^{(0)}}\right)_0\hat{A}^{(1)} + \frac{1}{2}\left(\frac{\partial^2\hat{\psi}}{\partial\nu^2}\right)_0 \tag{6}$$

etc., and where subscript zero means that the quantity is to be evaluated at $\nu = 0$ and hence also with $\hat{A} = \hat{A}^{(0)}$. In writing these equations, and in writing (2) earlier, we have, for compactness of notation, suppressed the labels on the A's. Thus in detail

$$\begin{aligned} \frac{\partial\hat{\psi}^{(0)}}{\partial\hat{A}^{(0)}}\hat{A}^{(1)} &\equiv \sum_k \frac{\partial\hat{\psi}^{(0)}}{\partial\hat{A}_k^{(0)}}\hat{A}_k^{(1)} \\ \frac{\partial^2\hat{\psi}^{(0)}}{\partial\hat{A}^{(0)}\,\partial\hat{A}^{(0)}}\hat{A}^{(1)}\hat{A}^{(1)} &\equiv \sum_k\sum_l \frac{\partial^2\hat{\psi}^{(0)}}{\partial\hat{A}_k^{(0)}\,\partial\hat{A}_l^{(0)}}\hat{A}_k^{(1)}\hat{A}_l^{(1)} \end{aligned} \tag{7}$$

etc.

For $\delta\hat{\psi}$, we have, from its definition, Eq. (4-2),

$$\delta\hat{\psi} \equiv \frac{\partial\hat{\psi}}{\partial\hat{A}}\,\delta A = \frac{\partial\hat{\psi}}{\partial\hat{A}^{(0)}}\,\delta A \tag{8}$$

and where, in accord with the discussion following Eq. (4-3), δA may without loss in generality be assumed independent of ν. Then using Eqs. (3)–(6), we find as the expansion of $\delta\hat{\psi}$

$$\delta\hat{\psi} = \delta^0\hat{\psi} + \nu\,\delta^1\hat{\psi} + \nu^2\,\delta^2\hat{\psi} + \cdots \tag{9}$$

where

$$\delta^0\hat{\psi} \equiv \frac{\partial\hat{\psi}^{(0)}}{\partial\hat{A}^{(0)}}\,\delta A \tag{10}$$

$$\delta^1\hat{\psi} \equiv \frac{\partial^2\hat{\psi}^{(0)}}{\partial\hat{A}^{(0)}\,\partial\hat{A}^{(0)}}\,\delta A\,\hat{A}^{(1)} + \left(\frac{\partial^2\hat{\psi}}{\partial\hat{A}^{(0)}\,\partial\nu}\right)_0\delta A \tag{11}$$

etc.

We now insert these results into the basic equations of the variation method,

$$(\delta\hat{\psi}, (H - \hat{E})\hat{\psi}) + \text{comp. conj.} = 0 \tag{12}$$

and

$$(\hat{\psi}, (H - \hat{E})\hat{\psi}) = 0 \tag{13}$$

Then, writing $\hat{E}$ as

$$\hat{E} = \hat{E}^{(0)} + \nu\hat{E}^{(1)} + \nu^2\hat{E}^{(2)} + \cdots$$

and equating separately to zero the terms of each order in ν, one readily finds that (12) yields

$$(\delta^0\hat{\psi}, (H^{(0)} - \hat{E}^{(0)})\hat{\psi}^{(0)}) + \text{comp. conj.} = 0 \tag{14}$$

$$\begin{aligned}\{(\delta^0\hat{\psi}, (H^{(0)} - \hat{E}^{(0)})\hat{\psi}^{(1)}) + (\delta^0\hat{\psi}, (H^{(1)} - \hat{E}^{(1)})\hat{\psi}^{(0)})\\ + (\delta^1\hat{\psi}, (H^{(0)} - \hat{E}^{(0)})\hat{\psi}^{(0)})\} + \text{comp. conj.} = 0\end{aligned} \tag{15}$$

etc., while (13) yields

$$(\hat{\psi}^{(0)}, (H^{(0)} - \hat{E}^{(0)})\hat{\psi}^{(0)}) = 0 \tag{16}$$

$$\begin{aligned}(\hat{\psi}^{(1)}, (H^{(0)} - \hat{E}^{(0)})\hat{\psi}^{(0)}) + (\hat{\psi}^{(0)}, (H^{(0)} - \hat{E}^{(0)})\hat{\psi}^{(1)})\\ + (\hat{\psi}^{(0)}, (H^{(1)} - \hat{E}^{(1)})\hat{\psi}^{(0)}) = 0\end{aligned} \tag{17}$$

$$\begin{aligned}(\hat{\psi}^{(2)}, (H^{(0)} - \hat{E}^{(0)})\hat{\psi}^{(0)}) + (\hat{\psi}^{(0)}, (H^{(0)} - \hat{E}^{(0)})\hat{\psi}^{(2)})\\ + (\hat{\psi}^{(1)}, (H^{(0)} - \hat{E}^{(0)})\hat{\psi}^{(1)}) + (\hat{\psi}^{(1)}, (H^{(1)} - \hat{E}^{(1)})\hat{\psi}^{(0)})\\ + (\hat{\psi}^{(0)}, (H^{(1)} - \hat{E}^{(1)})\hat{\psi}^{(1)}) + (\hat{\psi}^{(0)}, (H^{(2)} - \hat{E}^{(2)})\hat{\psi}^{(0)}) = 0\end{aligned} \tag{18}$$

etc. We will discuss these equations in detail only through first order in the wave function and second order in the energy, since this will suffice to illustrate the essential features.

Equations (14) and (16) need little comment. They are just the familiar equations for determining $\hat{\psi}^{(0)}$ and $\hat{E}^{(0)}$. Let us now consider (17). The essential observation here is that the first term on the right in (5) is of the form $\delta^0\hat{\psi}$ with $\delta A = \hat{A}^{(1)}$ and hence, from (14), makes no contribution to (17), which therefore becomes

$$\begin{aligned}\left(\left(\frac{\partial\hat{\psi}}{\partial\nu}\right)_0, (H^{(0)} - \hat{E}^{(0)})\hat{\psi}^{(0)}\right) + \left(\hat{\psi}^{(0)}, (H^{(0)} - \hat{E}^{(0)})\left(\frac{\partial\hat{\psi}}{\partial\nu}\right)_0\right)\\ + (\hat{\psi}^{(0)}, (H^{(1)} - \hat{E}^{(1)})\hat{\psi}^{(0)}) = 0\end{aligned} \tag{19}$$

Thus we can calculate $\hat{E}^{(1)}$ from $\hat{A}^{(0)}$ alone, we do not need to know $\hat{A}^{(1)}$. We will return to this equation in a little bit, but now, having determined $\hat{E}^{(1)}$ let us consider Eq. (15).

From the formulas (10) and (11) for $\delta^0\hat{\psi}$ and $\delta^1\hat{\psi}$ and from the formula (5) for $\hat{\psi}^{(1)}$ it follows that in general (15) will yield coupled linear inhomogeneous equations for the $\hat{A}^{(1)}$ and $\hat{A}^{(1)*}$. Further, it is easy to see that the same pattern will repeat in higher orders with only the structure of the inhomogeneity changing from order to order.

If the A_i are independent, arbitrary complex numbers, then one readily finds that these sets of equations can be written as

$$\sum_j (\gamma_{ij}^{(0)} \hat{A}_j^{(n)} + \eta_{ij}^{(0)} \hat{A}_j^{(n)*}) + C_j^{(n)} = 0, \qquad n > 0 \tag{20}$$

where $\gamma_{ij}^{(0)}$ and $\eta_{ij}^{(0)}$ are the γ_{ij} and η_{ij} of (4-13) but evaluated at $\nu = 0$, and where $C_j^{(n)}$ represents the inhomogeneity which involves only lower-order quantities and hence can be considered as known. ($C_j^{(n)}$ also involves $\hat{E}^{(n)}$, but as we have seen by example, and will prove in general, this can be calculated from lower-order results.) If we introduce the "two-component" vectors

$$\hat{\mathbb{A}}^{(n)} \equiv \begin{pmatrix} \hat{A}^{(n)} \\ \hat{A}^{(n)*} \end{pmatrix}; \qquad \mathbb{C}^{(n)} \equiv \begin{pmatrix} C^{(n)} \\ C^{(n)*} \end{pmatrix} \tag{21}$$

where $\hat{A}^{(n)}$ is itself a vector whose components are the $\hat{A}_j^{(n)}$, etc., then (20) and its complex conjugate can be combined into

$$\Lambda^{(0)}\hat{\mathbb{A}}^{(n)} + \mathbb{C}^{(n)} = 0, \qquad n > 0 \tag{22}$$

where $\Lambda^{(0)}$ is the Λ of (4-15) but evaluated at $\nu = 0$. In particular

$$C_j^{(1)} = \left(\left(\frac{\partial^2 \hat{\psi}}{\partial \hat{A}_j^{(0)}\, \partial \nu} \right)_0, (H^{(0)} - \hat{E}^{(0)})\hat{\psi}^{(0)} \right) + \left(\frac{\partial \hat{\psi}^{(0)}}{\partial \hat{A}_j^{(0)}}, (H^{(0)} - \hat{E}^{(0)}) \left(\frac{\partial \hat{\psi}}{\partial \nu} \right)_0 \right)$$
$$+ \left(\frac{\partial \hat{\psi}^{(0)}}{\partial \hat{A}_j^{(0)}}, (H^{(1)} - \hat{E}^{(1)})\hat{\psi}^{(0)} \right) \tag{23}$$

If we denote the orthonormal eigenvectors of $\Lambda^{(0)}$ by $Y_{(i)}^{(0)}$ and the eigenvalues by $\lambda_i^{(0)}$ [recall (4-18)–(4-19)], then the solution of (22) can be written as

$$\hat{\mathbb{A}}^{(n)} = -\sum_i{}' \, Y_{(i)}^{(0)} \frac{(Y_{(i)}^{(0)+} \cdot \mathbb{C}^{(n)})}{\lambda_i^{(0)}}, \qquad n > 0 \tag{24}$$

the prime meaning that the sum goes over only those $Y_{(i)}^{(0)}$ for which $\lambda_i^{(0)} \neq 0$. We could, of course, add to this solution arbitrary real multiples of any $Y_{(i)}^{(0)}$ [of the correct form—recall the discussion following (4-22)] for which $\lambda_i^{(0)} = 0$. [Note that (22) implies that the $\mathbb{C}^{(n)}$ must of necessity be orthogonal to such $Y_{(i)}^{(0)}$.] However, assuming that there are no real degeneracy problems, this ambiguity, if it exists, can only reflect a corresponding ambiguity in the $\hat{A}$'s of the sort which we discussed preceding Eq. (8-4), and so can be ignored.

To see this, we note that if there are ambiguities, then this means there are some variations of the A_i, call them $\bar{\delta} A_i$, which either leave $\psi(A)$ unchanged or turn it into a multiple of itself. Thus

$$\bar{\delta}\psi \equiv \sum_j \frac{\partial \psi}{\partial A_j} \bar{\delta} A_j = a\psi \tag{25}$$

where a is some number, possibly zero.

We will now show that

$$\bar{\delta}\mathbb{A}^{(0)} \equiv \begin{pmatrix} \bar{\delta}\hat{A}^{(0)} \\ \bar{\delta}\hat{A}^{(0)*} \end{pmatrix}$$

where $\bar{\delta}\hat{A}^{(0)}$ is $\bar{\delta}\hat{A}$ for $A = \hat{A}^{(0)}$ and $\nu = 0$, is an (in general unnormalized) eigenvector of $\Lambda^{(0)}$ with eigenvalue zero.

First we have that

$$\sum_j \gamma_{ij}^{(0)} \bar{\delta}\hat{A}_i^{(0)} = \sum_j \left(\frac{\partial \hat{\psi}^{(0)}}{\partial \hat{A}_i^{(0)}}, (H^{(0)} - \hat{E}^{(0)}) \frac{\partial \hat{\psi}^{(0)}}{\partial \hat{A}_j^{(0)}} \bar{\delta}\hat{A}_j^{(0)} \right)$$

Therefore, from (25) evaluated at $A = \hat{A}^{(0)}$ and $\nu = 0$,

$$\sum_j \gamma_{ij}^{(0)} \bar{\delta}\hat{A}_j^{(0)} = \left(\frac{\partial \hat{\psi}^{(0)}}{\partial \hat{A}_i^{(0)}}, (H^{(0)} - \hat{E}^{(0)})\hat{\psi}^{(0)} \right) \hat{a}^{(0)}$$

and hence

$$\sum_j \gamma_{ij}^{(0)} \bar{\delta}\hat{A}_j^{(0)} = 0$$

since

$$\left(\frac{\partial \hat{\psi}^{(0)}}{\partial \hat{A}_i^{(0)}}, (H^{(0)} - \hat{E}^{(0)})\hat{\psi}^{(0)} \right) = 0$$

are the defining equations for $\hat{\psi}^{(0)}$.

We now consider

$$\sum_j \eta_{ij}^{(0)}\, \bar{\delta}\hat{A}_j^{(0)*} \equiv \sum_j \left(\frac{\partial^2 \hat{\psi}^{(0)}}{\partial \hat{A}_i^{(0)}\, \partial \hat{A}_j^{(0)}}\, \bar{\delta}\hat{A}_j^{(0)}, (H^{(0)} - \hat{E}^{(0)})\hat{\psi}^{(0)} \right)$$

If we differentiate (25) with respect to A_i and then put $A = \hat{A}^{(0)}$ and $\nu = 0$, we find that

$$\sum_j \frac{\partial^2 \hat{\psi}^{(0)}}{\partial \hat{A}_i^{(0)}\, \partial \hat{A}_j^{(0)}}\, \bar{\delta}\hat{A}_j^{(0)} = \frac{\partial \hat{a}^{(0)}}{\partial \hat{A}_i^{(0)}} \hat{\psi}^{(0)} + \hat{a}^{(0)} \frac{\partial \hat{\psi}^{(0)}}{\partial \hat{A}_i^{(0)}} - \sum_j \frac{\partial \hat{\psi}^{(0)}}{\partial \hat{A}_j^{(0)}} \left(\frac{\partial \bar{\delta}\hat{A}_j^{(0)}}{\partial \hat{A}_i^{(0)}} \right)$$

and therefore

$$\begin{aligned} \sum_j \eta_{ij}^{(0)}\, \bar{\delta}\hat{A}_j^{(0)*} = {} & \left(\frac{\partial \hat{a}^{(0)}}{\partial \hat{A}_i^{(0)}} \right)^* (\hat{\psi}^{(0)}, (H^{(0)} - \hat{E}^{(0)})\hat{\psi}^{(0)}) \\ & + \hat{a}^{(0)*} \left(\frac{\partial \hat{\psi}^{(0)}}{\partial \hat{A}_j^{(0)}}, (H^{(0)} - \hat{E}^{(0)})\hat{\psi}^{(0)} \right) \\ & - \sum_j \left(\frac{\partial \bar{\delta}\hat{A}_j^{(0)}}{\partial \hat{A}_i^{(0)}} \right)^* \left(\frac{\partial \hat{\psi}^{(0)}}{\partial \hat{A}_j^{(0)}}, (H^{(0)} - \hat{E}^{(0)})\hat{\psi}^{(0)} \right) = 0 \end{aligned}$$

In summary, then,

$$\sum_j \{\gamma_{ij}^{(0)}\, \bar{\delta}\hat{A}_j^{(0)} + \eta_{ij}^{(0)}\, \bar{\delta}\hat{A}_j^{(0)*}\} = 0$$

and hence, as claimed,

$$\Lambda\, \bar{\delta}\mathscr{A}^{(0)} = 0$$

Further if we *do* add a multiple $\alpha^{(n)}$ of $\bar{\delta}\hat{A}^{(0)}$ to a given $\hat{A}^{(n)}$, then, since

$$\hat{\psi}^{(n)} = \sum_j \frac{\partial \hat{\psi}^{(0)}}{\partial \hat{A}^{(0)}} \hat{A}_j^{(n)} + \cdots$$

we see from (25) that $\hat{\psi}^{(n)} \to \hat{\psi}^{(n)} + \alpha^{(n)}\hat{a}^{(0)}\hat{\psi}^{(0)}$. Therefore, through nth order, as expected, $\hat{\psi}$ at most changes in overall scale and phase, namely

$$\hat{\psi} \to \hat{\psi} + \nu^n \alpha^{(n)} \hat{a}^{(0)} \hat{\psi}^{(0)} \simeq [\exp(\nu^n \alpha^{(n)} \hat{a}^{(0)})]\hat{\psi}$$

thereby justifying our claim that the omission of the $\bar{\delta}\hat{\mathscr{A}}^{(0)}$ type eigenvectors from the right hand side of (24) meant no real loss of generality.

On the other hand, if there are nontrivial eigenvectors with zero eigenvalues, and let us denote these eigenvectors collectively by Z, then some of our earlier statements need modification. Thus the lower-order $\hat{\mathscr{A}}^{(m)}$ and hence $\mathbb{C}^{(n)}$ may well not be completely known even when one imposes the further consistency condition that $\mathbb{C}^{(n)}$ be orthogonal to all the Z's. However, as in Rayleigh–Schrödinger degenerate perturbation theory, presumably one will still have enough information to compute $\hat{E}^{(n)}$. Further, one should also now really write (24) as

$$\hat{A}_{\perp}^{(n)} = -\sum_i{}' Y_{(i)}^{(0)} \frac{(Y_{(i)}^{(0)+} \cdot \mathbb{C}^{(n)})}{\lambda_i^{(0)}}$$

the $\perp$ meaning that it is only the part of $\hat{A}^{(n)}$ which is orthogonal to the Z's. The rest of $\hat{A}^{(n)}$, and also $\mathbb{C}^{(n)}$ if it is not completely known at this stage, then remain to be determined by higher-order equations and consistency conditions.

Turning now to Eq. (18) for $\hat{E}^{(2)}$, at first sight this seems to require a knowledge of $\hat{A}^{(2)}$. However, just as we showed that (17) does not involve $\hat{A}^{(1)}$, we will now show that (18) does not involve $\hat{A}^{(2)}$. The point is simply that from (6), the $\hat{A}^{(2)}$ in $\hat{\varphi}^{(2)}$ appear only in the combination $(\partial\hat{\varphi}^{(0)}/\partial\hat{A}^{(0)})\hat{A}^{(2)}$ which is of the form $\delta^0\hat{\varphi}$ with $\delta A = \hat{A}^{(2)}$ and hence makes no contribution to (18). Thus we can replace (18) by

$$\begin{aligned}(\bar{\varphi}^{(2)}, (H^{(0)} - \hat{E}^{(0)})\hat{\varphi}^{(0)}) &+ (\hat{\varphi}^{(0)}, (H^{(0)} - \hat{E}^{(0)})\bar{\varphi}^{(2)}) \\ &+ (\hat{\varphi}^{(1)}, (H^{(0)} - \hat{E}^{(0)})\hat{\varphi}^{(1)}) + (\hat{\varphi}^{(1)}, (H^{(1)} - \hat{E}^{(1)})\hat{\varphi}^{(0)}) \\ &+ (\hat{\varphi}^{(0)}, (H^{(1)} - \hat{E}^{(1)})\hat{\varphi}^{(1)}) + (\hat{\varphi}^{(0)}, (H^{(2)} - \hat{E}^{(2)})\hat{\varphi}^{(0)}) = 0 \end{aligned} \tag{26}$$

where

$$\bar{\varphi}^{(2)} \equiv \frac{1}{2}\frac{\partial^2\hat{\varphi}^{(0)}}{\partial\hat{A}^{(0)}\,\partial\hat{A}^{(0)}}\hat{A}^{(1)}\hat{A}^{(1)} + \left(\frac{\partial^2\hat{\varphi}}{\partial\nu\,\partial\hat{A}^{(0)}}\right)_0 \hat{A}^{(1)} + \frac{1}{2}\left(\frac{\partial^2\hat{\varphi}}{\partial\nu^2}\right)_0 \tag{27}$$

from which, knowing $\hat{A}^{(0)}$ and $\hat{A}^{(1)}$, we can calculate $\hat{E}^{(2)}$. Then, given $\hat{E}^{(2)}$, we can turn to the next equation in the sequence (14)-(15), an equation which we have not written out, and determine $\hat{A}^{(2)}$. Similarly, since in general $\hat{\varphi}^{(n)}$ involves $\hat{A}^{(n)}$ only in the combination $(\partial\hat{\varphi}^{(0)}/\partial\hat{A}^{(0)})\hat{A}^{(n)}$, it is clear that we can calculate $\hat{E}^{(n)}$ without knowing $\hat{A}^{(n)}$, use it to determine $\hat{A}^{(n)}$, etc. Further, we will show in Section 26 that in fact given $\hat{A}$ accurate through order ν^n, we can calculate $\hat{E}$ accurate through order ν^{2n+1}. Thus, in particular, given $\hat{A}^{(0)}$ and $\hat{A}^{(1)}$, we can calculate not only $\hat{E}^{(0)}$, $\hat{E}^{(1)}$, and $\hat{E}^{(2)}$, but also $\hat{E}^{(3)}$, although the argument of the preceding sentences would suggest that although we did not need $\hat{A}^{(3)}$ for the latter, we probably did need $\hat{A}^{(2)}$.

If the A_i are (independent, arbitrary) complex numbers, then one finds from the various definitions that (18) can be written as

$$(\hat{\psi}^{(0)}, (\hat{E}^{(2)} - H^{(2)})\hat{\psi}^{(0)}) = \sum_i \sum_j \{(\tfrac{1}{2}\hat{A}_i^{(1)*}\eta_{ij}^{(0)}\hat{A}_i^{(1)*} + \text{comp. conj.}) + \hat{A}_i^{(1)*}\gamma_{ij}^{(0)}\hat{A}_j^{(1)}\} + \sum_i (\hat{A}_i^{(1)*}C_i^{(1)} + C_i^{(1)*}\hat{A}_i^{(1)}) + \xi^{(2)} \tag{28}$$

where

$$\xi^{(2)} \equiv \left\{\frac{1}{2}\left(\left(\frac{\partial^2\hat{\psi}}{\partial\nu^2}\right)_0, (H^{(0)} - \hat{E}^{(0)})\hat{\psi}^{(0)}\right) + \left(\left(\frac{\partial\hat{\psi}}{\partial\nu}\right)_0, (H^{(1)} - \hat{E}^{(1)})\hat{\psi}^{(0)}\right) + \text{comp. conj.}\right\} + \left(\left(\frac{\partial\hat{\psi}}{\partial\nu}\right)_0, (H^{(0)} - \hat{E}^{(0)})\left(\frac{\partial\hat{\psi}}{\partial\nu}\right)_0\right) \tag{29}$$

Moreover, using (20) for $n = 1$ and the fact that $\gamma_{ij}^{(0)*} = \gamma_{ji}^{(0)}$, we can rewrite (28) as

$$(\hat{\psi}^{(0)}, (\hat{E}^{(2)} - H^{(2)})\hat{\psi}^{(0)}) = \tfrac{1}{2}\sum_i (\hat{A}_i^{(1)*}C_i^{(1)} + C_i^{(1)*}\hat{A}_i^{(1)}) + \xi^{(2)}$$

or

$$(\hat{\psi}^{(0)}, (\hat{E}^{(2)} - H^{(2)})\hat{\psi}^{(0)} = \tfrac{1}{2}(\mathbb{A}^{(1)+}\cdot \mathbb{C}^{(1)}) + \xi^{(2)} \tag{30}$$

Therefore using (24), we have the rather nice result that

$$(\hat{\psi}^{(0)}, (\hat{E}^{(2)} - H^{(2)})\hat{\psi}^{(0)}) = -\frac{1}{2}\sum_i{}' \frac{|(Y_{(i)}^{(0)+}\cdot \mathbb{C}^{(1)})|^2}{\lambda_i^{(0)}} + \xi^{(2)} \tag{31}$$

If there are two perturbations, then one simply makes the replacements

$$\nu H^{(1)} \to \nu_1 H^{(10)} + \nu_2 H^{(01)}$$
$$\nu^2 H^{(2)} \to \nu_1^2 H^{(20)} + \nu_1\nu_2 H^{(11)} + \nu_2^2 H^{(02)}$$

and

$$\nu\left(\frac{\partial\hat{\psi}}{\partial\nu}\right)_0 \to \nu_1\left(\frac{\partial\hat{\psi}}{\partial\nu_1}\right)_0 + \nu_2\left(\frac{\partial\hat{\psi}}{\partial\nu_2}\right)_0$$
$$\nu^2\left(\frac{\partial^2\hat{\psi}}{\partial\nu^2}\right)_0 \to \nu_1^2\left(\frac{\partial^2\hat{\psi}}{\partial\nu_1^2}\right)_0 + 2\nu_1\nu_2\left(\frac{\partial^2\hat{\psi}}{\partial\nu_1\,\partial\nu_2}\right)_0 + \nu_2^2\left(\frac{\partial^2\hat{\psi}}{\partial\nu_2^2}\right)_0$$

from which it follows that

$$\nu\mathbb{C}^{(1)} \to \nu_1\mathbb{C}^{(10)} + \nu_2\mathbb{C}^{(01)}$$

and

$$\nu^2 \xi^{(2)} \rightarrow \nu_1{}^2 \xi^{(20)} + \nu_1 \nu_2 \xi^{(11)} + \nu_2{}^2 \xi^{(20)}$$

the definition of the various quantities being obvious. Finally from (31) we have that

$$\nu^2 \hat{E}^{(2)} \rightarrow \nu_1{}^2 \hat{E}^{(20)} + \nu_1 \nu_2 \hat{E}^{(11)} + \nu_2{}^2 \hat{E}^{(02)}$$

where

$$(\hat{\psi}^{(0)}, (\hat{E}^{(20)} - H^{(20)})\hat{\psi}^{(0)}) = -\frac{1}{2} \sum_i{}' \frac{|\, (Y_{(i)}^{(0)+} \cdot \mathbb{C}^{(10)}\, |^2}{\lambda_i^{(0)}} + \xi^{(20)}$$

$$(\hat{\psi}^{(0)}, (\hat{E}^{(02)} - H^{(02)})\hat{\psi}^{(0)}) = -\frac{1}{2} \sum_i{}' \frac{|\, (Y_{(i)}^{(0)+} \cdot \mathbb{C}^{(01)}\, |^2}{\lambda_i^{(0)}} + \xi^{(02)}$$

and

$$(\hat{\psi}^{(0)}, (\hat{E}^{(11)} - H^{(11)})\hat{\psi}^{(0)})$$
$$= -\frac{1}{2} \sum_i{}' \frac{(\mathbb{C}^{(10)+} \cdot Y_{(i)}^{(0)})(Y_{(i)}^{(0)+} \cdot \mathbb{C}^{(01)}) + (\mathbb{C}^{(01)+} \cdot Y_{(i)}^{(0)})(Y_{(i)}^{(0)+} \cdot \mathbb{C}^{(10)})}{\lambda_i^{(0)}}$$
$$+ \xi^{(11)} \tag{32}$$

If all quantities are explicitly real, so that $\hat{A}_j^{(n)*} = \hat{A}_j^{(n)}$ and so that $\eta_{ij}^{(0)}$ is Hermitian, or if all the A_i are linear parameters so that $\eta_{ij}^{(0)}$ vanishes, then, as in Section 4, there is no need to introduce the two-component notation. Thus if we denote by $y_{(i)}^{(0)}$ the orthonormal eigenvectors of $\gamma^{(0)} + \eta^{(0)}$, then the solution of (20) can be written as

$$\hat{A}^{(n)} = -\sum_i{}' y_{(i)}^{(0)} \frac{(y_{(i)}^{(0)+} \cdot C^{(n)})}{\lambda_i^{(0)}}, \qquad n > 0 \tag{33}$$

whence, from (30), we have

$$(\hat{\psi}^{(0)}, (\hat{E}^{(2)} - H^{(2)})\hat{\psi}^{(0)}) = -\sum_i{}' \frac{|\, (y_{(i)}^{(0)+} \cdot C^{(1)})\, |^2}{\lambda_i^{(0)}} + \xi^{(2)} \tag{34}$$

with analogous formulas in the case of two perturbations. In particular

$$(\hat{\psi}^{(0)}, (\hat{E}^{(11)} - H^{(11)})\hat{\psi}^{(0)})$$
$$= -\sum_i{}' \frac{(C^{(10)+} \cdot y_{(i)}^{(0)})(y_{(i)}^{(0)+} \cdot C^{(01)}) + (C^{(01)+} \cdot y_{(i)}^{(0)})(y_{(i)}^{(0)+} \cdot C^{(10)})}{\lambda_i^{(0)}} + \xi^{(11)} \tag{35}$$

If the set of trial functions is independent of ν so that the generalized Hellmann–Feynman theorem for $\sigma = \nu$ is satisfied, then the various formulas simplify considerably, and also they can be written in other interesting ways. Thus from our discussion in Section 15, we know that under these conditions, $(\partial\hat{\psi}/\partial\nu)_0$ will be of the form $\delta^0\hat{\psi}$ for some δA and that therefore the sum of the first two terms on the left in (19) will vanish. Therefore when the set of trial functions is independent of ν, we will have

$$\hat{E}^{(1)} = \frac{(\hat{\psi}^{(0)}, H^{(1)}\hat{\psi}^{(0)})}{(\hat{\psi}^{(0)}, \hat{\psi}^{(0)})} \tag{36}$$

which of course also follows directly from the Hellmann–Feynman theorem

$$\left(\hat{\psi}, \left(\frac{\partial\hat{E}}{\partial\nu} - \frac{\partial H}{\partial\nu}\right)\hat{\psi}\right) = 0 \tag{37}$$

on putting ν equal to zero. Similarly the terms of first order in ν in (37) are readily seen to yield a compact formula for $\hat{E}^{(2)}$,

$$\begin{aligned}(\hat{\psi}^{(0)}, &(\hat{E}^{(2)} - H^{(2)})\hat{\psi}^{(0)}) \\ &= \tfrac{1}{2}[(\hat{\psi}^{(1)}, (H^{(1)} - \hat{E}^{(1)})\hat{\psi}^{(0)}) + (\hat{\psi}^{(0)}, (H^{(1)} - \hat{E}^{(1)})\hat{\psi}^{(1)})]\end{aligned} \tag{38}$$

By comparison of Eqs. (38) and (26), we see further that when the set of trial functions is independent of ν, we also have that

$$\begin{aligned}(\hat{\psi}^{(0)}, (\hat{E}^{(2)} - H^{(2)})\hat{\psi}^{(0)}) = -\{&(\bar{\psi}^{(2)}, (H^{(0)} - \hat{E}^{(0)})\hat{\psi}^{(0)}) \\ &+ (\hat{\psi}^{(1)}, (H^{(0)} - \hat{E}^{(0)})\hat{\psi}^{(1)}) \\ &+ (\hat{\psi}^{(0)}, (H^{(0)} - \hat{E}^{(0)})\bar{\psi}^{(2)})\}\end{aligned} \tag{39}$$

The right hand side of this equation evidently has a structure reminiscent of that of (4-10). We will now show that if $\hat{E}^{(0)}$ is a local minimum or, more precisely, and less restrictively, if $\delta_2\hat{E}^{(0)}$ is ≥ 0 for a certain δA to be defined later in Eq. (46), then [31, 31a]

$$(\hat{\psi}^{(0)}, (\hat{E}^{(2)} - H^{(2)})\hat{\psi}^{(0)}) \leq 0 \tag{40}$$

That is, when the Hellmann–Feynman theorem for $\sigma = \nu$ is satisfied, we will have the variational analog of the result in Rayleigh–Schrödinger perturbation theory for E' whose simplest and best-known corollary is that, in second order, $H^{(1)}$ always acts so as to lower the ground state energy.

To derive (40), we first note that from (5) and (27) it follows that

$$\hat{\psi}^{(0)} + \nu\hat{\psi}^{(1)} + \nu^2\bar{\psi}^{(2)} = \psi(\hat{A}^{(0)} + \nu\hat{A}^{(1)}, \nu) + \cdots \tag{41}$$

where the dots represent terms of third and higher order in ν. However, the fact that the set of trial functions is independent of ν implies that we can write the right hand side of (41) as

$$\psi(\hat{A}^{(0)} + \nu\hat{A}^{(1)} + \bar{A}, 0) + \cdots \tag{42}$$

for some $\bar{A}$ which, in line with our general assumptions, we will assume can be written as

$$\bar{A} = \nu\bar{A}^{(1)} + \nu^2\bar{A}^{(2)} + \cdots \tag{43}$$

If we combine these observations, then we see that they imply that we can write $\hat{\psi}^{(1)}$ as

$$\hat{\psi}^{(1)} = \frac{\partial\hat{\psi}^{(0)}}{\partial\hat{A}^{(0)}}\,\delta A \tag{44}$$

and $\bar{\psi}^{(2)}$ as

$$\bar{\psi}^{(2)} = \frac{1}{2}\,\frac{\partial^2\hat{\psi}^{(0)}}{\partial\hat{A}^{(0)}\,\partial\hat{A}^{(0)}}\,\delta A\,\delta A + \frac{\partial\hat{\psi}^{(0)}}{\partial\hat{A}^{(0)}}\,\bar{A}^{(2)} \tag{45}$$

where

$$\delta A \equiv \hat{A}^{(1)} + \bar{A}^{(1)} \tag{46}$$

We now use (44) and (45) in (39). Then, making use of the fact that by the standard argument the $\bar{A}^{(2)}$ term in (45) will make no contribution, we find that

$$\begin{aligned}(\hat{\psi}^{(0)}, (\hat{E}^{(2)} - H^{(2)})\hat{\psi}^{(0)}) = -\frac{1}{2}\Bigg\{&\left(\frac{\partial^2\hat{\psi}^{(0)}}{\partial\hat{A}^{(0)}\,\partial\hat{A}^{(0)}}\,\delta A\,\delta A, (H^{(0)} - \hat{E}^{(0)})\hat{\psi}^{(0)}\right)\\ &+ 2\left(\frac{\partial\hat{\psi}^{(0)}}{\partial\hat{A}^{(0)}}\,\delta A, (H^{(0)} - \hat{E}^{(0)})\,\frac{\partial\hat{\psi}^{(0)}}{\partial\hat{A}^{(0)}}\,\delta A\right)\\ &+ \left(\hat{\psi}^{(0)}, (H^{(0)} - \hat{E}^{(0)})\,\frac{\partial^2\hat{\psi}^{(2)}}{\partial\hat{A}^{(0)}\,\partial\hat{A}^{(0)}}\,\delta A\,\delta A\right)\Bigg\}\end{aligned}$$

However, from (4-2), (4-3), and (4-10), this can be written as

$$(\hat{\psi}^{(0)}, (\hat{E}^{(2)} - H^{(2)})\hat{\psi}^{(0)}) = -\tfrac{1}{2}\,\delta_2\hat{E}^{(0)}(\hat{\psi}^{(0)}, \hat{\psi}^{(0)})$$

with δA given by (46), and thus we see that if this $\delta_2\hat{E}^{(0)}$ is nonnegative, then (40) will follow, which proves the point.

PROBLEMS

1. Derive (19) as a special case of the result in Problem 2 of Section 15.

2. To illustrate the Λ technique, consider the equation $x + x^* = 14$. Show that the associated Λ matrix is $\binom{1\ 1}{1\ 1}$, and using it, obtain the obvious result that $x = 7 + i\alpha$, where α is an arbitrary real number.

25. Variation Methods within the Variation Method

Let us denote by $\delta\hat{\psi}^{(1)}$ and $\delta\overline{\psi}^{(2)}$ the results of an arbitrary variation δA of $\hat{A}^{(1)}$ in $\hat{\psi}^{(1)}$ and $\overline{\psi}^{(2)}$, respectively. That is, from (24-5) and (24-27),

$$\delta\hat{\psi}^{(1)} \equiv \frac{\partial\hat{\psi}^{(0)}}{\partial\hat{A}^{(0)}}\,\delta A \tag{1}$$

$$\delta\overline{\psi}^{(2)} \equiv \frac{\partial^2\hat{\psi}^{(0)}}{\partial\hat{A}^{(0)}\,\partial\hat{A}^{(0)}}\,\hat{A}^{(1)}\,\delta A + \left(\frac{\partial^2\hat{\psi}}{\partial v\,\partial\hat{A}^{(0)}}\right)_0 \delta A \tag{2}$$

Comparing (1) and (2) with (24-10) and (24-11) then shows that

$$\delta\hat{\psi}^{(1)} = \delta^0\hat{\psi} \tag{3}$$

$$\delta\overline{\psi}^{(2)} = \delta^1\hat{\psi} \tag{4}$$

Therefore with this understanding as to the meaning of δ, Eq. (24-15) can be written

$$\delta\hat{J}^{(2)} = 0 \qquad \text{all} \quad \delta A \tag{5}$$

where

$$\begin{aligned}\hat{J}^{(2)} = {} & (\hat{\psi}^{(1)}, (H^{(0)} - \hat{E}^{(0)})\hat{\psi}^{(1)}) \\ & + (\hat{\psi}^{(0)}, (H^{(0)} - \hat{E}^{(0)})\overline{\psi}^{(2)}) + (\overline{\psi}^{(2)}, (H^{(0)} - \hat{E}^{(0)})\hat{\psi}^{(0)}) \\ & + (\hat{\psi}^{(0)}, (H^{(1)} - \hat{E}^{(1)})\hat{\psi}^{(1)}) + (\hat{\psi}^{(1)}, (H^{(1)} - \hat{E}^{(1)})\hat{\psi}^{(0)})\end{aligned} \tag{6}$$

and where in the variation we keep $\hat{E}^{(0)}$ and $\hat{E}^{(1)}$ fixed (this is appropriate since, as we have seen, they do not involve $\hat{A}^{(1)}$). Further, from (24-26) we see that the numerical value of $\hat{J}^{(2)}$ is

$$\hat{J}^{(2)} = (\hat{\psi}^{(0)}, (\hat{E}^{(2)} - H^{(2)})\hat{\psi}^{(0)}) \tag{7}$$

Also, if we wish, we can clearly replace $\overline{\psi}^{(2)}$ by $\hat{\psi}^{(2)}$ and then extend the definition of δ to also include variations of $\hat{A}^{(2)}$ since, by a familiar argument, the extra contributions to $\hat{J}^{(2)}$ and to $\delta\hat{J}^{(2)}$ will vanish identically.

Now to the immediate purpose of this exercise. As we have seen, given $\hat{\psi}^{(0)}$, Eq. (24-15) yields equations to determine the $\hat{A}^{(1)}$. However, in many practical situations, one cannot actually solve these equations. The present considerations then suggest a *variation technique for approximating their solution.* Namely calculate $J^{(2)}$ defined by

$$J^{(2)} = (\psi^{(1)}, (H^{(0)} - \hat{E}^{(0)})\psi^{(1)}) + (\hat{\psi}^{(0)}, (H^{(0)} - \hat{E}^{(0)})\tilde{\psi}^{(2)}) + (\tilde{\psi}^{(2)}, (H^{(0)} - \hat{E}^{(0)})\psi^{(0)}) + (\hat{\psi}^{(0)}, (H^{(1)} - \hat{E}^{(1)})\psi^{(1)}) + (\psi^{(1)}, (H^{(1)} - \hat{E}^{(1)})\hat{\psi}^{(0)}) \tag{8}$$

where

$$\psi^{(1)} = \frac{\partial\hat{\psi}^{(0)}}{\partial\hat{A}^{(0)}} A^{(1)} + \left(\frac{\partial\hat{\psi}}{\partial\nu}\right)_0 \tag{9}$$

$$\tilde{\psi}^{(2)} = \frac{1}{2}\,\frac{\partial^2\hat{\psi}^{(0)}}{\partial\hat{A}^{(0)}\,\partial\hat{A}^{(0)}} A^{(1)}A^{(1)} + \left(\frac{\partial^2\hat{\psi}}{\partial\nu\,\partial\hat{A}^{(0)}}\right)_0 A^{(1)} + \frac{1}{2}\left(\frac{\partial^2\hat{\psi}}{\partial\nu^2}\right)_0 \tag{10}$$

and where the $A^{(1)}$ are to be drawn from some limited subset of the original A's, which is no larger than one can handle. Then require

$$\delta\check{J}^{(2)} = 0 \qquad \text{“all”} \quad \delta\check{A}^{(1)} \tag{11}$$

where we use the inverted caret to denote the optimal quantities which will emerge from this variation calculation. The $\check{A}^{(1)}$ are then used as approximations to the $\hat{A}^{(1)}$.

For example, within the linear variation method, this means that certain of the $A_i^{(1)}$ would be given definite values (for example, zero) while the rest are arbitrary. Similarly, within USCF, it means freezing certain of the $b_{\sigma j}$, the rest being arbitrary. Within UHF, there are more possibilities; however, the normal procedure is to restrict the $\varphi_i^{(1)}$ by assuming that they can be expanded in a finite basis (possibly containing nonlinear parameters) much as in USCF.

In support of this procedure, let us first note that it yields a variational approximation to $\hat{E}^{(2)}$, that is, an approximation involving only a second-order error. Namely, since $\delta\hat{J}^{(2)} = 0$ for all δA, it must be that

$$\check{J}^{(2)} = \hat{J}^{(2)} + O(\check{A}^{(1)} - \hat{A}^{(1)})^2 \tag{12}$$

Therefore if we define $\check{E}^{(2)}$ by

$$\check{J}^{(2)} = (\hat{\psi}^{(0)}, (\check{E}^{(2)} - H^{(2)})\hat{\psi}^{(0)}) \tag{13}$$

then from (7), (12), and (13), it follows that

$$\check{E}^{(2)} = \hat{E}^{(2)} + O(\check{A}^{(1)} - \hat{A}^{(1)})^2 \tag{14}$$

which proves the point. Also, by making the obvious approximation in Eq. (26-9) of the next section, one can produce a nonvariational approximation to $\hat{E}^{(3)}$.

Further, we will show that if $\hat{E}^{(0)}$ is itself a local minimum with respect to $\delta A = A^{(1)} - \hat{A}^{(1)}$, then $\check{E}^{(2)}$ will be a guaranteed, and improvable, upper bound to $\hat{E}^{(2)}$,[10] so that in such cases this variation method is, at least from the theoretical point of view, very well founded indeed, and would also be very well founded practically if one knew a priori that $\hat{E}^{(2)}$ is a guaranteed upper bound to $E'^{(2)}$. However, as we mentioned in Section 23, this latter will usually not be the case except possibly on an empirical basis. [Note that if the perturbation is a uniform electric or magnetic field, then, because of the minus sign in (15-9), an upper bound to $\hat{E}^{(2)}$ yields a *lower* bound to the approximate polarizabilities and susceptibilities provided by $\hat{E}^{(2)}$.]

If we write

$$\psi^{(1)} = \hat{\psi}^{(1)} + \Delta^{(1)} \tag{15}$$

$$\tilde{\psi}^{(2)} = \overline{\psi}^{(2)} + \Delta^{(2)} \tag{16}$$

then from (9) and (24-5), and from (10) and (24-27), it follows that

$$\Delta^{(1)} = \frac{\partial\hat{\psi}^{(0)}}{\partial\hat{A}^{(0)}}(A^{(1)} - \hat{A}^{(1)}) \tag{17}$$

and that

$$\Delta^{(2)} = \frac{1}{2}\,\frac{\partial^2\hat{\psi}^{(0)}}{\partial\hat{A}^{(0)}\,\partial\hat{A}^{(0)}}(A^{(1)}A^{(1)} - \hat{A}^{(1)}\hat{A}^{(1)}) + \left(\frac{\partial^2\hat{\psi}}{\partial\nu\,\partial\hat{A}^{(0)}}\right)_0(A^{(1)} - \hat{A}^{(1)})$$

which can also be written as

$$\begin{aligned}\Delta^{(2)} = {} & \frac{1}{2}\,\frac{\partial^2\hat{\psi}^{(0)}}{\partial\hat{A}^{(0)}\,\partial\hat{A}^{(0)}}(A^{(1)} - \hat{A}^{(1)})(A^{(1)} - \hat{A}^{(1)}) \\ & + \frac{\partial^2\hat{\psi}^{(0)}}{\partial\hat{A}^{(0)}\,\partial\hat{A}^{(0)}}\hat{A}^{(1)}(A^{(1)} - \hat{A}^{(1)}) + \left(\frac{\partial^2\hat{\psi}}{\partial\nu\,\partial\hat{A}^{(0)}}\right)_0(A^{(1)} - \hat{A}^{(1)})\end{aligned} \tag{18}$$

[10] See also Moccia [32].

However, $(J^{(2)} - \hat{J}^{(2)})$ can contain no terms of first order in $(A^{(1)} - \hat{A}^{(1)})$. Therefore, from the form of $J^{(2)}$ as given in (8), we must have

$$J^{(2)} - \hat{J}^{(2)} = (\hat{\psi}^{(0)}, (H^{(0)} - \hat{E}^{(0)})\,\bar{\Delta}^{(2)}) + (\Delta^{(1)}, (H^{(0)} - \hat{E}^{(0)})\,\Delta^{(1)}) + (\bar{\Delta}^{(2)}, (H^{(0)} - \hat{E}^{(0)})\hat{\psi}^{(0)}) \tag{19}$$

where $\bar{\Delta}^{(2)}$ is the first term on the right in (18),

$$\bar{\Delta}^{(2)} \equiv \frac{1}{2}\,\frac{\partial^2 \hat{\psi}^{(0)}}{\partial \hat{A}^{(0)}\,\partial \hat{A}^{(0)}}\,(A^{(1)} - \hat{A}^{(1)})(A^{(1)} - \hat{A}^{(1)})$$

We now note that $\Delta^{(1)}$ and $\bar{\Delta}^{(2)}$ are $\delta\hat{\psi}^{(0)}$ and $\frac{1}{2}\,\delta_2\hat{\psi}^{(0)}$ for $\delta A = (A^{(1)} - \hat{A}^{(1)})$, so that we can write (19) in the form

$$\begin{aligned} 2(J^{(2)} - \hat{J}^{(2)}) &= (\hat{\psi}^{(0)}, (H^{(0)} - \hat{E}^{(0)})\,\delta_2\hat{\psi}^{(0)}) \\ &\quad + 2(\delta\hat{\psi}^{(0)}, (H^{(0)} - \hat{E}^{(0)})\,\delta\hat{\psi}^{(0)}) + (\delta_2\hat{\psi}^{(0)}, (H^{(0)} - \hat{E}^{(0)})\hat{\psi}^{(0)}) \\ &= \delta_2\hat{E}^{(0)}(\hat{\psi}^{(0)}, \hat{\psi}^{(0)}) \end{aligned} \tag{20}$$

Therefore $\delta_2\hat{E}^{(0)} \geq 0$ implies that $(J^{(2)} - \hat{J}^{(2)}) \geq 0$ and hence

$$E^{(2)} \geq \hat{E}^{(2)} \tag{21}$$

Thus we have proven that if $\hat{E}^{(0)}$ is a local minimum with respect to the variation $A^{(1)} - \hat{A}^{(1)}$, then $E^{(2)}$ is an upper bound to $\hat{E}^{(2)}$. Conversely, $\hat{E}^{(2)}$ is a lower bound to $J^{(2)}$, which implies, pathological cases aside, that $J^{(2)}$ has a minimum in the calculus sense. Therefore, since $\delta\check{J}^{(2)} = 0$, $\check{J}^{(2)}$ [or the smallest $\check{J}^{(2)}$ if (11) has more than one solution, which can happen if $A^{(1)}$ involves nonlinear parameters] must be the absolute minimum of $J^{(2)}$, which in turn means that it is an improvable upper bound since it can only decrease (or at any rate not increase) as one increases the flexibility of the $A^{(1)}$.

In the approach which we have followed, the functional $J^{(2)}$ did not emerge in a very natural way and the derivations of (20) and (21) were a bit involved. However, there is another way of proceeding, which we will now describe, in which $J^{(2)}$ appears quite automatically and in which its stationary, minimal, and bounding properties are much more obvious. We first note that whether they are numerical parameters and/or functions, a general A could certainly itself be parameterized according to

$$A = A^{(0)} + \nu A^{(1)} + \nu^2 A^{(2)} + \cdots$$

with the $A^{(n)}$ *and their optimal values* independent of ν but otherwise a priori quite arbitrary. Suppose now that we restrict these A's by requiring that $A^{(0)} = \hat{A}^{(0)}$. Evidently we would thereby be considering only a subset of the possible A's. However, the important point is that since this is a subset which contains $\hat{A}$, it must lead to the same results as we found earlier.

Our new procedure, then, is to use as trial functions in the variation method functions of the form

$$\psi = \psi(A, \nu) \tag{22}$$

where, in accord with the preceding discussion,

$$A = \hat{A}^{(0)} + \nu A^{(1)} + \nu^2 A^{(2)} + \cdots \tag{23}$$

Expanding in powers of ν, we then find

$$\psi = \hat{\psi}^{(0)} + \nu\psi^{(1)} + \nu^2\psi^{(2)} + \cdots \tag{24}$$

where

$$\psi^{(1)} = \frac{\partial\hat{\psi}^{(0)}}{\partial\hat{A}^{(0)}} A^{(1)} + \left(\frac{\partial\hat{\psi}}{\partial\nu}\right)_0 \tag{25}$$

$$\psi^{(2)} = \frac{1}{2}\frac{\partial^2\hat{\psi}^{(0)}}{\partial\hat{A}^{(0)}\,\partial\hat{A}^{(0)}} A^{(1)}A^{(1)} + \left(\frac{\partial^2\hat{\psi}}{\partial\nu\,\partial\hat{A}^{(0)}}\right)_0 A^{(1)} + \frac{1}{2}\left(\frac{\partial^2\hat{\psi}}{\partial\nu^2}\right)_0 + \frac{\partial\hat{\psi}^{(0)}}{\partial\hat{A}^{(0)}} A^{(2)} \tag{26}$$

$$= \tilde{\psi}^{(2)} + \frac{\partial\hat{\psi}^{(0)}}{\partial\hat{A}^{(0)}} A^{(2)} \tag{27}$$

etc.

If we now insert these results into

$$(\psi, (H - E)\psi) = 0 \tag{28}$$

then we find

$$E = \hat{E}^{(0)} + \nu\hat{E}^{(1)} + \nu^2 E^{(2)} + \cdots \tag{29}$$

with $E^{(2)}$ given by

$$J^{(2)} + (\hat{\psi}^{(0)}, (H^{(2)} - E^{(2)})\hat{\psi}^{(0)}) = 0 \tag{30}$$

where $J^{(2)}$ is defined as in (8). [In deriving (29) and (30), we have made use of (24-14) with δA variously $A^{(1)}$ and $A^{(2)}$.]

We now require that

$$\delta\hat{E} = 0, \qquad \text{all} \quad \delta A^{(1)}, \delta A^{(2)}, \ldots \tag{31}$$

which will in turn imply the sequence of equations

$$\delta \hat{E}^{(n)} = 0 \qquad \text{all} \quad \delta A^{(1)}, \delta A^{(2)}, \ldots, \delta A^{(n)} \tag{32}$$

In particular, then, we see from (30) that

$$\delta \hat{E}^{(2)} = 0 \qquad \text{all} \quad \delta A^{(1)} \tag{33}$$

is identical to our earlier result (5) that

$$\delta \hat{J}^{(2)} = 0 \qquad \text{all} \quad \delta A \tag{34}$$

Also, it immediately follows from (30) that the numerical value of $\hat{J}^{(2)}$ is $(\hat{\psi}^{(0)}, (\hat{E}^{(2)} - H^{(2)})\hat{\psi}^{(0)})$, as we found before. Finally, the minimal properties of $\hat{J}^{(2)}$ when $\hat{E}^{(0)}$ is a local minimum with respect to the appropriate variation follows easily. First of all, we can argue by continuity that if $\hat{E}^{(0)}$ is a local minimum with respect to

$$\delta A = A^{(1)} - \hat{A}^{(1)} \tag{35}$$

[and therefore, from (4-10), with respect to any real multiple of this δA], then $\hat{E}$ will remain a minimum with respect to such variations for at least a small range of ν values around $\nu = 0$. Now E is derived from $\hat{E}$ by replacing $\hat{A}^{(0)} + \nu \hat{A}^{(1)} + \cdots$ by $\hat{A}^{(0)} + \nu A^{(1)} + \cdots$. That is,

$$A = \hat{A} + \delta' A \tag{36}$$

where

$$\delta' A = \nu(A^{(1)} - \hat{A}^{(1)}) + \cdots \tag{37}$$

Therefore it follows that if $\hat{E}^{(0)}$ is a local minimum with respect to $\delta A = (A^{(1)} - \hat{A}^{(1)})$ and hence with respect to $\delta' A$, then the terms of second order in ν in $E - \hat{E}$, which are evidently the terms of second order in $\delta' A$, must be nonnegative. That is, we must have, as before,

$$E^{(2)} \geq \hat{E}^{(2)} \tag{38}$$

What we have done for $\hat{A}^{(1)}$ and $\hat{E}^{(2)}$ given the $\hat{A}^{(0)}$ can also be done for $\hat{A}^{(2)}$ and $\hat{E}^{(4)}$ if one in fact knows the $\hat{A}^{(1)}$ exactly. Thus most simply, given the exact $\hat{A}^{(0)}$ and $\hat{A}^{(1)}$, one can start again with A's of the form

$$A = \hat{A}^{(0)} + \nu \hat{A}^{(1)} + \nu^2 A^{(2)} + \cdots \tag{39}$$

and derive a functional of $A^{(2)}$ which will yield a variational approximation to $\hat{E}^{(4)}$. Further, it follows, by the same argument which led to (38), that this functional will yield an improvable upper bound to $\hat{E}^{(4)}$ if $\hat{E}^{(0)}$ is a local minimum with respect to $\delta A = A^{(2)} - \hat{A}^{(2)}$. Similarly, if one knows the exact $\hat{A}^{(0)}$, $\hat{A}^{(1)}$, and $A^{(2)}$, one can start again and do the same for $\hat{A}^{(3)}$ and $\hat{E}^{(6)}$, etc.

Thus far, true to our original purpose, we have been discussing perturbation theory within a variation calculation. However, often one is confronted with rather a different problem. Namely one is simply given a function $\chi^{(0)}$ (derived perhaps from a variation calculation) as an approximation to an eigenfunction of $H^{(0)}$, and the problem is to perturb $\chi^{(0)}$ in some reasonable way by $\nu H^{(1)} + \nu^2 H^{(2)} + \cdots$ in order to approximate the response of the actual system to the same perturbations. In such cases, $H^{(0)}$ is usually the "accurate" Hamiltonian of an isolated system, while $\nu H^{(1)} + \nu^2 H^{(2)} + \cdots$ represents the effect of external fields and/or changes of parameters in $H^{(0)}$.

Several methods have been used to deal with such situations. One, of course, is to try and imbed $\chi^{(0)}$ in some consistent variation calculation of the type which we have been discussing. Another, which we will discuss to some extent here, is to pursue a path which is suggested by the discussion following (21). Some others will be described in Section 31.

Namely, suppose that we were to use in the variation method trial functions of the form

$$\psi = \chi^{(0)} + \nu\chi^{(1)} + \nu^2\chi^{(2)} + \cdots \tag{40}$$

with the $\chi^{(n)}$ containing various variational parameters, and where we require that the $\chi^{(n)}$ and their optimal values be independent of ν so as to ensure that $\hat{\psi}^{(0)} = \chi^{(0)}$. Then, expanding

$$(\psi, (H - E)\psi) = 0 \tag{41}$$

in powers of ν and writing

$$E = E^{(0)} + \nu E^{(1)} + \nu^2 E^{(2)} + \cdots \tag{42}$$

we are led to the sequence of equations

$$(\chi^{(0)}, (H^{(0)} - E^{(0)})\chi^{(0)}) = 0 \tag{43}$$

$$\begin{aligned}(\chi^{(0)}, (H^{(0)} - E^{(0)})\chi^{(1)}) + (\chi^{(1)}, (H^{(0)} - E^{(0)})\chi^{(0)})\\ + (\chi^{(0)}, (H^{(1)} - E^{(1)})\chi^{(0)}) = 0\end{aligned} \tag{44}$$

$$\begin{aligned}(\chi^{(0)}, (H^{(0)} - E^{(0)})\chi^{(2)}) + (\chi^{(2)}, (H^{(0)} - E^{(0)})\chi^{(0)})\\ + (\chi^{(0)}, (H^{(1)} - E^{(1)})\chi^{(1)}) + (\chi^{(1)}, (H^{(1)} - E^{(1)})\chi^{(0)})\\ + (\chi^{(1)}, (H^{(0)} - E^{(0)})\chi^{(1)}) + (\chi^{(0)}, (H^{(2)} - E^{(2)})\chi^{(0)}) = 0\end{aligned} \tag{45}$$

etc., and let us suppose that we want to determine only $\hat{E}^{(0)}$, $\hat{E}^{(1)}$, and $\hat{E}^{(2)}$.

Equation (43) tells us immediately that

$$\hat{E}^{(0)} = (\chi^{(0)}, H^{(0)}\chi^{(0)})/(\chi^{(0)}, \chi^{(0)}) \tag{46}$$

More interesting is Eq. (44). Requiring

$$\delta\hat{E}^{(1)} = 0 \qquad \text{"all"} \quad \delta\hat{\chi}^{(1)} \tag{47}$$

then yields

$$(\chi^{(0)}, (H^{(0)} - \hat{E}^{(0)})\,\delta\hat{\chi}^{(1)}) + (\delta\hat{\chi}^{(1)}, (H^{(0)} - \hat{E}^{(0)})\chi^{(0)}) = 0 \quad \text{"all"} \;\; \delta\hat{\chi}^{(1)} \tag{48}$$

and here we meet for the first time in all our discussions the possibility of inconsistent equations. That is, thus far, since there were no obvious inconsistencies in the formalism, we have simply implicitly assumed that (at least in principle) all our equations had solutions. However, suppose, for example, that $\chi^{(1)}$ and $\chi^{(2)}$ are linear variation functions

$$\chi^{(1)} = \sum_{k=1}^{M_1} A_k^{(1)}\phi_k \tag{49}$$

$$\chi^{(2)} = \sum_{k=1}^{M_2} A_k^{(2)}\phi_k{}' \tag{50}$$

where we have allowed for the possibility that one may use different basis sets for the two functions. Then (48) evidently yields

$$(\phi_k, (H^{(0)} - \hat{E}^{(0)})\chi^{(0)}) = 0, \qquad k = 1, \ldots, M_1 \tag{51}$$

That is, it implies conditions on the basis set ϕ_k and if these conditions are not met, we have an inconsistency at this point.

To go on, let us therefore suppose that (51) is satisfied (this will often be the case for reasons of symmetry). Then it also follows that

$$(\hat{\chi}^{(1)}, (H^{(0)} - E^{(0)})\chi^{(0)}) = 0 \tag{52}$$

so that from (44) we will have

$$\hat{E}^{(1)} = (\chi^{(0)}, H^{(1)}\chi^{(0)})/(\chi^{(0)}, \chi^{(0)}) \tag{53}$$

Turning to Eq. (45), $\delta\hat{E}^{(1)} = \delta\hat{E}^{(2)} = 0$ for all $\delta A_l^{(1)}$ then yields as equations to determine the $\hat{A}_k^{(1)}$

$$(\phi_k, (H^{(0)} - \hat{E}^{(0)})\hat{\chi}^{(1)} + (H^{(1)} - \hat{E}^{(1)})\chi^{(0)}) = 0 \tag{54}$$

Also, $\delta\hat{E}^{(1)} = \delta\hat{E}^{(2)} = 0$ for all $\delta A_k^{(2)}$ yields the consistency equations

$$(\phi_k', (H^{(0)} - \hat{E}^{(0)})\chi^{(0)}) = 0$$

which we can, however, ignore since they also imply that $\hat{\chi}^{(2)}$ will make no contribution to $\hat{E}^{(2)}$. Thus *to this order*, all is well if the consistency conditions (51) are met.

Things become a bit more complicated if nonlinear parameters are involved. Thus, for example, suppose $\chi^{(1)}$ and $\chi^{(2)}$ each involve one nonlinear real numerical parameter, which we will denote by A_1 and A_2, respectively. Then unless for some reason

$$\left(\frac{\partial\hat{\chi}^{(1)}}{\partial\hat{A}_1}, (H^{(0)} - \hat{E}^{(0)})\chi^{(0)}\right) + \left(\chi^{(0)}, (H^{(0)} - \hat{E}^{(0)})\frac{\partial\hat{\chi}^{(1)}}{\partial\hat{A}_1}\right) \tag{55}$$

vanishes identically, it is clear that Eq. (48) will determine $\hat{A}_1$, from which we can then calculate $\hat{E}_1$. However, now the problem is that if using (45) we require that $\partial\hat{E}^{(1)}/\partial\hat{A}_1 = \partial\hat{E}^{(2)}/\partial\hat{A}_1 = 0$, then we are led to

$$\left(\frac{\partial\hat{\chi}^{(1)}}{\partial\hat{A}^{(1)}}, (H^{(0)} - \hat{E}^{(0)})\hat{\chi}^{(1)} + (H^{(1)} - \hat{E}^{(1)})\hat{\chi}^{(0)}\right) + \text{comp. conj.} = 0 \tag{56}$$

which is another equation for $\hat{A}_1$, and one which will usually be inconsistent with (55). Of course, if (49) is satisfied identically for reasons of symmetry, then there is no problem and (56) will be *the* equation to determine $\hat{A}_1$.

Assuming consistency, then, to complete the calculation, we need $\hat{\psi}^{(2)}$. Therefore using (45), we require $\partial\hat{E}^{(1)}/\partial\hat{A}_2 = \partial\hat{E}^{(2)}/\partial\hat{A}_2 = 0$, to find

$$(\chi^{(0)}, (H^{(0)} - \hat{E}^{(0)})\,\partial\hat{\chi}^{(2)}/\partial\hat{A}_2) + \text{comp. conj.} = 0 \tag{57}$$

which, if the left hand side does not vanish identically, will determine $\hat{A}_2$. If it does vanish identically, then we must go on to the third-order equation and require $\partial\hat{E}^{(3)}/\partial\hat{A}_2 = \partial\hat{E}^{(2)}/\partial\hat{A}_2 = \partial\hat{E}^{(1)}/\partial\hat{A}_2 = 0$, etc.

The main lesson we would point to in all of this is that if one tries to "piece together" a variation-perturbation calculation, then one should be aware, if one worries about such things, that consistency is by no means automatic.[11]

A topic we have not yet discussed is the following: We know that, depending on the nature of the set of trial functions, the $\hat{\psi}$ and therefore the

[11] See, for example, Hambro [33], and Gutschick and McKoy [34].

$\hat{\psi}^{(n)}$ may satisfy various theorems. A natural question is then to what extent $\check{\psi}^{(1)}$, say, will satisfy theorems analogous to those satisfied by $\hat{\psi}^{(1)}$. However, we will postpone a discussion of this until Section 35.

PROBLEMS

1. Referring to Eq. (11), denote the variational parameters in the $A^{(1)}$ by B. Show that

$$\delta\check{\psi}^{(1)} = \frac{\partial\hat{\psi}^{(0)}}{\partial\hat{A}^{(0)}} \frac{\partial\check{A}^{(1)}}{\partial\check{B}}\,\delta B$$

and therefore that

$$\delta\check{A}^{(1)} = \frac{\partial\check{A}^{(1)}}{\partial\check{B}}\,\delta B$$

As usual, the δB form a real linear space. Show that the $\delta\check{A}^{(1)}$, however, form a linear space only if the B's are linear parameters.

2. If the $\delta\check{A}^{(1)}$ of Problem 1 do form a linear space and if $\hat{E}^{(0)}$ is an absolute minimum with respect to all δA (one can weaken this), show directly that, in accord with the conclusions in the text, $\check{E}^{(2)} \geq \hat{E}^{(2)}$. [Hint: Show first that under these conditions, $A^{(1)} - \check{A}^{(1)}$ is a possible $\delta\check{A}^{(1)}$ and then, with appropriate changes, follow the pattern of (15) et seq.]

26. The ν^{2n+1} Theorem and Interchange Theorems

As we have shown in Section 22, if one knows an eigenfunction ψ' correctly through order ν^n, then one can calculate the eigenvalue E' correctly through order ν^{2n+1}. We will now derive a generalization of that theorem within the variation method.[12] (An analogous theorem for certain time-dependent problems is given in Appendix C.)

Consider the function $\mathring{\psi}$ defined by

$$\mathring{\psi} = \psi(\mathring{A}, \nu) \tag{1}$$

where

$$\mathring{A} = \hat{A}^{(0)} + \nu\hat{A}^{(1)} + \cdots + \nu^n\hat{A}^{(n)} \tag{2}$$

[12] See also Silverman and van Leuvan [29]; also (for UHF) Rebane [35], and Langhoff *et al.* [36].

If we now define $\mathring{\Delta}$ and $\mathring{\hat{E}}$ by

$$\mathring{\psi} = \hat{\psi} + \mathring{\Delta} \tag{3}$$

and

$$\mathring{\hat{E}} = (\mathring{\psi}, H\mathring{\psi})/(\mathring{\psi}, \mathring{\psi}) \tag{4}$$

then one readily finds [recall (2-5)] that

$$\mathring{\hat{E}} - \hat{E} = \{(\mathring{\Delta}, (H - \hat{E})\hat{\psi}) + (\hat{\psi}, (H - \hat{E})\mathring{\Delta}) + (\mathring{\Delta}, (H - \hat{E})\mathring{\Delta})\}/(\mathring{\psi}, \mathring{\psi}) \tag{5}$$

To determine the order of magnitude of $\mathring{\hat{E}} - \hat{E}$, we first note that from (2), $(\mathring{A} - \hat{A})$, and hence $\mathring{\Delta}$, is of order ν^{n+1}, so that the last term on the right hand side of (5) is certainly of order ν^{2n+2}. Further, by expanding $\mathring{\psi}$ in powers of $\mathring{A} - \hat{A}$, one sees in more detail that $\mathring{\Delta}$ can be written as

$$\mathring{\Delta} = \frac{\partial \hat{\psi}}{\partial \hat{A}} (\mathring{A} - \hat{A}) + O(\nu^{2n+2}) \tag{6}$$

Therefore since the first term on the right in (6) is of the form $\delta\hat{\psi}$ with $\delta A = (\mathring{A} - \hat{A})$, its contribution to the sum of the first two terms on the right in (5) will vanish identically, and so the sum of these two terms is also of order ν^{2n+2}.

These results can be summarized in the "ν^{2n+1} theorem." Most generally, with a view toward the case of several perturbations, the theorem is evidently that

$$\mathring{\hat{E}} - \hat{E} = O((\mathring{A} - \hat{A})^2) \tag{7}$$

More specifically, it is the statement that if we know $\hat{A}$ correctly through order ν^n, then we can calculate $\hat{E}$ correctly through order ν^{2n+1} by expanding $\mathring{\psi}$ and then $\mathring{\hat{E}}$ through order ν^{2n+1}. We now give some examples.

Consider first $n = 0$, so that

$$\mathring{\psi} = \psi(\hat{A}^{(0)}, \nu) = \hat{\psi}^{(0)} + \nu\left(\frac{\partial \hat{\psi}}{\partial \nu}\right)_0 + \cdots$$

Then, writing (4) as

$$(\mathring{\psi}, (H - \mathring{\hat{E}})\mathring{\psi}) = 0$$

and expanding in powers of ν, we can determine $\hat{E}^{(0)}$ and $\hat{E}^{(1)}$ since from the ν^{2n+1} theorem

$$\mathring{\hat{E}} = \hat{E}^{(0)} + \nu\hat{E}^{(1)} + \cdots$$

Doing this, we obviously recover (24-16) and (24-19).

Similarly, if we go on to $n = 1$, we find

$$\mathring{\psi} = \hat{\psi}^{(0)} + \nu\hat{\psi}^{(1)} + \nu^2\bar{\psi}^{(2)} + \nu^3\bar{\psi}^{(3)} + \cdots$$

where $\bar{\psi}^{(2)}$ is as in (24-27) and where

$$\bar{\psi}^{(3)} = \frac{1}{6}\frac{\partial^3\hat{\psi}^{(0)}}{\partial\hat{A}^{(0)^3}}\hat{A}^{(1)^3} + \frac{1}{2}\left(\frac{\partial^3\hat{\psi}}{\partial\nu\,\partial\hat{A}^{(0)^2}}\right)_0\hat{A}^{(1)^2} + \frac{1}{2}\left(\frac{\partial^3\hat{\psi}}{\partial\nu^2\,\partial\hat{A}^{(0)}}\right)_0\hat{A}^{(1)} + \frac{1}{6}\left(\frac{\partial^3\hat{\psi}}{\partial\nu^3}\right)_0 \tag{8}$$

Then from

$$\mathring{E} = \hat{E}^{(0)} + \nu\hat{E}^{(1)} + \nu^2\hat{E}^{(2)} + \nu^3\hat{E}^{(3)} + \cdots$$

we recover (24-26) and gain a formula for $\hat{E}^{(3)}$:

$$\begin{aligned}\{(\bar{\psi}^{(3)}, (H^{(0)} - \hat{E}^{(0)})\hat{\psi}^{(0)}) + (\bar{\psi}^{(2)}, (H^{(0)} - \hat{E}^{(0)})\hat{\psi}^{(1)})\\ + (\bar{\psi}^{(2)}, (H^{(1)} - \hat{E}^{(1)})\hat{\psi}^{(0)}) + (\hat{\psi}^{(1)}, (H^{(2)} - \hat{E}^{(2)})\hat{\psi}^{(0)}) + \text{comp. conj.}\}\\ + (\hat{\psi}^{(1)}, (H^{(1)} - \hat{E}^{(1)})\hat{\psi}^{(1)}) + (\hat{\psi}^{(0)}, (H^{(3)} - \hat{E}^{(3)})\hat{\psi}^{(0)}) = 0\end{aligned} \tag{9}$$

Going on to $n = 2$, one derives formulas for $\hat{E}^{(4)}$ and $\hat{E}^{(5)}$, in terms of $\hat{A}^{(0)}$, $\hat{A}^{(1)}$, and $\hat{A}^{(2)}$; etc.

In all of our considerations, we have assumed that the $\hat{A}$ can be expanded in integral powers of ν. However, it has been shown [37, 38] that $\mathscr{Z}^{-1}$ expansions of certain extended HF approximations require that the $\hat{A}$ be expanded in powers of $\mathscr{Z}^{-1/2}$, though still $\hat{\psi}$ and $\hat{E}$ involve only integral powers. In this case, it would appear that to determine $\hat{E}$ through order ν^{2n+1}, we would have to include the term $\nu^{n+\frac{1}{2}}\hat{A}^{(n+\frac{1}{2})}$ in the definition of $\mathring{A}$, in addition, of course, to including the lower-order half-integral terms. However, in the examples mentioned, accuracy through order ν^n still suffices, for reasons which we will now give [38].

If we write

$$\mathring{A} = \hat{A}^{(0)} + \nu^{1/2}\hat{A}^{(1/2)} + \cdots + \nu^n\hat{A}^{(n)} \tag{10}$$

then instead of (6) we will have

$$\mathring{\Delta} = \frac{\partial\hat{\psi}}{\partial\hat{A}}(\mathring{A} - \hat{A}) + O(\nu^{2n+1}) \tag{11}$$

where, as it stands, the first term on the right in (11) appears to be of order $\nu^{n+\frac{1}{2}}$. However, in these examples, it turns out that $\mathring{\psi}$, like $\hat{\psi}$, can also be expanded in integral powers of ν $(=\mathscr{Z}^{-1})$ only, and so $\mathring{\Delta}$ is actually of order ν^{n+1}. Therefore the last term in (5) is again of order ν^{2n+2}. Further, the first term in $\mathring{\Delta}$ still makes no contribution to the sum of the first two terms in (4). Therefore we have the result that if in addition

$$(H - \hat{E})\hat{\varphi} = O(\nu) \tag{12}$$

then we will again have

$$\mathring{\hat{E}} - \hat{E} = O(\nu^{2n+2})$$

and indeed it is the case in these examples that the $\hat{\varphi}$ do become exact eigenfunctions of H when $\nu = 0$, so that (12) is satisfied.

Returning to situations in which only integral powers appear in $\hat{A}$, we will now extend our discussion to the case of two perturbations (the further extension to more than two then being straightforward.) The main point of interest here lies in what in a general way are now usually called interchange theorems.[13] Namely we will show that there is more than one path to calculating $\hat{E}$ to a given order in ν_1 and ν_2, one perhaps requiring the calculation of $\hat{A}$ to a low order in ν_1 and a higher order in ν_2 while the other (the interchanged form) requires just the reverse. Clearly such theorems are of interest if, as is often the case, one perturbation is more difficult to deal with than the other. Thus we have

$$H = H^{(0)} + \nu_1 H^{(10)} + \nu_2 H^{(01)} + \nu_1{}^2 H^{(20)} + \nu_1\nu_2 H^{(11)} + \nu_2{}^2 H^{(02)} + \cdots \tag{13}$$

$$\hat{E} = \hat{E}^{(0)} + \nu_1 \hat{E}^{(10)} + \nu_2 \hat{E}^{(01)} + \nu_1{}^2 \hat{E}^{(20)} + \nu_1\nu_2 \hat{E}^{(11)} + \nu_2{}^2 \hat{E}^{(02)} + \cdots \tag{14}$$

and

$$\hat{A} = \hat{A}^{(0)} + \nu_1 \hat{A}^{(10)} + \nu_2 \hat{A}^{(01)} + \nu_1{}^2 \hat{A}^{(20)} + \nu_1\nu_2 \hat{A}^{(11)} + \nu_2{}^2 \hat{A}^{(02)} + \cdots \tag{15}$$

and we wish to discuss ways of calculating $\hat{E}^{(n_1, n_2)}$.

We will consider in detail only the case $n_1 = 1$, n_2 arbitrary. The general case $n_1 > 1$ can be discussed in a similar way. However, since many more possibilities present themselves, we will not enter into the details.[14] The ν^{2n+1} theorem implies that we should be able to calculate $\hat{E}$ exactly through first order in ν_1 by knowing $\hat{A}$ only through zeroth order in ν_1, and indeed this is the case. Namely if we use

$$\mathring{A} = \hat{A}^{(0)} + \nu_2 \hat{A}^{(01)} + \nu_2{}^2 \hat{A}^{(02)} + \cdots + \nu_2^{n_2} \hat{A}^{(0,n_2)} \tag{16}$$

[13] Used, for example, by Sternheimer and Foley [39], but introduced systematically by Dalgarno and Stewart [40]. Name due to Hirschfelder *et al.* [28].

[14] See, for example, Tuan [41]. For the case of three pertubations, see Tuan [42], and Schulman and Tobin [43].

then

$$\mathring{A} - \hat{A} = O(\nu_1, \nu_2^{n_2+1}) \tag{17}$$

the notation meaning that each term in the difference is at least of order ν_1 and/or of order $\nu_2^{n_2+1}$. Therefore from (7) we will have

$$\mathring{E} - \hat{E} = O(\nu_1^2, \nu_1\nu_2^{n_2+1}, \nu_2^{2n_2+2}) \tag{18}$$

and so, with this $\mathring{A}$, we can, as desired, calculate $\hat{E}^{(1,n_2)}$.

The virtue of this $\mathring{A}$ is that we do not have to apply the first perturbation explicitly. However, if we are willing to calculate $\hat{A}$ to first order in ν_1, then we can reduce the necessary accuracy in ν_2. Consider first the case n_2 odd:

$$n_2 = 2m_2 + 1 \tag{19}$$

Then the ν^{2n+1} theorem implies that we ought to be able to calculate $\hat{E}$ accurately through order $\nu_2^{n_2}$ by knowing $\hat{A}$ accurately only through order $\nu_2^{m_2}$, and indeed this is the case. Namely if we use

$$\begin{aligned}\mathring{A} = \hat{A}^{(0)} + \nu_2\hat{A}^{(01)} + \cdots + \nu_2^{m_2}\hat{A}^{(0m_2)} + \nu_1\hat{A}^{(10)} + \nu_1\nu_2\hat{A}^{(11)} + \cdots \\ + \nu_1\nu_2^{m_2}\hat{A}^{(1m_2)}\end{aligned} \tag{20}$$

then

$$\mathring{A} - \hat{A} = O(\nu_1^2, \nu_1\nu_2^{m_2+1}, \nu_2^{m_2+1}) \tag{21}$$

and therefore

$$\mathring{E} - \hat{E} = O(\ldots, \nu_1\nu_2^{2m_2+2}, \ldots) \tag{22}$$

and so with this $\mathring{A}$ we can, as desired, calculate $\hat{E}^{(1,2m_2+1)}$.

The case

$$n_2 = 2m_2 \tag{23}$$

is a bit more subtle in that one does not need complete accuracy to order $\nu_2^{m_2}$. Namely it suffices to use

$$\begin{aligned}\mathring{A} = \hat{A}^{(0)} + \nu_2\hat{A}^{(01)} + \cdots + \nu_2^{m_2}\hat{A}^{(0m_2)} + \nu_1\hat{A}^{(10)} + \nu_1\nu_2\hat{A}^{(11)} + \cdots \\ + \nu_1\nu_2^{m_2-1}\hat{A}^{(1,m_2-1)}\end{aligned} \tag{24}$$

since then

$$\mathring{A} - \hat{A} = O(\nu_1^2, \nu_1\nu_2^{m_2}, \nu_2^{m_2+1}) \tag{25}$$

and therefore

$$\mathring{E} - \hat{E} = O(\ldots, \nu_1\nu_2^{2m_2+1}, \ldots) \tag{26}$$

which proves the point.

Thus, as examples, to calculate $\hat{E}^{(11)}$ we can use either, from (16)

$$\mathring{A} = \hat{A}^{(0)} + \nu_2\hat{A}^{(01)} \tag{27}$$

or from (20)

$$\mathring{A} = \hat{A}^{(0)} + \nu_1\hat{A}^{(10)} \tag{28}$$

(this is obviously "interchange" in its purest form), while to calculate $\hat{E}^{(12)}$ we can use either, from (16)

$$\mathring{A} = \hat{A}^{(0)} + \nu_2\hat{A}^{(01)} + \nu_2{}^2\hat{A}^{(02)} \tag{29}$$

or from (24)

$$\mathring{A} = \hat{A}^{(0)} + \nu_2\hat{A}^{(01)} + \nu_1\hat{A}^{(10)} \tag{30}$$

Returning to $\hat{E}^{(11)}$, we will now write out some of the formulas that arise from the use of these $\mathring{A}$'s. Using (27), we have

$$\mathring{\psi} = \psi(\hat{A}^{(0)} + \nu_2\hat{A}^{(01)}, \nu_1, \nu_2) \tag{31}$$

which when expanded in powers of ν_1 and ν_2 yields

$$\mathring{\psi} = \hat{\psi}^{(0)} + \nu_2\hat{\psi}^{(01)} + \nu_1\bar{\psi}^{(10)} + \nu_1\nu_2\bar{\psi}^{(11)} + \cdots \tag{32}$$

where

$$\hat{\psi}^{(01)} = \frac{\partial\hat{\psi}^{(0)}}{\partial\hat{A}^{(0)}}\hat{A}^{(01)} + \left(\frac{\partial\hat{\psi}}{\partial\nu_2}\right)_0 \tag{33}$$

$$\bar{\psi}^{(10)} = \left(\frac{\partial\hat{\psi}}{\partial\nu_1}\right)_0 \tag{34}$$

and

$$\bar{\psi}^{(11)} = \left(\frac{\partial^2\hat{\psi}}{\partial\hat{A}^{(0)}\,\partial\nu_1}\right)_0 \hat{A}^{(01)} + \left(\frac{\partial^2\hat{\psi}}{\partial\nu_1\,\partial\nu_2}\right)_0 \tag{35}$$

and where the subscript zero means $\nu_1 = \nu_2 = 0$. Then writing

$$(\mathring{\psi}, (H - \mathring{E})\mathring{\psi}) = 0 \tag{36}$$

and picking out the terms of order $\nu_1\nu_2$, one finds, since from our theorem $\mathring{E}^{(11)} = \hat{E}^{(11)}$, that

$$(\hat{\psi}^{(0)}, (H^{(11)} - \hat{E}^{(11)})\hat{\psi}^{(0)}) + \{(\bar{\psi}^{(11)}, (H^{(0)} - \hat{E}^{(0)})\hat{\psi}^{(0)}) + (\bar{\psi}^{(10)}, (H^{(0)} - \hat{E}^{(0)})\hat{\psi}^{(01)}) + (\bar{\psi}^{(10)}, (H^{(01)} - \hat{E}^{(01)})\hat{\psi}^{(0)}) + (\hat{\psi}^{(01)}, (H^{(10)} - \hat{E}^{(10)})\hat{\psi}^{(0)}) + \text{comp. conj.}\} = 0 \tag{37}$$

In writing down (37), we have also used the ν^{2n+1} theorem to infer for this $\mathring{\psi}$ that, since $\mathring{A}$ is accurate through zeroth order in both ν_1 and ν_2, then $\mathring{E}^{(0)} = \hat{E}^{(0)}$, $\mathring{E}^{(01)} = \hat{E}^{(01)}$ and $\mathring{E}^{(10)} = \hat{E}^{(10)}$. The interchanged formula then of course follows in this case simply by interchanging the roles of ν_1 and ν_2.

Since it is quite simple, let us also write out the formula one gets for $\hat{E}^{(12)}$ using (29), an analogous formula for $\hat{E}^{(21)}$ then following by interchanging the roles of ν_1 and ν_2. The formula derived from (30) is given, for a special case, later in (54). From (29), we evidently have

$$\mathring{\psi} = \hat{\psi}^{(0)} + \nu_2\hat{\psi}^{(01)} + \nu_2{}^2\hat{\psi}^{(02)} + \cdots$$

whence, picking out the terms of order $\nu_1\nu_2{}^2$ in (36), we readily find that

$$(\hat{\psi}^{(0)}, (\hat{E}^{(12)} - H^{(12)})\hat{\psi}^{(0)}) = (\hat{\psi}^{(02)}, (H^{(10)} - \hat{E}^{(10)})\hat{\psi}^{(0)}) + (\hat{\psi}^{(01)}, (H^{(10)} - \hat{E}^{(10)})\hat{\psi}^{(01)}) + (\hat{\psi}^{(0)}, (H^{(10)} - \hat{E}^{(10)})\hat{\psi}^{(02)}) + (\hat{\psi}^{(0)}, (H^{(11)} - \hat{E}^{(11)})\hat{\psi}^{(01)}) + (\hat{\psi}^{(01)}, (H^{(11)} - \hat{E}^{(11)})\hat{\psi}^{(0)}) \tag{38}$$

where we have also used the fact that, for this $\mathring{\psi}$, again $\mathring{E}^{(10)} = \hat{E}^{(10)}$ and $\mathring{E}^{(11)} = \hat{E}^{(11)}$.

If ν_1 and ν_2 represent external fields, then quantities like $\hat{E}^{(11)}$, $\hat{E}^{(21)}$, etc., describe effects involving the two fields simultaneously, for example, shielding phenomena of the sort which we described in Section 15. Another common situation is one in which $H^{(0)}$ is an approximate Hamiltonian, ν_2 is some external field, and ν_1 is simply an order parameter with numerical value one (the $\mathscr{Z}^{-1}$ expansion is an exception) with $\nu_1 H^{(10)} + \cdots$ representing the corrections to $H^{(0)}$. Also, in such cases, there will usually be no "cross terms" in H; that is, H will have the form

$$H = H^{(0)} + \nu_1 H^{(10)} + \nu_1{}^2 H^{(20)} + \cdots + \nu_2 H^{(01)} + \nu_2{}^2 H^{(02)} + \cdots \tag{39}$$

When $H^{(0)}$ is an approximate Hamiltonian, then, for example, $\hat{E}^{(11)}$ might be the first-order correction, within whatever overall variation approximation one is using, to a permanent moment calculated using the approximate Hamiltonian alone to represent the isolated system. Similarly, $\hat{E}^{(12)}$ will yield the first-order correction to a polarizability, or a susceptibility, or a force constant, etc. Of course the moments, polarizabilities, etc., that we have been mentioning are those calculated from energy derivatives as described in Section 15. Also, by force constant we evidently meant the Born–Oppenheimer force constant. However, if the generalized Hellmann–Feynman theorem for $\sigma = \nu_2$ is satisfied, then, as we have discussed, the "other definition" will yield the same results.

Also, if this Hellmann–Feynman theorem is satisfied, then the fact that, with H given by (39),

$$\left(\frac{\partial \hat{E}}{\partial \nu_2}\right)_{\nu_2=0} = \left[\frac{(\hat{\psi}, H^{(01)}\hat{\psi})}{(\hat{\psi}, \hat{\psi})}\right]_{\nu_2=0} \tag{40}$$

is especially interesting for yet another reason, which we will now explain. With $H^{(0)}$ still an approximate Hamiltonian, let us return to the case of a single perturbation, that perturbation being the correction to the approximate Hamiltonian. Then

$$\langle W \rangle_0 \equiv (\hat{\psi}^{(0)}, W\hat{\psi}^{(0)})/(\hat{\psi}^{(0)}, \hat{\psi}^{(0)}) \tag{41}$$

provides a zeroth approximation to the "exact" (that is, exact within whatever variation approximation one is using) average value of W, which quantity we will denote by $[W]$; thus

$$[W] \equiv \frac{((\hat{\psi}^{(0)} + \nu\hat{\psi}^{(1)} + \cdots), W(\hat{\psi}^{(0)} + \nu\hat{\psi}^{(1)} + \cdots))}{(\hat{\psi}^{(0)} + \nu\hat{\psi}^{(1)} + \cdots, \hat{\psi}^{(0)} + \nu\hat{\psi}^{(1)} + \cdots)} \tag{42}$$

(Normally we would denote this $\langle W \rangle$, but this would cause confusion with what comes next.)

Now one way to calculate the corrections, to $\langle W \rangle_0$, is, as implied in (42), to calculate $\hat{\psi}^{(1)}$, etc. However, (40) suggests another approach. Namely suppose we add $\nu_2 W$ as a further perturbation in the Hamiltonian [40] and make the identifications

$$\nu \equiv \nu_1, \qquad H^{(n)} \equiv H^{(n,0)}, \qquad W = H^{(01)} \tag{43}$$

all other

$$H^{(nm)} = 0 \tag{44}$$

If, further, we use the same set of trial functions with this extended Hamiltonian as we did for the original one, then, assuming no degeneracy problems, evidently we will find that

$$(\hat{\psi})_{\nu_2=0} \equiv \hat{\psi}^{(0)} + \nu\hat{\psi}^{(10)} + \nu^2\hat{\psi}^{(20)} + \cdots$$

will just equal

$$\hat{\psi}^{(0)} + \nu\hat{\psi}^{(0)} + \nu^2\hat{\psi}^{(2)} + \cdots \tag{45}$$

that is, we will find $\hat{\psi}^{(n0)} = \hat{\psi}^{(n)}$. Also, since the set of trial functions is certainly independent of ν_2, (40) will hold, whence we see that under these conditions, we will have

$$(\partial\hat{E}/\partial\nu_2)_{\nu_2=0} = [W]$$

that is,

$$[W] = \hat{E}^{(01)} + \nu\hat{E}^{(11)} + \nu^2\hat{E}^{(21)} + \cdots \tag{46}$$

Thus in particular, the first order in ν correction to $\langle W\rangle_0$ is just $\nu\hat{E}^{(11)}$. Therefore from the interchange theorem we see that we can calculate it by treating $W = H^{(01)}$ as the perturbation rather than, as earlier suggested, $H^{(1)} \equiv H^{(10)}$, and the point is that this may be much easier to do. Similarly, higher-order corrections, $\nu^2 E^{(21)}$, etc., can be evaluated by solving equations to lower order in ν_1 than (42) would suggest.

When the generalized Hellmann–Feynman theorems for $\sigma = \nu_1$ and/or ν_2 are satisfied, then the resultant formulas for the $\hat{E}^{(nm)}$ become somewhat simplified, and since this is a common case in practice, it will be useful to record some of the results.

For simplicity, we will assume that the relevant Hellmann–Feynman theorem is satisfied because the individual trial functions do not depend explicitly on the corresponding ν. If there is no explicit ν_1 dependence, then from their definitions (34) and (35), $\bar{\psi}^{(10)}$ and $\bar{\psi}^{(11)}$ will vanish, whence (37) will become

$$(\hat{\psi}^{(0)}, (\hat{E}^{(11)} - H^{(11)})\hat{\psi}^{(0)}) = (\hat{\psi}^{(01)}, (H^{(10)} - \hat{E}^{(10)})\hat{\psi}^{(0)}) + (\hat{\psi}^{(0)}, (H^{(10)} - \hat{E}^{(10)})\hat{\psi}^{(01)}) \tag{47}$$

where, from (24-36) with obvious changes in notation,

$$\hat{E}^{(10)} = (\hat{\psi}^{(0)}, H^{(10)}\hat{\psi}^{(0)})/(\hat{\psi}^{(0)}, \hat{\psi}^{(0)}) \tag{48}$$

Similarly if the theorem for $\sigma = \nu_2$ is satisfied, then we have as the interchanged form

$$(\hat{\psi}^{(0)}, (\hat{E}^{(11)} - H^{(11)})\hat{\psi}^{(0)}) = (\hat{\psi}^{(10)}, (H^{(01)} - \hat{E}^{(01)})\hat{\psi}^{(0)}) + (\hat{\psi}^{(0)}, (H^{(01)} - \hat{E}^{(01)})\hat{\psi}^{(10)}) \tag{49}$$

where

$$\hat{E}^{(01)} = (\hat{\psi}^{(0)}, H^{(01)}\hat{\psi}^{(0)})/(\hat{\psi}^{(0)}, \hat{\psi}^{(0)}) \tag{50}$$

We have already discussed the use of (29) to yield a formula for $\hat{E}^{(12)}$ the result being (38), and this is its simplest form. If instead we use (30), then assuming that there is neither explicit ν_1 nor ν_2 dependence, we have

$$\psi = \hat{\psi}^{(0)} + \nu_1\hat{\psi}^{(10)} + \nu_2\hat{\psi}^{(01)} + \nu_2{}^2\underline{\psi}^{(02)} + \nu_1\nu_2\underline{\psi}^{(11)} + \nu_1\nu_2{}^2\underline{\psi}^{(12)} + \cdots \tag{51}$$

where

$$\underline{\psi}^{(02)} = \frac{1}{2}\,\frac{\partial^2\hat{\psi}^{(0)}}{\partial\hat{A}^{(0)}\,\partial\hat{A}^{(0)}}\,\hat{A}^{(01)}\,\hat{A}^{(01)} \tag{52}$$

$$\underline{\psi}^{(11)} = \frac{\partial^2\hat{\psi}^{(0)}}{\partial\hat{A}^{(0)}\,\partial\hat{A}^{(0)}}\,\hat{A}^{(01)}\,\hat{A}^{(10)} \tag{53}$$

and

$$\underline{\psi}^{(12)} = \frac{1}{2}\,\frac{\partial^3\hat{\psi}^{(0)}}{\partial\hat{A}^{(0)}\,\partial\hat{A}^{(0)}\,\partial\hat{A}^{(0)}}\,\hat{A}^{(10)}\,\hat{A}^{(01)}\,\hat{A}^{(01)} \tag{54}$$

Then one readily finds the still not very pretty result

$$\begin{aligned}
(\hat{\psi}^{(0)}, (\hat{E}^{(12)} - H^{(12)})\hat{\psi}^{(0)}) &= \{(\underline{\psi}^{(12)}, (H^{(0)} - \hat{E}^{(0)})\hat{\psi}^{(0)}) \\
&+ (\underline{\psi}^{(11)}, (H^{(0)} - \hat{E}^{(0)})\hat{\psi}^{(01)}) + (\underline{\psi}^{(02)}, (H^{(0)} - \hat{E}^{(0)})\hat{\psi}^{(10)}) \\
&+ (\underline{\psi}^{(11)}, (H^{(01)} - \hat{E}^{(01)})\hat{\psi}^{(0)}) + (\underline{\psi}^{(02)}, (H^{(10)} - \hat{E}^{(10)})\hat{\psi}^{(0)}) + \text{comp. conj.}\} \\
&+ \{(\hat{\psi}^{(01)}, (H^{(01)} - \hat{E}^{(01)})\hat{\psi}^{(10)}) + (\hat{\psi}^{(10)}, (H^{(02)} - \hat{E}^{(02)})\hat{\psi}^{(0)}) \\
&+ (\hat{\psi}^{(01)}, (H^{(11)} - \hat{E}^{(11)})\hat{\psi}^{(0)}) + \text{comp. conj.}\} \\
&+ (\hat{\psi}^{(01)}, (H^{(10)} - \hat{E}^{(10)})\hat{\psi}^{(01)})
\end{aligned} \tag{55}$$

In deriving (55), we have also used the facts that with this $\mathring{\psi}$, $\mathring{E}^{(01)} = \hat{E}^{(01)}$, $\mathring{E}^{(10)} = \hat{E}^{(10)}$, $\mathring{E}^{(02)} = \hat{E}^{(02)}$, and $\mathring{E}^{(11)} = \hat{E}^{(11)}$, where $\hat{E}^{(11)}$ is given by either (47) or (49), and where $E^{(02)}$ is given by (24-38) with obvious changes in notation ($\hat{\psi}^{(n)} \to \hat{\psi}^{(0n)}$, etc.). An analogous formula for $\hat{E}^{(21)}$ can of course be derived simply by interchanging the roles of ν_1 and ν_2.

PROBLEM

1. Show that the assumption of "no explicit ν dependence" can be dropped by deriving (47) and (49) directly from the relevant Hellmann–Feynman theorems.

References

1. R. A. Silverman, *Phys. Rev.* **85**, 227 (1952).
2. A. Dalgarno and A. L. Stewart, *Proc. Roy. Soc. Ser. A* **238**, 269 (1956).
3. F. Dupont-Bourdelet, J. Tillieu, and J. Guy, *J. Phys. Radium* **21**, 776 (1960).
4. T. K. Kikuta, *Progr. Theor. Phys.* **12**, 10 (1954); **14**, 457 (1955).
5. L. Biedenharn and J. M. Blatt, *Phys. Rev.* **93**, 230 (1954).
6. M. D. Kostin and H. Brooks, *J. Math. Phys.* (*N.Y.*) **5**, 1691 (1964).
7. J. O. Hirschfelder, *J. Chem. Phys.* **39**, 2099 (1963).
8. P. J. Turinsky, *J. Math. Anal. Appl.* **33**, 605 (1971).
9. G. C. Pomraning, *J. Math. Phys.* (*N.Y.*) **8**, 149 (1967), and references therein.
10. R. E. Watson and A. J. Freeman, *Phys. Rev.* **131**, 250 (1963).
11. H. D. Cohen and C. C. J. Roothaan, *J. Chem. Phys.* **43**, S34 (1965).
12. A. D. McLean and M. Yoshimine, *J. Chem. Phys.* **46**, 3682 (1967).
13. M. N. Grasso, K. T. Chung, and R. P. Hurst, *Phys. Rev.* **167**, 1 (1968).
14. J. A. Pople, J. W. McIver, Jr., and N. S. Ostlund, *J. Chem. Phys.* **49**, 2960 (1968).
15. F. P. Billingsley II and M. Krauss, *Phys. Rev. A* **6**, 855 (1972).
16. D. M. Bishop, and A. Macias, *J. Chem. Phys.* **53**, 3515 (1970), and references therein.
17. C. J. H. Schutte, *Struct. Bonding* (*Berlin*) **9**, 213 (1971).
18. P. D. Yde, K. Thomsen, and P. Swanstrom, *Mol. Phys.* **23**, 691 (1972), and references therein.
19. T. L. Barr, *Int. J. Quantum Chem. Symp.* **4**, 239 (1971); E. J. Brändas and R. J. Bartlett, *Chem. Phys. Lett.* **8**, 153 (1971); E. J. Brändas and O. Goscinski, *Phys. Rev. A* **1**, 552 (1970).
20. S. Diner, J. P. Malrieu, and P. Claverie, *Theor. Chim. Acta* **13**, 1 (1969), and references therein.
21. Z. Gershgorn and I. Shavitt, *Int. J. Quantum Chem.* **2**, 751 (1968).
22. B. Roos, *Chem. Phys. Lett.* **15**, 153 (1972).
23. B. Huron, J. P. Malrieu, and P. Rancurel, *J. Chem. Phys.* **S8**, 5745 (1973).
24. K. Aashamar, G. Lyslo, and J. Midtdal, *J. Chem. Phys.* **52**, 3324 (1970); *Phys. Norv.* **6**, 21 (1972).
25. H. Doyle, M. Oppenheimer, and G. W. F. Drake, *Phys. Rev. A* **5**, 26 (1972).
26. M. Cohen, *Proc. Roy. Soc. Ser. A* **293**, 365 (1966).
27. C. S. Sharma, *Proc. Phys. Soc. London* (*At. Mol. Phys.*) **2**, 1010 (1969).
28. J. O. Hirschfelder, W. Byers Brown, and S. T. Epstein, *Advan. Quantum Chem.* **1** (1964).
29. J. N. Silverman and J. L. van Leuvan, *Phys. Rev.* **162**, 1175 (1967).
30. J. E. Harriman, *J. Chem. Phys.* **54**, 902 (1971).
31. J. N. Silverman and J. L. van Leuvan, *Chem. Phys. Lett.* **7**, 37; Erratum **7**, 640 (1970).

31a. E. R. Davidson, *J. Chem. Phys.* **36**, 2527 (1962).

32. R. Moccia, *Chem. Phys. Lett.* **5**, 260, 265 (1970).
33. L. Hambro, *Phys. Rev. A* **7**, 479 (1973).
34. V. P. Gutschick and V. McKoy, *J. Chem. Phys.* **58**, 2397 (1973).
35. T. K. Rebane, *Opt. Spectrosc.* (*USSR*) **19**, 179 (1965).
36. P. W. Langhoff, J. D. Lyons, and R. P. Hurst, *Phys. Rev.* **148**, 18 (1966).
37. A. L. Stewart, *Proc. Phys. Soc. London* **83**, 1033 (1964); W. Byers Brown and G. V. Nazaroff, *Int. J. Quantum Chem.* **1**, 463 (1967).

38. C. A. Coulson and A. Hibbert, *Proc. Phys. Soc. London* **91**, 33 (1967), and references there; also A. Hibbert, *Proc. Phys. Soc. London Sect. B* **1**, 1048 (1968).
39. R. M. Sternheimer and H. M. Foley, *Phys. Rev.* **92**, 1460 (1953).
40. A. Dalgarno and A. L. Stewart, *Proc. Roy. Soc. Ser. A* **238**, 269, 276 (1956); **247**, 245 (1958).
41. D. F.-T. Tuan, *J. Chem. Phys.* **46**, 2435 (1967).
42. D. F.-T. Tuan, *J. Chem. Phys.* **54**, 4631 (1971).
43. J. M. Schulman and F. L. Tobin, *J. Chem. Phys.* **53**, 3662 (1970).

Chapter VI / Perturbation Theory and the Variation Method: Applications

27. Perturbation Analysis of the Linear Variation Method

Within the linear variation method, perturbation theory has been traditionally used to calculate polarizabilities, etc. Moreover, in recent years, perturbation theory, with $H^{(0)}$ an approximate Hamiltonian, has often been carried to very high order within the linear variation method. For example various $\hat{E}$ for the helium atom isoelectronic sequence have been expanded to very high orders in $\mathscr{Z}^{-1}$ [1] (although in the original literature it has not always been made clear that these were expansions *within* the linear variation method; we will elaborate on this rather cryptic comment in Section 30). Also perturbation theory with some usually rather formal choice of $H^{(0)}$ has been frequently used as a means of carrying out CI-type calculations.[2]

The $\hat{\psi}$ and $\hat{E}$ whose perturbation expansion we will be discussing are of course one of the $\hat{\psi}_k$, $\hat{E}_k$ of Section 6, say $\hat{\psi}_l$ and $\hat{E}_l$. However, to avoid complicating the notation, we will, until the last part of this section, refrain from introducing the extra subscript l. Nevertheless, one should keep in mind that

$$\hat{\psi}^{(0)} \equiv \hat{\psi}_l^{(0)}, \qquad \hat{E}^{(0)} \equiv \hat{E}_l^{(0)}, \qquad \hat{A}^{(0)} \equiv \hat{A}_{(l)}^{(0)} \tag{1}$$

and in accord with this, we will therefore assume that $\hat{\psi}^{(0)}$ is normalized,

$$(\hat{\psi}^{(0)}, \hat{\psi}^{(0)}) = 1 \tag{2}$$

Actually, to say that we are concerned with $\hat{\psi}_l$ and $\hat{E}_l$ is not quite a trivial statement since it requires that the labeling be unambiguous. That

[1] Many papers by Midtdal and collaborators, for example, Aashamar *et al.* [1]. Also many papers by Dalgarno, Drake, and collaborators, for example, Doyle *et al.* [2].

[2] For example, Diner *et al.* [3], Gershgorn and Shavitt [4], Roos [5], and Huron *et al.* [6].

is, there should be no crossing of energy levels (of appropriate symmetry type, if that is relevant) as ν changes away from zero, before the ν value of interest is reached. Now in fact there does not appear to be any simple guarantee of no crossing [7], but chances are it will not happen (the famous "no crossing theorem") [8]. It is also of interest to note that if there is no crossing (and if the perturbation series converges), then it follows from the discussion in Section 7 that if $\hat{E}^{(0)}$ is the lth root at $\nu = 0$, then $\hat{E}$ will be an upper bound to E_l'.[3]

The basic formulas that we need are all in Sections 24 and 26, but for convenience we will repeat some of them here. They are of course also identical to what one gets by a direct perturbation expansion of (6-3), a point to which we will return later. Since $\eta \equiv 0$, we have [(24-33), (24-34)]

$$\hat{A}^{(n)} = -\sum_i{}' y_{(i)}^{(0)} \frac{(y_{(i)}^{(0)+} \cdot C^{(n)})}{\lambda_i^{(0)}}, \qquad n > 0 \tag{3}$$

$$\hat{E}^{(2)} = (\hat{\varphi}^{(0)}, H^{(2)}\hat{\varphi}^{(0)}) - \sum_i{}' \frac{|(y_{(i)}^{(0)+} \cdot C^{(1)})|^2}{\lambda_i^{(0)}} + \xi^{(2)} \tag{4}$$

with $y_{(i)}^{(0)}$ and $\lambda_i^{(0)}$ the orthonormal eigenvectors and eigenvalues, respectively, of $\gamma^{(0)}$, the latter, from (4-13), being given by

$$\gamma_{ij}^{(0)} = (\phi_i^{(0)}, (H^{(0)} - \hat{E}^{(0)})\phi_j^{(0)}) \tag{5}$$

or in the notation of (6-3),

$$\gamma^{(0)} = \mathscr{H}^{(0)} - \hat{E}^{(0)} S^{(0)} \tag{6}$$

(Note that with the normalization convention implied by (3), $\hat{\varphi}$ is probably not normalized. Thus we have departed from the conventions of Section 6 in this regard.)

Also, from (24-23) or (6-3)

$$C_j^{(1)} = (\phi_j^{(1)}, (H^{(0)} - \hat{E}^{(0)})\hat{\varphi}^{(0)}) + \left(\phi_j^{(0)}, (H^{(0)} - \hat{E}^{(0)})\left(\frac{\partial \hat{\varphi}}{\partial \nu}\right)_0\right) + (\phi_j^{(0)}, (H^{(1)} - \hat{E}^{(1)})\hat{\varphi}^{(0)}) \tag{7}$$

where

$$\left(\frac{\partial \hat{\varphi}}{\partial \nu}\right)_0 = \sum_{k=1}^{M} \hat{A}_k^{(0)} \phi_k^{(1)} \tag{8}$$

[3] For a weaker result of this kind, see Lyslo *et al.* [9].

Further, from (24-29)

$$\xi^{(2)} = \left\{\frac{1}{2}\left(\left(\frac{\partial^2\hat{\psi}}{\partial\nu^2}\right)_0, (H^{(0)} - \hat{E}^{(0)})\hat{\psi}^{(0)}\right) + \left(\left(\frac{\partial\hat{\psi}}{\partial\nu}\right)_0, (H^{(1)} - \hat{E}^{(1)})\hat{\psi}^{(0)}\right)\right.$$

$$\left. + \text{ comp. conj.}\right\} + \left(\left(\frac{\partial\hat{\psi}}{\partial\nu}\right)_0, (H^{(0)} - \hat{E}^{(0)})\left(\frac{\partial\hat{\psi}}{\partial\nu}\right)_0\right) \tag{9}$$

with

$$\tfrac{1}{2}(\partial^2\hat{\psi}/\partial\nu^2)_0 = \sum_{k=1}^{M} A_k^{(0)}\phi_k^{(2)} \tag{10}$$

Finally, (26-9) provides a formula for $\hat{E}^{(3)}$ given $\hat{A}^{(1)}$. However, we will not write it out in detail except to note that in the present context (24-27) and (26-8) become

$$\bar{\psi}^{(2)} = \sum_{k=1}^{M} \hat{A}_k^{(1)}\phi_k^{(1)} + \sum_{k=1}^{M} \hat{A}_k^{(0)}\phi_k^{(2)}$$

$$\bar{\psi}^{(3)} = \sum_{k=1}^{M} \hat{A}_k^{(1)}\phi_k^{(2)} + \sum_{k=1}^{M} \hat{A}_k^{(0)}\phi_k^{(3)} \tag{11}$$

As we have discussed, in the linear variation method, the only ambiguity, barring degeneracies, is that the overall scale of the $\hat{A}_i$ is not determined. Thus, in the notation of (24-25), $\overline{\delta A}_i = \hat{A}_i$ evidently yields

$$\sum_i \frac{\partial\hat{\psi}}{\partial\hat{A}_i}\,\overline{\delta A}_i = \hat{\psi}$$

Therefore from the analysis following Eq. (24-25) it follows that $\hat{A}^{(0)}$ must be an eigenvector of $\gamma^{(0)}$ with eigenvalue zero, and indeed, we see from (6-3) and Eq. (6) that

$$\gamma^{(0)}\hat{A}^{(0)} = 0 \tag{12}$$

is just the defining equation for $\hat{A}^{(0)}$. Further, we see, as expected, that (12) has no other solutions if $\hat{E}^{(0)}$, as we will assume for now, is nondegenerate. Also, by taking the scalar product of

$$\gamma^{(0)}\hat{A}^{(n)} + C^{(n)} = 0, \qquad n > 0 \tag{13}$$

with $\hat{A}^{(0)}$ it then follows that $C^{(n)}$ must of necessity satisfy

$$\hat{A}^{(0)+}\cdot C^{(n)} = 0, \qquad n > 0 \tag{14}$$

For $n = 1$, (14) is, as one readily verifies, just the formula (24-19) for $\hat{E}^{(2)}$ specialized to the linear variation method, while for $n = 2$, it yields the specialization of formula (24-26) for $\hat{E}^{(2)}$, etc. Thus in this case, we actually had no need to invoke the formulas (24-19) and (24-26) explicitly, and indeed this is what we should have expected. Namely in our general discussion in Section 24, we treated Eqs. (24-12) and (24-13) separately, the latter being the prime source of (24-19) and (24-26). However, the point is, we know that in the linear variation method, and indeed in any variation method in which the set of trial functions has no fixed overall scale, (24-12) already implies (24-13).

Although (3) is certainly a perfectly correct way of calculating $\hat{A}^{(n)}$, it is perhaps not the expected one. Namely since perturbation theory within the linear variation method must be equivalent to the Rayleigh–Schrödinger perturbation theory of $\bar{H}$, we would expect formulas in which the denominators are differences of zeroth-order energies, i.e., $(\hat{E}_k^{(0)} - \hat{E}^{(0)})$. However, unless $S^{(0)}$ is the unit matrix (a multiple of the unit matrix yields essentially the same results), $y_{(i)}^{(0)}$ will not be $\hat{A}_{(i)}^{(0)}$, and $\lambda_i^{(0)}$ will not equal $(\hat{E}_i^{(0)} - \hat{E}^{(0)})$. Indeed, unless $S^{(0)}$ is a multiple of the unit matrix, we will evidently have to compute a new set of eigenvectors and eigenvalues $y_{(i)}^{(0)}$ and $\lambda_i^{(0)}$ for each $\hat{\psi}^{(0)}$, that is, for each value of l.

Now in fact Rayleigh–Schrödinger perturbation theory of $\bar{H}$ is the most commonly used method, so we certainly want to discuss it. However, first we will develop the present method a little further, and exhibit one way of, in effect, eliminating the zero eigenvalue from the outset. To do this, we first select one of the $\phi_i^{(0)}$ for which $\hat{A}_i^{(0)} \neq 0$; let us suppose that it is $\phi_M^{(0)}$. Then evidently we can, with no loss in generality, write $\hat{A}^{(n)}$ for $n > 1$ as

$$\hat{A}_i^{(n)} = \bar{A}_i^{(n)} + \alpha^{(n)} \hat{A}_i^{(0)}, \qquad \hat{A}_M^{(n)} = \alpha^{(n)} \hat{A}_M^{(0)}, \qquad i = 1, 2, \ldots, M-1 \tag{15}$$

where $\alpha^{(n)}$ is some constant.

If now we denote by $\bar{\gamma}^{(0)}$ the $(M-1)\times(M-1)$ Hermitian matrix derived from $\gamma^{(0)}$ by deleting its Mth row and column, and by $\bar{C}^{(n)}$ the column vector formed from the first $M-1$ components of $C^{(n)}$, then from (12) and the first $M-1$ equations of (13), we find as the equation for $\bar{A}^{(n)}$ (the column vector formed from the $\bar{A}_i^{(n)}$)

$$\bar{\gamma}^{(0)} \bar{A}^{(n)} + \bar{C}^{(n)} = 0 \tag{16}$$

The solution of (16) is then

$$\bar{A}^{(n)} = -\sum_i \bar{y}_{(i)}^{(0)} \frac{(\bar{y}_{(i)}^{(0)+} \cdot \bar{C}^{(n)})}{\bar{\lambda}_i^{(0)}} \tag{17}$$

where the $\bar{y}_{(i)}^{(0)}$ are the eigenvectors of $\bar{\gamma}^{(0)}$ and the $\bar{\lambda}_i^{(0)}$ are the corresponding eigenvalues, none of the latter being zero since clearly $\bar{A}^{(n)}$ is an unambiguous quantity. Namely given $C^{(n)}$, (13) defines $\hat{A}^{(n)}$ only up to a multiple of $\hat{A}^{(0)}$. However, we see from (15) that under the transformation $\hat{A}^{(n)} \rightarrow \hat{A}^{(n)} + \beta\hat{A}^{(0)}$, $\bar{A}^{(n)}$ remains unchanged (while $\alpha^{(n)} \rightarrow \alpha^{(n)} + \beta$).

Having determined $\bar{A}^{(n)}$, one then chooses $\alpha^{(n)}$ as one wishes so as to complete the determination of $\hat{A}^{(n)}$; for example, one could choose it equal to zero. Further, we can still, for example, calculate $\hat{E}^{(2)}$ very simply since from (4)

$$\hat{E}^{(2)} = (\hat{\psi}^{(0)}, H^{(2)}\hat{\psi}^{(0)}) + \hat{A}^{(1)+}\cdot C^{(1)} + \xi^{(2)} \tag{18}$$

However, from (15)

$$\hat{A}^{(1)+}\cdot C^{(1)} = \bar{A}^{(1)+}\cdot \bar{C}^{(1)} + \alpha^{(1)}\hat{A}^{(0)+}\cdot C^{(1)} \tag{19}$$

and therefore from (14)

$$\hat{A}^{(1)+}\cdot C^{(1)} = \bar{A}^{(1)+}\cdot \bar{C}^{(1)} \tag{20}$$

Thus

$$\hat{E}^{(2)} = (\hat{\psi}^{(0)}, H^{(2)}\hat{\psi}^{(0)}) - \sum_i \frac{|\,\bar{y}_{(i)}^{(0)+}\cdot \bar{C}^{(1)}\,|^2}{\lambda_i^{(0)}} + \xi^{(2)} \tag{21}$$

Also, in a similar way, one finds, if there are two perturbations,

$$\hat{E}^{(11)} = (\hat{\psi}^{(0)}, H^{(11)}\hat{\psi}^{(0)}) - \sum_i \frac{(\bar{C}^{(10)+}\cdot \bar{y}_{(i)}^{(0)})(\bar{y}_{(i)}^{(0)+}\cdot \bar{C}^{(01)}) + \text{comp. conj.}}{\bar{\lambda}_i^{(0)}} + \xi^{(11)} \tag{22}$$

In addition, we can if we wish eliminate all mention of eigenvectors and eigenvalues and write (17), (21), and (22) as

$$\begin{aligned}
\bar{A}^{(n)} &= -(\bar{y}^{(0)})^{-1}\bar{C}^{(n)} \\
\hat{E}^{(2)} &= (\hat{\psi}^{(0)}, H^{(2)}\hat{\psi}^{(0)}) - \bar{C}^{(1)+}(\bar{y}^{(0)})^{-1}\bar{C}^{(1)} + \xi^{(2)} \\
\hat{E}^{(11)} &= (\hat{\psi}^{(0)}, H^{(11)}\hat{\psi}^{(0)}) - \bar{C}^{(10)+}(\bar{y}^{(0)})^{-1}\bar{C}^{(01)} - \bar{C}^{(01)+}(\bar{y}^{(0)})^{-1}\bar{C}^{(10)} + \xi^{(11)}
\end{aligned} \tag{23}$$

That is, all we really need is $(\bar{y}^{(0)})^{-1}$, however expressed.

Turning now to Rayleigh–Schrödinger perturbation theory, this is particularly simple if the space is independent of ν. Since Π is then independent of ν, it follows that

$$\bar{H}^{(n)} = \overline{H^{(n)}} \tag{24}$$

which in turn means that

$$(\hat{\psi}_j^{(0)}, \bar{H}^{(n)}\hat{\psi}_k^{(0)}) = (\hat{\psi}_j^{(0)}, H^{(n)}\hat{\psi}_k^{(0)}) \tag{25}$$

Thus in this case the perturbation theory for $\bar{H}$ will look much like that for H. One can simply take over the standard formulas of Rayleigh–Schrödinger perturbation theory for H (nondegenerate or degenerate as the case may be) but with all sums finite, since the $\bar{H}^{(n)}$ have no matrix elements outside the basis set, and with the zeroth-order functions and zeroth-order energies replaced by the $\hat{\psi}_k^{(0)}$ and $\hat{E}_k^{(0)}$, respectively. Calculations in the nondegenerate case can then be carried out to high order, in a simple iterative fashion.[4]

Indeed we can derive these formulas as a special case of the γ method. If the space is independent of ν, then the orthonormal $\hat{\psi}_k^{(0)}$ must already span the space. Therefore we may use them as the basis set instead of the ϕ_k, whence we will have that in this basis S is the unit matrix and that [recall Eq. (1)]

$$\hat{A}_i^{(0)} = \delta_{il} \tag{26}$$

and

$$\gamma_{ij}^{(0)} = (\hat{\psi}_i^{(0)}, (H^{(0)} - \hat{E}^{(0)})\hat{\psi}_j^{(0)}) = (\hat{E}_i^{(0)} - \hat{E}^{(0)})\,\delta_{ij} \tag{27}$$

Thus $\gamma^{(0)}$ is diagonal, its eigenvalues being $(\hat{E}_i^{(0)} - \hat{E}^{(0)})$, with the corresponding eigenvectors (the $\hat{A}_{(i)}^{(0)}$) having a one in the ith position and zero elsewhere. Therefore we have, since $\lambda_l^{(0)}$ is the zero eigenvalue, that

$$\hat{A}_j^{(n)} = -\frac{C_j^{(n)}}{(\hat{E}_j^{(0)} - \hat{E}^{(0)})}, \quad j \neq l, \qquad \hat{A}_l^{(n)} = 0, \quad n > 0 \tag{28}$$

In particular, since in this basis

$$C_j^{(1)} = (\hat{\psi}_j^{(0)}, (H^{(1)} - \hat{E}^{(1)})\hat{\psi}^{(0)})$$

and since now $\xi^{(2)} \equiv 0$, we have

$$\hat{A}_j^{(1)} = -\frac{(\hat{\psi}_j^{(0)}, H^{(1)}\hat{\psi}^{(0)})}{\hat{E}_j^{(0)} - \hat{E}^{(0)}}, \quad j \neq l, \qquad \hat{A}_l^{(1)} = 0, \tag{29}$$

and

$$\hat{E}^{(2)} = (\hat{\psi}^{(0)}, H^{(2)}\hat{\psi}^{(0)}) - \sum_j{}' \frac{|(\hat{\psi}_j^{(0)}, H^{(1)}\hat{\psi}^{(0)})|^2}{\hat{E}_j^{(0)} - \hat{E}^{(0)}} \tag{30}$$

[4] For example, Grimaldi [10], Dalgarno and Drake [11], Brändas and Goscinski [12], Diner *et al.* [3], Gershgorn and Shavitt [4], Roos [5], and Huron *et al.* [6].

Further, since in this basis $\bar{\psi}^{(2)} = \bar{\psi}^{(3)} = 0$, (26-9) yields

$$\hat{E}^{(3)} = (\hat{\psi}^{(0)}, H^{(3)}\hat{\psi}^{(0)}) + \{(\hat{\psi}^{(1)}, (H^{(2)} - \hat{E}^{(2)})\hat{\psi}^{(0)}) + \text{comp. conj.}\} + (\hat{\psi}^{(1)}, (H^{(1)} - \hat{E}^{(1)})\hat{\psi}^{(1)}) \tag{31}$$

and of course all of these results are just what we would expect on the basis of our remarks following Eq. (25). [Note, however, that (28) means that we are using the so-called "intermediate normalization" convention, that is, $(\hat{\psi}^{(0)}, \hat{\psi}) = 1$, rather than the more conventional normalization $(\hat{\psi}, \hat{\psi}) = 1$.]

Also, if the space is independent of ν, there is a sometimes useful corollary of these considerations. Suppose that an exact eigenfunction ψ' lies in the space through order ν^n. That is, suppose that $\psi'^{(0)}, \psi'^{(1)}, \ldots, \psi'^{(n)}$ all lie in the space. Then evidently ψ' will also be an eigenfunction of $\bar{H}$ through order ν^n and therefore there will be a $\hat{\psi}$ and $\hat{E}$ which will be exact, that is, agree with ψ' and E', through orders ν^n and ν^{2n+1}, respectively.

Even if the basis set depends on ν, the discussion following Eq. (25) shows that one must still be able to write the results in a Rayleigh–Schrödinger form. One way to do this is of course to write out $\bar{H}$ as a power series in ν and apply Rayleigh–Schrödinger perturbation theory to it. An equivalent and more transparent approach is to return to Eqs. (16) and write $\hat{A}^{(n)}$ not as a linear combination of the $y_{(i)}^{(0)}$, which are the eigenvectors of $\gamma^{(0)} \equiv (\mathscr{H}^{(0)} - \hat{E}^{(0)}S^{(0)})$, but rather as a linear combination of the $\hat{A}_k^{(0)}$, which satisfy

$$(\mathscr{H}^{(0)} - \hat{E}_k^{(0)}S^{(0)})\hat{A}_k^{(0)} = 0 \tag{32}$$

We will now indicate how this procedure goes [13] in the nondegenerate case using, as is customary, the notation of Eq. (6-3), which equation we now repeat:

$$(\mathscr{H} - \hat{E}S)\hat{A} = 0 \tag{33}$$

First one expands $\mathscr{H}$ and S in powers of ν:

$$\mathscr{H} = \mathscr{H}^{(0)} + \nu\mathscr{H}^{(1)} + \nu^2\mathscr{H}^{(2)} + \cdots, \qquad S = S^{(0)} + \nu S^{(1)} + \nu^2 S^{(2)} + \cdots \tag{34}$$

where

$$\mathscr{H}_{ij}^{(0)} = (\phi_i^{(0)}, H^{(0)}\phi_j^{(0)}); \qquad S_{ij}^{(0)} = (\phi_i^{(0)}, \phi_j^{(0)})$$

$$\mathscr{H}_{ij}^{(1)} = (\phi_i^{(0)}, H^{(1)}\phi_j^{(0)}) + (\phi_i^{(1)}, H^{(0)}\phi_j^{(0)}) + (\phi_i^{(0)}H^{(0)}\phi_j^{(1)})$$

$$\mathscr{H}_{ij}^{(2)} = (\phi_i^{(0)}, H^{(2)}\phi_j^{(0)}) + (\phi_i^{(1)}, H^{(1)}\phi_j^{(0)}) + (\phi_i^{(0)}, H^{(1)}\phi_j^{(1)}) + (\phi_i^{(1)}, H^{(0)}\phi_j^{(1)}) + (\phi_i^{(0)}, H^{(0)}\phi_j^{(2)}) + (\phi_i^{(2)}, H^{(0)}\phi_j^{(0)})$$

$$S_{ij}^{(1)} = (\phi_i^{(0)}, \phi_j^{(1)}) + (\phi_i^{(1)}, \phi_j^{(0)})$$

$$S_{ij}^{(2)} = (\phi_i^{(0)}, \phi_j^{(2)}) + (\phi_i^{(1)}, \phi_j^{(1)}) + (\phi_i^{(2)}, \phi_j^{(0)})$$

etc. Then putting (34) into (33), and also now explicitly introducing an extra subscript l, Eqs. (13) become in this notation

$$(\mathscr{H}^{(0)} - \hat{E}_l^{(0)} S^{(0)}) \hat{A}_{(l)}^{(0)} = 0 \tag{35}$$

$$(\mathscr{H}^{(0)} - \hat{E}_l^{(0)} S^{(0)}) \hat{A}_{(l)}^{(1)} + (\mathscr{H}^{(1)} - \hat{E}_l^{(0)} S^{(1)} - \hat{E}_l^{(1)} S^{(0)}) \hat{A}_{(l)}^{(0)} = 0 \tag{36}$$

$$\begin{aligned}(\mathscr{H}^{(0)} - \hat{E}_l^{(0)} S^{(0)}) \hat{A}_{(l)}^{(2)} &+ (\mathscr{H}^{(1)} - \hat{E}_l^{(0)} S^{(1)} - \hat{E}_l^{(1)} S^{(0)}) \hat{A}_{(l)}^{(1)} \\ &+ (\mathscr{H}^{(2)} - \hat{E}_l^{(0)} S^{(2)} - \hat{E}_l^{(1)} S^{(1)} - \hat{E}_l^{(2)} S^{(0)}) \hat{A}_{(l)}^{(0)} = 0\end{aligned} \tag{37}$$

etc.

If we take the scalar product of (36) with $\hat{A}_{(l)}^{(0)}$, then, using (35), we find that in this notation (14) for $n = 1$ becomes

$$\hat{A}_{(l)}^{(0)+} \cdot (\mathscr{H}^{(1)} - \hat{E}_l^{(0)} S^{(1)} - \hat{E}_l^{(1)} S^{(0)}) \hat{A}_{(l)}^{(0)} = 0 \tag{38}$$

and this in turn, as we mentioned earlier, is just Eq. (24-19) for $\hat{E}^{(1)}$ in this special case. Also, taking the scalar product of (37) with $\hat{A}_{(l)}^{(0)}$, we find that Eq. (14) for $n = 2$ becomes

$$\begin{aligned}\hat{A}_{(l)}^{(0)+} \cdot (\mathscr{H}^{(1)} - \hat{E}_l^{(0)} S^{(1)} - \hat{E}_l^{(1)} S^{(0)}) \hat{A}_{(l)}^{(1)} \\ + \hat{A}_{(l)}^{(0)+} \cdot (\mathscr{H}^{(2)} - \hat{E}_l^{(0)} S^{(2)} - \hat{E}_l^{(1)} S^{(1)} - \hat{E}_l^{(2)} S^{(0)}) \hat{A}_{(l)}^{(0)} = 0\end{aligned} \tag{39}$$

which in turn is just Eq. (24-26) for $\hat{E}^{(2)}$ in this special case. Finally, if we take the scalar product of (36) with $A_{(k)}^{(0)}$, then we find, using (32), that

$$\hat{A}_{(k)}^{(0)+} \cdot S^{(0)} \hat{A}_{(l)}^{(1)} = - \frac{\hat{A}_{(k)}^{(0)+} \cdot (\mathscr{H}^{(1)} - \hat{E}_l^{(0)} S^{(1)} - \hat{E}_l^{(1)} S^{(0)}) \hat{A}_{(l)}^{(0)}}{\hat{E}_k^{(0)} - \hat{E}_l^{(0)}}, \qquad k \neq l \tag{40}$$

As we said, we now want to expand $\hat{A}_{(l)}^{(1)}$ as a linear combination of the $\hat{A}_{(k)}^{(0)}$. To see how this is done, we note that from

$$(\hat{\psi}_j^{(0)}, \hat{\psi}_k^{(0)}) = \delta_{jk} \tag{41}$$

it follows that

$$\sum_{i=1}^{M} \sum_{m=1}^{M} \hat{A}_{(j)i}^{(0)*} (\phi_i^{(0)}, \phi_m^{(0)}) \hat{A}_{(k)m}^{(0)} = \delta_{jk} \tag{42}$$

or

$$\hat{A}_{(j)}^{(0)+} \cdot S^{(0)} \hat{A}_{(k)}^{(0)} = \delta_{jk} \tag{43}$$

Therefore the general expansion formula is

$$\hat{A}_{(l)}^{(m)} = \sum_{k=1}^{M} \hat{A}_{(k)}^{(0)} (\hat{A}_{(k)}^{(0)+} \cdot S^{(0)} \hat{A}_{(l)}^{(m)}) \tag{44}$$

and hence in particular from (40) and (43)

$$\hat{A}_{(l)}^{(1)} = -\sum_k{}' \hat{A}_{(k)}^{(0)} \frac{\hat{A}_{(k)}^{(0)+}\cdot(\mathscr{H}^{(1)} - \hat{E}_l^{(0)}S^{(1)})\hat{A}_l^{(0)}}{\hat{E}_k^{(0)} - \hat{E}_l^{(0)}} \tag{45}$$

where we have arbitrarily, but without any real loss in generality, chosen $(A_{(l)}^{(1)+}\cdot S^{(0)}\hat{A}_{(l)}^{(0)})$ to be zero. Also, from (39), (45), and (43) it follows that

$$\hat{E}_l^{(2)} = \hat{A}_{(l)}^{(0)+}\cdot(\mathscr{H}^{(2)} - \hat{E}^{(0)}S^{(2)} - \hat{E}^{(1)}S^{(1)})\hat{A}_{(l)}^{(0)+} - \sum_k{}' \frac{|\hat{A}_{(k)}^{(0)+}\cdot(\mathscr{H}^{(1)} - \hat{E}_l^{(0)}S^{(1)})\hat{A}_{(l)}^{(0)}|^2}{\hat{E}_k^{(0)} - \hat{E}_l^{(0)}} \tag{46}$$

The formulas in higher order follow in a similar fashion. Also, the formulas for two or more perturbations follow by making the obvious substitutions, for example,

$$\hat{E}_l^{(11)} = \hat{A}_{(l)}^{(0)+}\cdot(\mathscr{H}^{(11)} - \hat{E}_l^{(0)}S^{(11)} - \hat{E}_l^{(10)}S^{(01)} - \hat{E}_l^{(01)}S^{(10)})\hat{A}_{(l)}^{(0)} - \sum_k{}' \frac{(\hat{A}_{(l)}^{(0)+}\cdot(\mathscr{H}^{(10)} - \hat{E}_l^{(0)}S^{(10)})\hat{A}_{(k)}^{(0)})(\hat{A}_{(k)}^{(0)+}\cdot(\mathscr{H}^{(01)} - \hat{E}^{(0)}S^{(01)})\hat{A}_{(l)}^{(0)}) + \text{comp. conj.}}{\hat{E}_k^{(0)} - \hat{E}_{(l)}^{(0)}} \tag{47}$$

We have presented four alternative ways of writing the solution of the nth-order equation of the linear variation method, the results being given in (3), (17), (23), and (45). Since they are all equivalent, any choice of one form over another, or the use of some other form, must be based on more practical considerations of speed, cost, availability of one program or another, etc.[5] However, it should be emphasized that if one extends the discussion to time-dependent problems, for example, to the calculation of frequency-dependent electric dipole polarizabilities, then the Rayleigh–Schrödinger approach is singled out on physical grounds. The point is that in such problems one will have equations like (36) except that $(\mathscr{H}^{(0)} - \hat{E}_l^{(0)}S^{(0)})$ will be replaced by $(\mathscr{H}^{(0)} - \hat{E}_l^{(0)}S^{(0)} \pm \omega S^{(0)})$, where ω is the frequency. Therefore the only essential change from (45) will be that the denominator will be replaced by

$$\hat{E}_k^{(0)} - \hat{E}_l^{(0)} \pm \omega$$

thus clearly exhibiting the $(\hat{E}_k^{(0)} - \hat{E}_l^{(0)})$ as the excitation energies of the system in this approximation. Further, the rest of the coefficient of $\hat{A}_{(k)}^{(0)}$,

[5] See, for example, Sanders [14].

when squared and multiplied by $\hat{E}_k^{(0)} - \hat{E}_l^{(0)}$, then provides an approximate oscillator strength.

Our considerations in this section has been confined to the "pure" linear variation method. If the basis functions also involve nonlinear parameters, then the matrix η will not vanish; however, one can still use the general methods of Section 24. Also, though we will not discuss it, a more Rayleigh–Schrödinger-like procedure has also been described and used in the literature [15] to calculate $\hat{\psi}^{(1)}$.

PROBLEMS

1. In the discussion following (15), we ignored the Mth equation (13). Show that, not surprisingly, as a consequence of (12), (14), and (16), it is satisfied identically.

2. (Adapted from Kane [16].) Suppose that the ϕ_k are independent of ν. Suppose also that $H^{(n)} = 0$, $n > 1$. Then show that $\sum_{k=1}^{M} (\hat{E}_k)^p$ can be calculated exactly knowing only the $\hat{E}_k^{(n)}$ for $n \leq p$. [Hints: $(\mathscr{H}')^p$, where $\mathscr{H}'$ is as in Problem 3 of Section 6, involves ν only up to the pth power. The sum of the eigenvalues of a matrix is the sum of its diagonal elements.]

3. Suppose that $\hat{E}_l^{(0)}$ of (35) is degenerate. Assume that the degeneracy is completely broken in first order and derive explicit formulas for $\hat{A}$ and $\hat{E}$ through first order and third order, respectively.

28. Perturbation Analysis of USCF and UHF: One-Particle Perturbations

We will start by considering USCF. Here, as in the linear variation method, the variation parameters [the $\hat{b}_{\sigma j}$ of (9-1)] are simply numbers. In our discussion, we will concentrate on attempting to illuminate a few points of principle and not actually write down many detailed results. Hopefully, however, these can then be extracted from the extensive literature[6] by the interested reader. (The literature mostly deals with RSCF and SUSCF; however, the general principles are the same.)

Although the $b_{\sigma j}$ are numbers, we cannot just take over the general formulas (24-20), etc., because it will be recalled that in our discussion of

[6] See, for example, Stevens *et al.* [17], Lipscomb [18], Gerratt and Mills [19], Moccia [20]; for discussion in terms of density matrices, see Diercksen and McWeeny [21], and Mestechkin [22].

USCF, we found it convenient to suppose that the spin orbitals were orthonormal, thereby imposing constraints. In a general way, the effect of constraints is to add Lagrange multiplier terms here and there to Eq. (24-20), etc. However, for the case in hand, there is no need to redo the general formalism since the result would be simply the perturbation expansion of Eqs. (9-9) and (9-7), equations which we repeat here (as in Section 9, we will not explicitly consider the possibility that the basis spin orbitals themselves contain nonlinear parameters):

$$(\hat{f} - \varepsilon_j S)\hat{B}_{(j)} = 0 \tag{1}$$

$$\hat{B}^{+}_{(j)} \cdot S\hat{B}_{(k)} = \delta_{jk} \tag{2}$$

where from (9-4)

$$\begin{aligned}\hat{f}_{\sigma\tau} &= (u_\sigma, hu_\tau) + \sum_{l=1}^{N} \sum_{\varrho=1}^{M} \sum_{\delta=1}^{M} \hat{b}^*_{\varrho l}\hat{b}_{\delta l}[(u_\sigma u_\varrho, gu_\tau u_\delta) - (u_\sigma u_\varrho, gu_\delta u_\tau)] \\ &= (u_\sigma, hu_\tau) + \sum_{l=1}^{N} [(u_\sigma\hat{\varphi}_l, gu_\tau\hat{\varphi}_l) - (u_\sigma\hat{\varphi}_l, g\hat{\varphi}_l u_\tau)]\end{aligned} \tag{3}$$

and where

$$S_{\sigma\tau} = (u_\sigma, u_\tau) \tag{4}$$

Also, it will be recalled that if the Lagrange multipliers ε_j are nondegenerate, then the $\hat{B}_{(j)}$ are unique up to an irrelevant phase, and the orthogonality conditions, conditions (2) for $j \neq k$, are automatically fulfilled. (Actually, we showed this only for UHF, but the same arguments obviously carry over to USCF).

As announced in the title of this section, we will for now confine attention to one-electron perturbations only [in this context, USCF is often called coupled, or fully coupled, SCF. For what is called uncoupled SCF, see following Eq. (13)]. Thus we have

$$h = h^{(0)} + \nu h^{(1)} + \nu^2 h^{(2)} + \cdots \tag{5}$$

Also, we will allow for the possibility that the basis spin orbitals depend on ν:

$$u_\sigma = u_\sigma^{(0)} + \nu u_\sigma^{(1)} + \nu^2 u_\sigma^{(2)} + \cdots \tag{6}$$

Then, writing

$$\hat{B}_{(j)} = \hat{B}^{(0)}_{(j)} + \nu\hat{B}^{(1)}_{(j)} + \nu^2\hat{B}^{(2)}_{(j)} + \cdots$$

and

$$\varepsilon_j = \varepsilon_j^{(0)} + \nu\varepsilon_j^{(1)} + \nu^2\varepsilon_j^{(2)} + \cdots \tag{7}$$

we are led to a set of lengthy equations which we will not write out in complete detail. The general nth-order equation is of the form

$$(\hat{\mathscr{f}}^{(0)} - \varepsilon_j^{(0)} S^{(0)})\hat{B}_{(j)}^{(n)} + G_{(j)}(\hat{B}^{(n)}) - \varepsilon_j^{(n)} S^{(0)} \hat{B}_{(j)}^{(0)} + C_{(j)}^{(n)} = 0 \tag{8}$$

where $C_{(j)}^{(n)}$, which is the part we will not write out in detail, involves only $h^{(n)}$ and lower-order quantities and hence can be considered known, and where, explicitly,

$$\hat{\mathscr{f}}_{\sigma\tau}^{(0)} = (u_\sigma^{(0)}, h^{(0)} u_\tau^{(0)}) + \sum_{l=1}^{N} [(u_\sigma^{(0)} \hat{\varphi}_l^{(0)}, g u_\tau^{(0)} \hat{\varphi}_l^{(0)}) - (u_\sigma^{(0)} \hat{\varphi}_l^{(0)}, g \hat{\varphi}_l^{(0)} u_\tau^{(0)})] \tag{9}$$

and

$$S_{\sigma\tau}^{(0)} = (u_\sigma^{(0)}, u_\tau^{(0)}) \tag{10}$$

so that

$$(\hat{\mathscr{f}}^{(0)} - \varepsilon_k^{(0)} S^{(0)})\hat{B}_{(k)}^{(0)} = 0 \tag{11}$$

Also, explicitly,

$$\begin{aligned} G_{(j)\sigma}(\hat{B}^{(n)}) = & \sum_{l=1}^{N} \sum_{\varrho=1}^{M} \hat{b}_{\varrho l}^{(n)*} [(u_\sigma^{(0)} u_\varrho^{(0)}, g \hat{\varphi}_j^{(0)} \hat{\varphi}_l^{(0)}) - (u_\sigma^{(0)} u_\varrho^{(0)}, g \hat{\varphi}_l^{(0)} \hat{\varphi}_j^{(0)})] \\ & + \sum_{l=1}^{N} \sum_{\varrho=1}^{M} \hat{b}_{\varrho l}^{(n)} [(u_\sigma^{(0)} \hat{\varphi}_l^{(0)}, g \hat{\varphi}_j^{(0)} u_\varrho^{(0)}) - (u_\sigma^{(0)} \hat{\varphi}_l^{(0)}, g u_\varrho^{(0)} \hat{\varphi}_j^{(0)})] \end{aligned} \tag{12}$$

Further, the condition (2) evidently yields in nth order that

$$\hat{B}_{(j)}^{(n)+} \cdot S^{(0)} \hat{B}_{(k)}^{(0)} + \hat{B}_{(j)}^{(0)+} \cdot S^{(0)} \hat{B}_{(k)}^{(n)} + \theta_{jk}^{(n)} = 0 \tag{13}$$

where $\theta^{(n)}$, which we will not write out in detail, involves only $S^{(n)}$ and lower-order quantities, and hence can be considered known.

A straightforward approach at this point is simply to solve Eqs. (8) (and their complex conjugates) as a set of linear inhomogeneous equations in the $\hat{b}_{\varrho l}^{(n)}$ and $\hat{b}_{\varrho l}^{(n)*}$. The solution will obviously be linear in the $\varepsilon_k^{(n)}$ and the latter can then be determined by requiring that (13) hold for $j = k$. In practice, however, if N and M are at all large, this is a fairly unwieldy procedure because of the large number of coupled equations, and the usual approach is through an iterative process much like that used in determining $\hat{\varphi}^{(0)}$ itself. Namely the G term in (8) causes most of the complication because it couples the various orbitals to one another. (If one simply ignores it, one has the simplest version of what is called uncoupled SCF [23]. We will discuss a related approximation further in Section 36.) Therefore one starts by guessing some approximation to the $\hat{B}^{(n)}_{(k)}$ and replaces $G_{(j)}$ by the approximation one calculates from them. With G "known," the equations uncouple, and further, taking the scalar product of (8) with $\hat{B}^{(0)+}_{(j)}$ and using (11) for $k = j$, it follows that

$$\varepsilon^{(n)}_j = \hat{B}^{(0)+}_{(j)} \cdot G_{(j)} + \hat{B}^{(0)+}_{(j)} \cdot C^{(n)}_{(j)} \tag{14}$$

so that $\varepsilon^{(n)}_j$ is now also "known." One now solves the uncoupled equations to determine a new set of $\hat{B}^{(n)}_{(k)}$. The latter will evidently be arbitrary to within a multiple of $\hat{B}^{(0)}_{(k)}$, which can then be chosen so as to meet the normalization conditions (this still leaves a phase ambiguity to be disposed of as one wishes). Then one uses these solutions to compute a new $G_{(j)}$, etc., until hopefully things converge.

There is another way of writing the perturbation equations which has the additional virtue of, in effect, reducing the number of unknowns by $N(M + 1)$. Also, as we will see, it is well adapted to handling any problems arising from a possible degeneracy of the $\varepsilon^{(0)}_j$, a possibility which we have been ignoring, and which we will continue to ignore for a while longer.

We can certainly write

$$\hat{B}^{(n)}_{(j)} = \sum_{k=1}^{N} \hat{B}^{(0)}_{(k)} (\hat{B}^{(0)+}_{(k)} \cdot S^{(0)} \hat{B}^{(n)}_{(j)}) + \bar{B}^{(n)}_{(j)} \tag{15}$$

thereby defining $\bar{B}^{(n)}_{(j)}$ as the part of $\hat{B}^{(n)}_{(j)}$ orthogonal to the $\hat{B}^{(0)}_{(k)}$ in the sense that

$$\hat{B}^{(0)+}_{(k)} \cdot S^{(0)} \bar{B}^{(n)}_{(j)} = 0, \qquad j, k = 1, 2, \ldots, N \tag{16}$$

Then, as one readily sees, the contribution of $\bar{B}^{(n)}_{(j)}$ to $\hat{\psi}_j$, which is $\sum u^{(0)}_\sigma \bar{b}^{(n)}_{\sigma j}$, is orthogonal to all $\hat{\psi}^{(0)}_k$. We will now show that we can separate the determination of the $\hat{B}^{(n)}_{(j)}$ into two steps: First determine the $\bar{B}^{(n)}_{(j)}$ and then determine the rest. To this end, we first rewrite (15) as

$$\hat{B}^{(n)}_{(j)} = \bar{B}^{(n)}_{(j)} + \sum_{k=1}^{N} \hat{B}^{(0)}_{(k)} \cdot \tfrac{1}{2}(\hat{B}^{(0)+}_{(k)} \cdot S^{(0)} \hat{B}^{(n)}_{(j)} + \hat{B}^{(n)+}_{(k)} \cdot S^{(0)} \hat{B}^{(0)}_{(j)}) + \sum_{k=1}^{N} X^{(n)}_{jk} \hat{B}^{(0)}_{(k)} \tag{17}$$

where

$$X^{(n)}_{jk} \equiv \tfrac{1}{2}(\hat{B}^{(0)+}_{(k)} \cdot S^{(0)} \hat{B}^{(n)}_{(j)} - \hat{B}^{(n)+}_{(k)} \cdot S^{(0)} \hat{B}^{(0)}_{(j)}) \tag{18}$$

We now note that, from (13), we know the second part of (17) in terms of lower-order quantities. Thus when we substitute (17) into (8), its contribution can be put together with $C^{(n)}_{(j)}$ to give a new known quantity,

$\bar{C}_{(j)}^{(n)}$ say. Further, and equally important, we observe that since

$$X_{jk}^{(n)} = - X_{kj}^{(n)*} \tag{19}$$

it follows that the transformation

$$\hat{\varphi}_j \to \hat{\varphi}_j - \nu^n \sum_k X_{jk}^{(n)} \hat{\varphi}_j + \cdots = \hat{\varphi}_j - \nu^n \sum_k X_{jk}^{(n)} \hat{\varphi}_j^{(0)} + \cdots \tag{20}$$

where the $\cdots$ represent terms of $O(\nu^{n+1})$ and higher, is, to order ν^n, a unitary transformation of the $\hat{\varphi}_j$. Also, we note that since $\hat{b}_{\sigma j}^{(n)}$ is the coefficient of $u_\sigma^{(0)}$ in $\hat{\varphi}_k^{(n)}$, it follows that we can equally write the transformation as

$$\hat{B}_{(j)}^{(n)} \to \hat{B}_{(j)}^{(n)} - \sum_{k=1}^{N} X_{jk}^{(n)} \hat{B}_{(k)}^{(0)} \tag{21}$$

Therefore we see that the third part of (17) could be removed by a unitary transformation. This has two consequences of interest. First of all, it means that if one is going to stop at the nth order, then one really does not need to know the $X_{jk}^{(n)}$ at all since the sum of the first two parts of (17), call it $\underline{B}_{(j)}^{(n)}$, all by itself is now seen to yield a perfectly good solution of the USCF equations to this order. True, it is not the canonical one, but it is unitarily equivalent to the canonical one to order ν^n and that is all that really counts. Second, and of more immediate interest, we can infer that the contribution of the third term of (17) to (8) will simply be in the nature of Lagrange multiplier-like terms since we know in general that the only effect of unitary transformations of the spin orbitals on the structure of the USCF equations is to change the Lagrange multiplier terms.

Indeed, in more detail, since the unitary transformation in question is

$$C_{kj} = \delta_{kj} + \nu^n X_{kj}^{(n)} + \cdots \tag{22}$$

we expect from (8-21) that the term of order ν^n in ε_{kj} will be changed from $\varepsilon_k^{(n)} \delta_{kj}$ by the amount

$$\sum_i (\delta_{ij} \varepsilon_i^{(0)} X_{ik}^{(n)} + X_{ij}^{(n)*} \varepsilon_i^{(0)} \delta_{ik}) = (\varepsilon_j^{(0)} - \varepsilon_k^{(0)}) X_{jk}^{(n)}$$

and since already

$$(\mathscr{f}^{(0)} - \varepsilon_j^{(0)} S^{(0)}) \sum_{k=1}^{N} X_{jk}^{(n)} \hat{B}_{(k)}^{(0)} = \sum_{k=1}^{N} (\varepsilon_k^{(0)} - \varepsilon_j^{(0)}) X_{jk}^{(n)} S^{(0)} \hat{B}_{(k)}^{(0)}$$

we can infer that

$$G_{(j)}(X^{(n)} \hat{B}^{(0)}) \equiv 0 \tag{23}$$

which we will now confirm. Thus consider the direct terms on the left in (23). From (12), they are

$$\sum_{l=1}^{N}\sum_{\varrho=1}^{M}\sum_{k=1}^{N} X_{lk}^{(n)*}\hat{b}_{\varrho k}^{(0)*}(u_\sigma^{(0)}u_\varrho^{(0)}, g\hat{\varphi}_j^{(0)}\hat{\varphi}_l^{(0)}) + X_{lk}^{(n)}\hat{b}_{\varrho k}^{(0)}(u_\sigma^{(0)}\hat{\varphi}_l^{(0)}, g\hat{\varphi}_j^{(0)}u_\varrho^{(0)})$$

$$= \sum_{l=1}^{N}\sum_{k=1}^{N} X_{lk}^{(n)*}(u_\sigma^{(0)}\hat{\varphi}_k^{(0)}, g\hat{\varphi}_j^{(0)}\hat{\varphi}_l^{(0)}) + X_{lk}^{(n)}(u_\sigma^{(0)}\hat{\varphi}_l^{(0)}, g\hat{\varphi}_j^{(0)}\hat{\varphi}_k^{(0)})$$

If we now interchange k and l in the second term and use (19), we see that the direct term vanishes. A similar discussion then shows that the contribution of the exchange term also vanishes, which proves the point.

In summary, if we introduce (17) into (8), we will be led to equations of the form

$$(\hat{\mathscr{f}}^{(0)} - \varepsilon_j^{(0)}S^{(0)})\bar{B}_{(j)}^{(n)} + G_{(j)}(\bar{B}^{(n)}) + \bar{C}_{(j)}^{(n)} - \varepsilon_j^{(n)}S^{(0)}\hat{B}_{(j)}^{(0)}$$
$$+ \sum_k (\varepsilon_k^{(0)} - \varepsilon_j^{(0)})X_{jk}^{(n)}S^{(0)}\hat{B}_{(k)}^{(0)} = 0 \qquad (24)$$

(Note that the $X_{jj}^{(n)}$ do not appear and so will be left arbitrary. However, this was to be expected since the pure imaginary number $X_{jj}^{(n)}$ contributes only to the arbitrary phase of $\hat{\varphi}_j$.)

We now introduce $M-N$ vectors $V_{(t)}$ which are orthogonal to all the $\hat{B}_{(k)}^{(0)}$ in the sense that

$$V_{(t)}^{+}\cdot S^{(0)}\hat{B}_{(k)}^{(0)} = 0, \qquad t = 1, 2, \ldots, M-N, \quad k = 1, 2, \ldots, N \qquad (25)$$

For example, in practice [recall the discussion of (10-27)] they are usually taken to be the virtual USCF vectors satisfying

$$(\hat{\mathscr{f}}^{(0)} - \varepsilon_t S^{(0)})V_{(t)} = 0$$

In any case, if we take the scalar product of (24) for each j with each $V_{(t)}^{+}$ in turn, we will evidently produce a set of equations involving neither the unknown Lagrange multipliers $\varepsilon_k^{(n)}$ nor the unknown X_{jk}, $j \neq k$. Further, since we can certainly write

$$\bar{B}_{(j)}^{(n)} = \sum_{t=1}^{M-N} V_{(t)}\bar{b}_{tj}^{(n)} \qquad (26)$$

where the $\bar{b}_{tj}^{(n)}$ are numbers, these equations can then be solved, perhaps directly, but in practice usually by iterating on the G terms, for the $\bar{b}_{tj}^{(n)}$.

Given the $\bar{b}_{tj}^{(n)}$ and hence the $\bar{B}_{(j)}^{(n)}$, we then return to (24). If we take its scalar product with $\hat{B}_k^{(0)+}$, we find, using (16), that

$$\hat{B}_{(k)}^{(0)+} \cdot (G_{(j)}(\bar{B}^{(n)}) + \bar{C}_{(j)}^{(n)}) - \varepsilon_j^{(n)} \delta_{jk} + (\varepsilon_k^{(0)} - \varepsilon_j^{(0)}) X_{jk}^{(n)} = 0 \qquad (27)$$

from which it follows that

$$\varepsilon_j^{(n)} = -B_{(j)}^{(0)+}(G_{(j)}(\bar{B}^{(n)}) + \bar{C}_{(j)}^{(n)}) \qquad (28)$$

and

$$X_{jk}^{(n)} = -\hat{B}_k^{(0)+} \cdot \frac{(G_{(j)}(\bar{B}^{(n)}) + \bar{C}_{(j)}^{(n)})}{\varepsilon_k^{(0)} - \varepsilon_j^{(0)}}, \qquad k \neq j \qquad (29)$$

which, together with

$$X_{kk}^{(n)} = i\alpha \qquad (30)$$

α any real number, for example, zero, completes the solution.

Our discussion to this point has both implicitly and explicitly [in the case of Eq. (29)] assumed that there are no complications because of possible degeneracy among the $\varepsilon_j^{(0)}$. Let us now discuss what happens in first order if, say, $\varepsilon_1^{(0)} = \varepsilon_2^{(0)} \neq \varepsilon_k^{(0)}$ for $k > 2$. Then for $j, k = 1, 2$, (27) yields

$$\hat{B}_{(k)}^{(0)+} \cdot (G_{(j)}(\bar{B}^{(1)}) + \bar{C}_{(j)}^{(1)}) - \varepsilon_j^{(1)} \delta_{jk} = 0 \qquad j, k = 1, 2 \qquad (31)$$

which are the sort of equations familiar from Rayleigh–Schrödinger degenerate perturbation theory for determining the correct zeroth-order functions. However, there does seem to be a complication in that the $G_{(j)}$ and $\bar{C}_{(j)}^{(1)}$ look very nonlinear. We will now show that in fact this is not a source of difficulty.

First of all, we note that if one were to abandon the canonical orbitals in first order so that $\varepsilon_j^{(1)} \delta_{jk} \to \varepsilon_{jk}^{(1)}$, then Eqs. (31) would become innocuous equations for $\varepsilon_{jk}^{(1)}$, $j, k = 1, 2$. Therefore if one is going to stop at first order, one can simply ignore (31) completely. That is, having selected some orthonormal set, $\hat{\varphi}_j^{(0)\prime}$ say, and calculated the corresponding $\underline{\hat{B}}_{(j)}^{(1)\prime}$ from Eq. (24) supplemented by primes here and there, one can simply use them to calculate $\hat{\varphi}$, since from what we have just said here and following (21), these will be solutions of some noncanonical set of USCF equations.

However, suppose that one wants to go on and that one does want the canonical spin orbitals. Proceeding as above, one can then calculate the 2×2 matrix $\mathscr{M}'$ whose components are

$$\mathscr{M}_{kj}' = \hat{B}_{(k)}^{(0)\prime+} \cdot (G_{(j)}(\bar{B}^{(1)\prime}) + \bar{C}_{(j)}^{(1)\prime}) \qquad (32)$$

and which one can easily show is Hermitian. Further, one can infer from the invariance of the form of the USCF equations to unitary transformations of the spin orbitals (we leave the detailed verification as a problem) that if one makes a ν-independent 2×2 unitary transformation of $\hat{B}_{(1)}^{(0)\prime}$ and $\hat{B}_{(2)}^{(0)\prime}$, then the $\bar{B}_{(k)}^{(1)\prime}$ and $\bar{C}_{(k)}^{(1)\prime}$ for $k = 1, 2$ will undergo the same unitary transformation, and that $\mathscr{M}'$ will undergo the corresponding matrix transformation. Therefore it follows that if, in the ordinary way, one finds a 2×2 unitary matrix which diagonalizes $\mathscr{M}'$, one will at the same time determine $\hat{B}_{(j)}^{(0)}$, $\bar{B}_{(j)}^{(1)}$, and $\varepsilon_j^{(1)}$ which satisfy (31).

At this stage, $X_{12}^{(1)}$ and possibly the $\hat{B}_{(k)}^{(0)}$ are as yet undetermined. However, if the degeneracy is removed in first order, then the $\hat{B}_{(k)}^{(0)}$ will have been determined from the preceding, and, as one can readily show, $X_{12}^{(1)}$ is determined by the second-order equations when one takes the scalar product of the $j = 1$ equation with $\hat{B}_{(2)}^{(0)}$.

Once the $B_{(j)}^{(n)}$ have been determined, then we can, if we wish, calculate $\hat{E}$ to order ν^{2n+1}. In particular, given the $B_{(j)}^{(1)}$ we can calculate $\hat{E}^{(1)}$, $\hat{E}^{(2)}$, and $\hat{E}^{(3)}$. Although the general formulas (24-19), (24-26), and (26-9) apply, it is perhaps simpler to start again, and in accord with (26-1), define

$$\mathring{\psi} \equiv | \mathring{\varphi}_1, \mathring{\varphi}_2, \ldots, \mathring{\varphi}_N | \tag{33}$$

where

$$\mathring{\varphi}_j = \sum_{\sigma=1}^{M} u_\sigma(\hat{b}_{\sigma j}^{(0)} + b_{\sigma j}^{(1)}) \tag{34}$$

and extract the terms up to third order from

$$(\mathring{\psi}, (H - \hat{E})\mathring{\psi}) = 0$$

Further, to simplify the calculation, it is helpful to replace the $\mathring{\varphi}_j$ by an orthonormal set so that one can readily use Slater's rules in the calculation. Since the $\mathring{\varphi}_j$ are already orthonormal through first order in ν, the orthonormality corrections will be of order $\nu^2, \nu^4, \ldots$, with only the former being needed to order ν^3.

If the u_σ do not depend on ν, then the calculation of $\hat{E}^{(1)}$ and $\hat{E}^{(2)}$ is quite simple since we can use (24-36) and (24-38). Thus from the former we immediately find

$$\hat{E}^{(1)} = \sum_{j=1}^{N} (\hat{\varphi}_j^{(0)}, h^{(1)}\hat{\varphi}_j^{(0)}) \tag{35}$$

Also, since, as one readily sees, in this case $\theta^{(1)}$ of Eq. (8) for $n = 1$ is identically zero, the second part of (17) vanishes for $n = 1$. Further, since, as we know, $X_{kj}^{(1)}$ does not contribute to $\hat{\psi}^{(1)}$, the latter takes the form

$$\hat{\psi}^{(1)} = \sum_{j=1}^{N} | \hat{\varphi}_1^{(0)}, \ldots, \bar{\varphi}_j^{(1)}, \ldots | \tag{36}$$

where

$$\bar{\varphi}_j^{(1)} = \sum_{\sigma} u_{\sigma} \bar{b}_{\sigma j}^{(1)}$$

and hence [from (16)] is orthogonal to all $\hat{\varphi}_k^{(0)}$. Therefore from Slater's rules, (24-38) immediately yields

$$\hat{E}^{(2)} = \sum_{j=1}^{N} (\hat{\varphi}_j^{(0)}, h^{(2)}\hat{\varphi}_j^{(0)}) + \tfrac{1}{2}\sum_{j=1}^{N} [(\hat{\varphi}_j^{(0)}, h^{(1)}\bar{\varphi}_j^{(1)}) + (\bar{\varphi}_j^{(1)}, h^{(1)}\hat{\varphi}_j^{(0)})] \tag{37}$$

which we note can also be written as

$$\hat{E}^{(2)} = \sum_{j=1}^{N} (\hat{\varphi}_j^{(0)}, h^{(2)}\hat{\varphi}_j^{(0)}) + \tfrac{1}{2}\sum_{j=1}^{N} [(\hat{\varphi}_j^{(0)}, h^{(1)}\hat{\varphi}_j^{(1)}) + (\hat{\varphi}_j^{(1)}, h^{(1)}\hat{\varphi}_j^{(0)})] \tag{38}$$

the $X^{(1)}$ terms making no contribution, because of (19).

This completes our discussion of USCF. As we said at the outset, our goal was not to arrive at detailed results but rather to explore some points of principle. It is in this same spirit that we turn to a consideration of UHF.[7] In addition, we confine attention to the nondegenerate case. Since the general structure of the equations is the same in each order, it will be sufficient to consider only the first-order equations in detail. From (8-20) and (8-9), they have the form

$$(\hat{\mathscr{f}}^{(0)} - \varepsilon_j^{(0)})\hat{\varphi}_j^{(1)} + G_j^{(1)} + (h^{(1)} - \varepsilon_j^{(1)})\hat{\varphi}_j^{(0)} = 0 \tag{39}$$

where, if χ is an arbitrary spin orbital,

$$\hat{\mathscr{f}}^{(0)}\chi \equiv h^{(0)}\chi + \sum_{l=1}^{N} [(\hat{\varphi}_l^{(0)}, g\hat{\varphi}_l^{(0)})\chi - (\hat{\varphi}_j^{(0)}, g\chi)\hat{\varphi}_l^{(0)}] \tag{40}$$

Hence in particular

$$(\hat{\mathscr{f}}^{(0)} - \varepsilon_k^{(0)})\hat{\varphi}_k^{(0)} = 0 \tag{41}$$

Also,

$$\begin{aligned} G_j^{(1)} = \sum_{l=1}^{N} \{ &(\hat{\varphi}_l^{(1)}, g\hat{\varphi}_l^{(0)})\hat{\varphi}_j^{(0)} - (\hat{\varphi}_l^{(1)}, g\hat{\varphi}_j^{(0)})\hat{\varphi}_l^{(0)} \\ &+ (\hat{\varphi}_l^{(0)}, g\hat{\varphi}_l^{(1)})\hat{\varphi}_j^{(0)} - (\hat{\varphi}_l^{(0)}, g\hat{\varphi}_j^{(0)})\hat{\varphi}_l^{(1)} \} \end{aligned} \tag{42}$$

[7] For more details, see, for example, Rebane [24], and Langhoff *et al.* [25].

If the problem is such that one can separate out angular variables and reduce (39) to coupled one-dimensional "radial equations," then one can attack them by direct numerical techniques, presumably in the process iterating on the $G_j^{(1)}$ so as to uncouple the equations. However, usually such an approach is not possible and the alternative is to use variation methods as described in Section 25.

In practice, as we mentioned following (25-11), this means restricting the $\varphi_i^{(1)}$ to be expandable in a finite basis set. The equations which result then have much the same character as the USCF equations, so we will not enter into any details except to mention a point connected with the Lagrange multipliers. Let us define the quantity $\hat{Q}_j$ by writing (39) as

$$\hat{Q}_j - \varepsilon_j^{(1)}\hat{\varphi}_j^{(0)} = 0 \tag{43}$$

Then with this definition, one finds for $\delta\check{J}^{(2)}$ [$J^{(2)}$ being given by (25-8)] a result which is quite analogous to (8-8), namely

$$\delta\check{J}^{(2)} = \sum_{j=1}^{N} [(\delta\check{\varphi}_j^{(1)}, \check{Q}_j) + (\check{Q}_j, \delta\check{\varphi}_j^{(1)})] \tag{44}$$

Further, to take care of the constraints

$$(\hat{\varphi}_j^{(1)}, \hat{\varphi}_k^{(0)}) + (\hat{\varphi}_j^{(0)}, \hat{\varphi}_k^{(1)}) = 0 \tag{45}$$

we introduce Lagrange multipliers, thereby in effect replacing $\check{J}^{(2)}$ by

$$\check{J}^{(2)} - \sum_j \sum_k \check{\varepsilon}_{jk}[(\check{\varphi}_j^{(1)}, \hat{\varphi}_k^{(0)}) + (\hat{\varphi}_j^{(0)}, \check{\varphi}_k^{(1)})] \tag{46}$$

Then we require that

$$\delta\check{J}^{(2)} - \sum_j \sum_k \check{\varepsilon}_{jk}[(\delta\check{\varphi}_j^{(1)}, \hat{\varphi}_k^{(0)}) + (\hat{\varphi}_j^{(0)}, \delta\check{\varphi}_k^{(1)})] = 0$$

and that

$$(\check{\varphi}_j^{(1)}, \hat{\varphi}_k^{(0)}) + (\hat{\varphi}_j^{(0)}, \check{\varphi}_k^{(1)}) = 0 \tag{47}$$

and proceed to draw the consequences.

However, instead of using all the machinery of Section 25, one can simply write down a functional which yields Eq. (43), and indeed this is usually what is done. However, it is clear from (44) that that functional will be just

$$\check{J}^{(2)} - \sum \check{\varepsilon}_j^{(1)}[(\check{\varphi}_j^{(1)}, \hat{\varphi}_j^{(0)}) + (\hat{\varphi}_j^{(0)}, \check{\varphi}_j^{(1)})] \tag{48}$$

which differs from (46) in that there are no off-diagonal Lagrange multiplier terms.

There are now two points to be made. First of all, if one uses (48), then in general, unless symmetry conditions intervene, the conditions (47) for $j \neq k$ will not be satisfied. Thus to be consistent one should in general use (46), although one might argue that since $\check{\varphi}_i^{(1)}$ will be only an approximation to $\hat{\varphi}_i^{(0)}$, one really need not worry specifically about this one source of error. The second point is that, as one might expect, if one makes the set of trial functions $\varphi_l^{(1)}$ in (48) flexible enough, then the conditions (47) for $k \neq j$ *will* be satisfied automatically just as they are in canonical UHF itself. Namely we will show that this is the case if one uses trial $\varphi_l^{(1)}$ of the form

$$\varphi_l^{(1)} = \sum_{i=1}^{N} A_{li}^{(1)} \hat{\varphi}_i^{(0)} + \cdots \tag{49}$$

where the $A_{li}^{(1)}$ are arbitrary numerical parameters and where the dots can be anything so long as it does not depend explicitly on the $A_{li}^{(1)}$. We first note that this means that $\delta A_{kj}^{(1)} \hat{\varphi}_j^{(0)}$ with $\delta A_{kj}^{(1)}$ an arbitrary complex number is a possible $\delta\check{\varphi}_k^{(1)}$. One then readily finds that for such a variation, use of (48) will yield [varying $\check{\varphi}_k^{(1)*}$ and not $\check{\varphi}_k^{(1)}$ and making use of (44)]

$$(\hat{\varphi}_j^{(0)}, (\hat{\mathscr{f}}^{(0)} - \varepsilon_j^{(0)})\check{\varphi}_k^{(1)}) + (\hat{\varphi}_j^{(0)}, (h^{(1)} - \check{\varepsilon}_k^{(1)})\hat{\varphi}_k^{(0)}) + \times\times\times = 0 \tag{50}$$

where $\times\times\times$ represents the $\check{G}$ terms. Similarly varying $\check{\varphi}_j^{(1)}$ but not $\check{\varphi}_j^{(1)*}$, with $\delta\check{\varphi}_j^{(1)} = \delta A_{jk}^{(1)} \hat{\varphi}_k^{(0)}$ one finds

$$(\check{\varphi}_j^{(1)}, (\hat{\mathscr{f}}^{(0)} - \varepsilon_j^{(0)})\hat{\varphi}_k^{(0)}) + (\hat{\varphi}_j^{(0)}(h^{(1)} - \check{\varepsilon}_j^{(1)})\hat{\varphi}_k^{(0)}) + \times\times\times = 0 \tag{51}$$

where the $\check{G}$ terms in (51) are identical to those in (50). If now one uses (41) and the orthogonality of $\hat{\varphi}_j^{(0)}$ and $\hat{\varphi}_k^{(0)}$, (50) and (51) can be written

$$(\varepsilon_j^{(0)} - \varepsilon_k^{(0)})(\hat{\varphi}_j^{(0)}, \check{\varphi}_k^{(1)}) + (\hat{\varphi}_j^{(0)}, h^{(1)}\hat{\varphi}_k^{(0)}) + \times\times\times = 0 \tag{52}$$

and

$$(\varepsilon_k^{(0)} - \varepsilon_j^{(0)})(\check{\varphi}_j^{(1)}, \hat{\varphi}_k^{(0)}) + (\hat{\varphi}_j^{(0)}, h^{(1)}\hat{\varphi}_k^{(0)}) + \times\times\times = 0 \tag{53}$$

respectively, whence subtraction and cancellation of a factor of $(\varepsilon_k^{(0)} - \varepsilon_j^{(0)})$ yields (47), as desired.

Thus far in our discussion, we have implicitly assumed that the $\hat{\varphi}_j^{(0)}$ are known exactly, or at least are known with sufficient precision so that, for example, integrals involving $(\hat{\mathscr{f}}^{(0)} - \varepsilon_j^{(0)})\hat{\varphi}_j^{(0)}$ vanish to the desired number of decimals. However, this is often not at all the case and yet the variational formalism in the form given here is still used but with the $\hat{\varphi}_j^{(0)}$ replaced by the approximation, and one may well ask, what is the status of such calculations? If the approximate $\hat{\varphi}_j^{(0)}$ are USCF functions, then if the set of $\varphi_i^{(1)}$ is chosen appropriately, it may be that one is really making a USCF calculation (or a variational approximation to it). However, in a general way the status, meaning, and consistency of such calculations are not immediately clear and (presuming that one is interested) would require separate investigations in each case. Indeed the situation is quite analogous to what which we discussed at the end of Section 25. Further, if one goes on to higher order, for example, using $J^{(4)}$ but with the $\hat{\varphi}_i^{(0)}$ *and* the $\hat{\varphi}_i^{(1)}$ replaced by approximations, then the situation becomes even more obscure. Therefore, having drawn attention to the problem, we will not attempt to discuss the matter further.

Given the $\hat{\varphi}_j^{(1)}$, we can calculate $\hat{E}^{(1)}$, $\hat{E}^{(2)}$, and $\hat{E}^{(3)}$, the formulas for $\hat{E}^{(1)}$ and $\hat{E}^{(2)}$ evidently being just (34) and (38) again. To determine $\hat{E}^{(3)}$, the simplest procedure, as in the USCF case, is to use

$$\mathring{\psi} = |\, (\hat{\varphi}_1^{(0)} + \nu\hat{\varphi}_1^{(1)}),\ (\hat{\varphi}_2^{(0)} + \nu\hat{\varphi}_2^{(1)}),\ \ldots \,|$$

orthogonalize the spin orbitals, and extract the third-order terms from $(\mathring{\psi}, (H - \mathring{E})\mathring{\psi}) = 0$. Also, $\check{E}^{(1)}$ and $\check{E}^{(2)}$ are given by analogous formulas. Further, one can, for example, write the formula for $\hat{E}^{(3)}$ in other ways by using (39).[7]

PROBLEMS

1. Be sure you understand in detail the argument which led to the inference (23) (an inference which we then verified directly).
2. The $\varepsilon_i^{(0)}$ for closed shell solutions have some obvious degeneracies. What are they? Do they cause any problems?
3. Derive formulas (2.20)–(2.26) of Langhoff *et al.* [25].

29. The $\mathscr{Z}^{-1}$ Expansion

We have several times alluded to the fact that if for an atom of nuclear charge $\mathscr{Z}$ one treats the Coulomb interaction between the electrons as a perturbation, then the result is an expansion in powers of $\mathscr{Z}^{-1}$. The proof is as follows [26]: The Hamiltonian for the system is

$$H = \sum_t \left\{ \frac{[\mathbf{p}_t + (1/2c)\mathscr{B}\times\mathbf{r}_t]^2}{2} - \frac{\mathscr{Z}}{|\,\mathbf{r}_t\,|} + \mathscr{E}\cdot\mathbf{r}_t \right\} + \sum_{s>t}\sum \frac{1}{|\,\mathbf{r}_s - \mathbf{r}_t\,|} \tag{1}$$

where we have taken the nucleus to be the coordinate origin and have included the possibility that uniform electric and magnetic fields are present. (The discussion is easily generalized to include any number of independent pure multipole fields.) We now change variables (or equivalently make a canonical transformation) according to

$$\mathbf{r}_s = \mathbf{r}_s'/\mathscr{Z}, \qquad \mathbf{p}_s = \mathscr{Z}\mathbf{p}_s' \tag{2}$$

and readily find that

$$H/\mathscr{Z}^2 = \sum_t \left\{ \frac{[\mathbf{p}_t' + (1/2c)\mathscr{B}' \times \mathbf{r}_t']^2}{2} - \frac{1}{|\mathbf{r}_t'|} + \mathscr{E}' \cdot \mathbf{r}_t' \right\} + \frac{1}{\mathscr{Z}} \sum_{s>t} \sum \frac{1}{|\mathbf{r}_s' - \mathbf{r}_t'|} \tag{3}$$

where

$$\mathscr{B}' \equiv \mathscr{B}/\mathscr{Z}^2 \tag{4}$$

and

$$\mathscr{E}' \equiv \mathscr{E}/\mathscr{Z}^3 \tag{5}$$

Therefore since on the right hand side of (3) $\mathscr{Z}$ appears explicitly only as a factor $\mathscr{Z}^{-1}$ multiplying the Coulomb interaction between the electrons, we see that, to give a precise statement of the theorem, if we treat that interaction as a perturbation, then for fixed $\mathscr{B}'$ and $\mathscr{E}'$, the result will be an expansion of the eigenvalues divided by $\mathscr{Z}^2$ in powers of $\mathscr{Z}^{-1}$, and an expansion of the eigenfunctions as functions of the $\mathbf{r}_s'$, also in powers of $\mathscr{Z}^{-1}$. Further, by similar arguments, the same kinds of expansions will presumably be possible for any variational $\hat{E}$ and $\hat{\psi}$ derived from a set of trial functions which (in $\mathbf{r}_s'$ variables and for fixed $\mathscr{B}'$ and $\mathscr{E}'$) is independent of $\mathscr{Z}$ or which depends on $\mathscr{Z}$ in an analytic way.

Having derived this general result, we will now prove the $\mathscr{Z}^{-1}$ theorems for UHF and SUHF which were quoted in Section 10. To do this most simply, it is helpful to continue to use $\mathbf{r}_s'$ variables and to measure energies in units of $\mathscr{Z}^2$ so that we can think of the right hand side of (3) as *the* Hamiltonian which then is of our standard form

$$H^{(0)} + \nu H^{(1)} \tag{6}$$

with

$$\nu = \mathscr{Z}^{-1} \tag{7}$$

and can directly take over our earlier general results without making any change in notation. In particular, let us recall Eq. (24-15), which, since there are no reality restrictions in UHF or SUHF, we can write as

$$(\delta^0\hat{\psi}, (H^{(0)} - \hat{E}^{(0)})\hat{\psi}^{(1)}) + (\delta^0\hat{\psi}, (H^{(1)} - \hat{E}^{(1)})\hat{\psi}^{(0)}) \\ + (\delta^1\hat{\psi}, (H^{(0)} - \hat{E}^{(0)})\hat{\psi}^{(0)}) = 0 \tag{8}$$

Now $H^{(0)}$ is a spin-free one-electron Hamiltonian. Therefore the $\hat{\psi}^{(0)}$ of UHF and of SUHF will certainly be eigenfunctions of it, so we can drop the last term in (8) and write instead

$$(\delta^0\hat{\psi}, (H^{(0)} - \hat{E}^{(0)})\hat{\psi}^{(1)}) + (\delta^0\hat{\psi}, (H^{(1)} - \hat{E}^{(1)})\hat{\psi}^{(0)}) = 0 \tag{9}$$

Further, $\delta^0\hat{\psi}$ consists in a general way of a multiple of $\hat{\psi}^{(0)}$ plus (possibly sums of) one-electron excitations of $\hat{\psi}^{(0)}$. However, the former will contribute nothing to the left hand side of (8), either because $(\hat{\psi}^{(0)}, (H^{(0)} - \hat{E}^{(0)})\hat{\psi}^{(0)}) = 0$ or because $(\hat{\psi}^{(0)}, (H^{(1)} - \hat{E}^{(1)})\hat{\psi}^{(0)}) = 0$. As for the one-electron excitations, from the discussion of their actual form which we gave in Section 10, it should be clear that they can be assumed without loss of generality to be also eigenfunctions of $H^{(0)}$, call them $\psi_k^{(0)}$ with eigenvalues $E_k^{(0)}$ say, so that for them (9) yields

$$(\psi_k^{(0)}, \hat{\psi}_k^{(1)}) = -\frac{(\psi_k^{(0)}, H^{(1)}\hat{\psi}^{(0)})}{E_k^{(0)} - \hat{E}^{(0)}}, \qquad E_k^{(0)} \neq \hat{E}^{(0)} \tag{10}$$

So far, we have made no assumptions beyond those of UHF and SUHF. However, now let us suppose further that $\psi'^{(0)}$, where as usual ψ' is the eigenfunction, is also a single determinant. Hence in such cases,

$$\hat{\psi}^{(0)} = \psi'^{(0)} \tag{11}$$

One immediate consequence of this is of course that

$$\hat{E}^{(0)} = E'^{(0)}, \qquad \hat{E}^{(1)} = E'^{(1)} \tag{12}$$

and therefore that (returning to ordinary energy units for the moment) the so-called correlation energy $(E' - \hat{E})$ is in first approximation independent of $\mathscr{Z}$. Of more immediate interest, however, is that if now, in accord with (11) and (12), we write (10) as

$$(\psi_k^{(0)}, \hat{\psi}^{(1)}) = -\frac{(\psi_k^{(0)}, H^{(1)}\psi'^{(0)})}{E_k^{(0)} - E'^{(0)}}, \qquad E_k^{(0)} \neq E'^{(0)} \tag{13}$$

then, by comparing with the familiar result of first-order Rayleigh–Schrödinger perturbation theory for $\psi'^{(1)}$, it follows that

$$(\psi_k^{(0)}, \hat{\varphi}^{(1)}) = (\psi_k^{(0)}, \psi'^{(1)}), \qquad E_k^{(0)} \neq E'^{(0)} \tag{14}$$

Finally, we make one further assumption. Namely we will suppose that if there are any $\psi_k^{(0)}$ which are degenerate with $\hat{\varphi}^{(0)}$, then (invariably for reasons of symmetry) they are orthogonal to $\hat{\varphi}^{(1)}$ and to $\psi'^{(1)}$, so that we can remove the qualification in (14) and write simply

$$(\psi_k^{(0)}, \hat{\varphi}^{(1)}) = (\psi_k^{(0)}, \psi'^{(1)}) \tag{15}$$

and this is the essential result. For UHF, the $\psi_k^{(0)}$ can be any one-electron excitation of $\hat{\varphi}^{(0)}$, whence, from (15) it follows that under the stated conditions, the one-electron excitations in $\hat{\varphi}^{(1)}$ and $\psi'^{(1)}$ will be identical. (Of course, $\hat{\varphi}^{(1)}$ is simply a linear combination of $\hat{\varphi}^{(0)}$ and one-electron excitations of $\psi^{(0)}$. However, $\psi'^{(1)}$ involves two-electron excitations as well.)

One corollary of these considerations is the following: Consider the ψ', which in the $\mathscr{Z} \to \infty$ limit become single determinants which are one-electron excitations of one another, for example, 1s*n*s ^{3}S ($M_s = 1$) for various n or $(1s)^2 ns$ ^{2}S. Then, although, as we mentioned at the outset of Section 21, unless symmetry conditions intervene, the corresponding $\hat{\varphi}$ of UHF are not in general exactly orthogonal; still, as we will now show, they are orthogonal through order $\mathscr{Z}^{-1}$ [27]. Let one of the states be labeled k, the other l. Then in obvious notation it follows from (10) and the fact that they are orthogonal in zeroth order that

$$\begin{aligned}(\hat{\varphi}_k, \hat{\varphi}_l) &= \mathscr{Z}^{-1}\{(\hat{\varphi}_k^{(0)}, \hat{\varphi}_l^{(1)}) + (\hat{\varphi}_k^{(1)}, \hat{\varphi}_l^{(0)})\} + \cdots \\ &= -\mathscr{Z}^{-1}(\hat{\varphi}_k^{(0)}, H^{(1)}\hat{\varphi}_l^{(0)})\left(\frac{1}{\hat{E}_k^{(0)} - \hat{E}_l^{(0)}} + \frac{1}{\hat{E}_l^{(0)} - \hat{E}_k^{(0)}}\right) + \cdots = 0 + \cdots\end{aligned}$$

which proves the point.

Turning to the theorem of Section 10, let W be a Hermitian one-electron operator. We can write its UHF expectation value, correct through relative order $\mathscr{Z}^{-1}$, as

$$\begin{aligned}\langle W\rangle &= \langle W\rangle_0 + \mathscr{Z}^{-1}\,\frac{(\hat{\varphi}^{(1)}, (W - \langle W\rangle_0)\hat{\varphi}^{(0)}) + \text{comp. conj.}}{(\hat{\varphi}^{(0)}, \hat{\varphi}^{(0)})} + \cdots \\ &\equiv \langle W\rangle_0 + \mathscr{Z}^{-1}\langle W\rangle_1 + \cdots\end{aligned} \tag{16}$$

where

$$\langle W\rangle_0 \equiv (\hat{\varphi}^{(0)}, W\hat{\varphi}^{(0)})/(\hat{\varphi}^{(0)}, \hat{\varphi}^{(0)}) \tag{17}$$

However, from (11), it follows in obvious notation that

$$\langle W \rangle_0 = \langle W \rangle_0' \tag{18}$$

Further, because $(W - \langle W \rangle_0)\hat{\varphi}^{(0)}$ involves only one-electron excitations of $\hat{\varphi}^{(0)}$, we see from (15) that

$$\langle W \rangle_1 = \langle W \rangle_1' \tag{19}$$

Therefore, as stated in Section 10, we have the result that $\langle W \rangle$ is exact through first order in $\mathscr{Z}^{-1}$ [28]. In particular since the theorem, when applicable, holds in the presence of external fields, we can conclude, [recalling the discussion of (15-6), etc.] that UHF polarizabilities, susceptibilities, etc., will all be accurate through relative order $\mathscr{Z}^{-1}$.

For SUHF, one can make essentially the same argument except that because of the somewhat restricted nature of the $\psi_k^{(0)}$ (one-electron excitation with no spin change, or paired one-electron excitations with no spin change) one must add the further proviso that W be spin free [29]. Of course, if SUHF agrees with UHF, then one may invoke the more general UHF theorem. However, since this will be the case only if $\hat{\varphi}^{(0)}$ has spin zero, this extra generality is only a formal one since, because W is a one-electron operator, the average value of any spin-dependent part of W will then anyway (correctly and exactly) vanish identically.

Having proven the theorem, it is now appropriate to point to cases in which it does and does not apply. First of all, it obviously does not apply in situations in which $\psi'^{(0)}$ is a vector-coupled sum of several determinants. On the other hand, it certainly does apply to the ground state of rare gas-like atoms. However, there can be some surprises in the case of nominally closed (sub) shell systems because of the peculiar degeneracy of hydrogenic energy levels. Thus consider the ground state of the isolated Be atom. Nominally one says that this is $(1s)^2(2s)^2\ {}^2S$. However, in the $\mathscr{Z} = \infty$ limit (the hydrogenic limit), this becomes degenerate with $(1s)^2(2p)^2\ {}^2S$ and indeed $\psi'^{(0)}$ is a certain linear combination of these, and so the theorem is not applicable.[8]

As another example, consider the ground state of an isolated Li atom. This is $(1s)^2 2s\ {}^2S$. However, the spin degeneracy clearly causes no problem, ψ' and $\hat{\varphi}$ both being eigenfunctions of the z component of spin. Also, the fact that in the hydrogenic limit it becomes degenerate with $(1s)^2 2p\ {}^2P$ is no problem because ψ' and $\hat{\varphi}$ both have even parity. Thus $\psi'^{(0)}$ *is* a single determinant and so the theorem applies.

[8] See Cohen and Dalgarno [30].

As a final example, let us again consider the ground state of Li but now in a uniform electric field. Here for $\mathscr{Z} \to \infty$ one will have states which for $\mathscr{E}' = 0$ become $(1s)^2 2p_+$, $(1s)^2 2p_-$, and a linear combination of the degenerate hydrogenic pair $(1s)^2 2s$ and $(1s)^2 2p_0$ (the subscript on the p in each case denoting the component of orbital angular momentum along $\mathscr{E}'$) and each is a single determinant, and so the theorem would seem to apply. Nevertheless, though it applies, the theorem is not very interesting for the linear combination state. Namely the theorem tells us what happens as $\mathscr{Z} \to \infty$ with $\mathscr{E}'$ fixed. However, usually the electric field perturbation is much weaker than the Coulomb perturbation, so that what one actually wants to do is to first expand in powers of the field and then to treat the Coulomb interaction as a perturbation in each term of that expansion. On the other hand, the theorem has reference to a situation in which one first expands in powers of $\mathscr{Z}^{-1}$ and then, possibly, further treats $\mathscr{E}'$ as a perturbation in each term of that series. Now if there is no degeneracy as $\mathscr{Z} \to \infty$ and $\mathscr{E}' \to 0$, then the order in which one does things is irrelevant, the result in either case being the same double power series in $\mathscr{E}'$ and $\mathscr{Z}^{-1}$. However, when there is degeneracy, then the double power series may not exist and the order may matter. Thus in the example in question, evidently for $\mathscr{Z} \equiv \infty$ and $\mathscr{E}' \to 0$, $\psi'^{(0)}$ becomes the linear combination state, while for $\mathscr{E}' \equiv 0$ and $\mathscr{Z} \to \infty$, it becomes simply $(1s)^2 2s$. This then leaves the question, is there an interesting theorem for such cases? However, we will not attempt to answer this.[9]

Although we wrote $\langle W \rangle_1$, and by implication $\langle W \rangle_1'$, in terms of Coulomb perturbed functions, naturally one would want to use the interchange method of Eq. (26-40), etc., to actually calculate them and higher corrections since, because it is a one-electron operator, W will usually be a simpler perturbation to deal with than the Coulomb interaction. In fact one can solve the first-order equations analytically in closed form.[10] However, as we noted preceding (26-45), to employ this method, one must have no complications due to degeneracy, and the present discussion might therefore lead one to worry about using the technique in the case of, say, the isolated Li ground state for W's which can mix the ^2S and the ^2P, and/or the two spin components of the doublets. We will now show that in fact there is no difficulty. First of all, only odd parity parts of W can mix the S and P, but since the expectation values (UHF, SUHF, or exact) of these are identically zero, they are not interesting, and so in effect we can confine

[9] Some discussion can be found in Cohen [31].

[10] See, for example, Cohen and Dalgarno [29] and other references there.

attention to even parity W's. Second, turning to the spin degeneracy, since W is a one-particle operator, one readily sees that it can involve the s_{tx}, s_{ty}, and s_{tz} only linearly, where the $\mathbf{s}_t$ are the spin operators for the tth electron. However, in a state of definite z component of spin, the averages of s_{tx} and s_{ty} vanish identically, so that only the s_{tz} terms are of interest and these do not mix the degenerate states.

Once one recognizes that under the conditions of the $\mathscr{Z}^{-1}$ theorem interchange theorems apply, one can prove the former theorem almost trivially, the point being simply that with neglect of the Coulomb interaction and in the absence of degeneracy problems UHF (SUHF) perturbed by any one-electron (spin-free one-electron) W obviously yields exact eigenfunctions and eigenvalues, and hence in particular one will have, in the notation of (26-44), etc., that $\hat{\varphi}^{(01)} = \psi'^{(01)}$.

Numerical results and additional discussion of all aspects of the $\mathscr{Z}^{-1}$ perturbation theory of Hartree–Fock theories can be found in a series of papers by Dalgarno and collaborators, notably Cohen and Stewart.[10] Indeed these papers are the source of much of the information which we have been discussing (though if carried through in all details, our treatment, especially of spin-dependent W's for a system like lithium, would differ somewhat from theirs. However, the end result would of course be the same).

These authors also introduce and use various "shielding approximations," the simplest of which, returning to $\mathbf{r}_s$ coordinates and the usual energy units, consists in replacing

$$-\sum_s \mathscr{Z}/|\,\mathbf{r}_s\,|$$

in $H^{(0)}$ by

$$-\sum_s (\mathscr{Z} - \sigma)/|\,\mathbf{r}_s\,|$$

with the idea that for proper choice of the shielding constant σ, one may be able to accelerate the convergence of the resultant perturbation series, the perturbation now consisting of the Coulomb interaction plus $-\sum_s \sigma/|\,\mathbf{r}_s\,|$. (Actually, they use this simple method only for the helium ground state. When there is more than one orbital, they in effect introduce different shielding constants for each orbital.) In this connection, it should be noted that $\hat{\varphi}^{(1)}$ here has a similar relationship to $\psi'^{(1)}$ as in the $\mathscr{Z}^{-1}$ expansion since the considerations which lead to (15) hold for any one-electron $H^{(0)}$ given the further requirements about degeneracy, etc.

Other zeroth-order potentials have also been studied to some extent,[11] again with the hope of speeding convergence. In all such cases, the UHF perturbation equations through first order for an isolated atom are (in ordinary units and $\mathbf{r}_s$ coordinates) since g is now a first-order quantity

$$(h^{(0)} - \varepsilon_j^{(0)})\hat{\varphi}_j^{(0)} = 0, \qquad (\hat{\varphi}_j^{(0)}, \hat{\varphi}_j^{(0)}) = 1 \tag{20}$$

$$(h^{(0)} - \varepsilon_j^{(0)})\hat{\varphi}_j^{(1)} + \sum_k \{(\hat{\varphi}_k^{(0)}, g\hat{\varphi}_k^{(0)})\hat{\varphi}_j^{(0)} - (\hat{\varphi}_k^{(0)}, g\hat{\varphi}_j^{(0)})\hat{\varphi}_k^{(0)}\}$$
$$+ (h^{(1)} - \varepsilon_j^{(1)})\hat{\varphi}_j^{(0)} = 0 \tag{21}$$

$$(\hat{\varphi}_j^{(1)}, \hat{\varphi}_j^{(0)}) + (\hat{\varphi}_j^{(0)}, \hat{\varphi}_j^{(1)}) = 0 \tag{22}$$

where

$$h^{(0)} = \tfrac{1}{2}p^2 + v \qquad \text{and} \qquad h^{(1)} = -(\mathscr{Z}/|\,\mathbf{r}\,|) - v \tag{23}$$

v being the zeroth-order potential.

In the $\mathscr{Z}^{-1}$ expansion or the simple shielding approximation, the solutions of (20) are of course hydrogenic functions, and, for appropriate choice of v, may also be known analytically in other cases. Also, equations (21) and (22) and their analogs in SUHF and various other Hartree–Fock type theories[12] are simpler than, say, Eq. (28–39) in that they involve no coupling of the $\hat{\varphi}_j^{(1)}$ (or coupling only through Lagrange multiplier terms). Indeed their solutions can often be reduced to quadratures [34], though in any case variational techniques are available.[13]

As a final point, we will now derive a nice formula [36] which relates the $\varepsilon_i^{(n)}$ to the $\hat{E}^{(n)}$ in the $\mathscr{Z}^{-1}$ expansion of UHF. We will then infer a similar formula for SUHF. First we write down a series of relations between the various parts of $\hat{E}$ (we revert to ordinary units and $\mathbf{r}_s$ coordinates). From the virial theorem (20-18) we have

$$2\langle T\rangle + \langle V_{\mathrm{N}}\rangle + \langle V_e\rangle = 0 \tag{24}$$

or

$$2\hat{E} = \langle V_{\mathrm{N}}\rangle + \langle V_e\rangle \tag{25}$$

where $\langle V_{\mathrm{N}}\rangle$ is the average electron–nucleus potential energy and $\langle V_e\rangle$ is the average electron–electron potential energy. Further, from the general-

[11] For example, Weber *et al.* [32].

[12] For a general survey, see Sharma [33].

[13] For example, Dalgarno [35].

ized Hellmann–Feynman theorem for $\sigma = \mathscr{Z}$, we have

$$\mathscr{Z}\, \partial\hat{E}/\partial\mathscr{Z} = \langle V_{\mathrm{N}} \rangle \tag{26}$$

and in addition from (8-22) it follows that

$$\hat{E} = \sum_i \varepsilon_i - \langle V_c \rangle \tag{27}$$

If now we combine these results, we have that

$$3\hat{E} - \mathscr{Z}(\partial\hat{E}/\partial\mathscr{Z}) = \sum_i \varepsilon_i \tag{28}$$

Therefore if we expand $\hat{E}$ and ε_i according to

$$\hat{E} = \mathscr{Z}^2 \sum_n \hat{E}^{(n)} \mathscr{Z}^{-n} \tag{29}$$

and

$$\varepsilon_i = \mathscr{Z}^2 \sum_n \varepsilon_i^{(n)} \mathscr{Z}^{-n} \tag{30}$$

then (28) immediately yields the desired formula

$$(n+1)\hat{E}^{(n)} = \sum_i \varepsilon_i^{(n)} \tag{31}$$

To derive the formula for SUHF, we note [recall Eq. (8-42)] that there

$$\delta\hat{E} = \sum_{\mathrm{o}} (\delta\hat{\xi}_{\mathrm{o}}, \hat{f}_{\mathrm{o}}\, \hat{\xi}_{\mathrm{o}}) + \sum_{\mathrm{d}} (\delta\hat{\xi}_{\mathrm{d}}, \hat{f}_{\mathrm{d}}\, \hat{\xi}_{\mathrm{d}}) + \text{comp. conj.} \tag{32}$$

which therefore suggests, and one can check by detailed calculation, that since $\langle V_c \rangle$ is quadratic in the ξ_i^* and quadratic in the ξ_i while the rest of $\hat{E}$ is linear in each, then

$$\hat{E} = \sum_{\mathrm{o}} (\hat{\xi}_{\mathrm{o}}, \hat{f}_{\mathrm{o}}\, \hat{\xi}_{\mathrm{o}}) + \sum_{\mathrm{d}} (\hat{\xi}_{\mathrm{d}}, \hat{f}_{\mathrm{d}}\, \hat{\xi}_{\mathrm{d}}) - \langle V_c \rangle \tag{33}$$

If now we use (8-43)–(8-45), this becomes

$$\hat{E} = \sum_{\mathrm{o}} \varepsilon_{\mathrm{o}} + \sum_{\mathrm{d}} 2\varepsilon_{\mathrm{d}}' - \langle V_c \rangle \tag{34}$$

and therefore, since (25)-(26) hold in SUHF as well as in UHF, we find

$$(n+1)\hat{E}^{(n)} = \sum_{\mathrm{o}} \varepsilon_{\mathrm{o}}^{(n)} + \sum_{\mathrm{d}} 2\varepsilon_{\mathrm{d}}^{(n)} \tag{35}$$

Whether or not the analogous theorems hold in USCF and SUSCF will of course depend on the nature of the basis sets which are used.

PROBLEMS

1. Discuss the nature of the perturbation expansion for molecules when one treats the Coulomb interaction between the electrons as a perturbation. (See, for example, Matcha, and Byers Brown [37].)

2. Let $\psi_1^{(0)}$ and $\psi_2^{(0)}$ be orthogonal eigenfunctions of some $H^{(0)}$ with eigenvalues $E_1^{(0)}$ and $E_2^{(0)}$, respectively, and where $E_1^{(0)} \neq E_2^{(0)}$. Show that a solution of

$$(H^{(0)} - E_1^{(0)})\Phi = \alpha\psi_2^{(0)}$$

where α is a constant, is

$$\Phi = \frac{\alpha}{E_2^{(0)} - E_1^{(0)}}\,\psi_2^{(0)}$$

Show that if $E_1^{(0)} = E_2^{(0)}$, then this equation has a solution only if $\alpha = 0$.

3. In the text, we mentioned that for spin-dependent W's our treatment would differ in some details from that of Cohen and Dalgarno [29]. To illustrate this point, use the procedures of the text to derive a formula analogous to their (110) and then show that the result is, as it must be, identical to their (110) (in this latter step, the result found in Problem 2 will be helpful).

References

1. K. Aashamar, G. Lyslo, and J. Midtdal, *J. Chem. Phys.* **52**, 3324 (1970); *Phys. Norv.* **6**, 21 (1972).
2. H. Doyle, M. Oppenheimer, and G. W. F. Drake, *Phys. Rev. A* **5**, 26 (1972).
3. S. Diner, J. P. Malrieu, and P. Claverie, *Theor. Chem. Acta* **13**, 1 (1969), and references therein.
4. Z. Gershgorn and I. Shavitt, *Int. J. Quantum Chem.* **2**, 751 (1968).
5. B. Roos, *Chem. Phys. Lett.* **15**, 153 (1972).
6. B. Huron, J. P. Malrieu, and P. Rancurel, *J. Chem. Phys.* **58**, 5745 (1973).
7. T. P. Valkering, and W. J. Caspers, *Physica (Utrecht)* **63**, 113 (1973), and references there to earlier work.
8. J. von Neumann and E. Wigner, *Z. Phys.* **30**, 467 (1929).
9. G. Lyslo, K. Aashamar, and J. Midtdal, *Int. J. Quantum Chem.* **5**, 583 (1971), and references therein.
10. F. Grimaldi, *J. Chem. Phys.* **43**, S59 (1965).
11. A. Dalgarno and G. W. F. Drake, *Chem. Phys. Lett.* **3**, 349 (1969).
12. E. Brändas and O. Goscinski, *Phys. Rev. A* **1**, 552 (1970).

13. S. Bratoz, *Colloq. Int. Cent. Nat. Rech. Sci.* **82**, 287 (1958); R. Carbo, *Theor. Chim. Acta* **17**, 74 (1970).
14. F. C. Sanders, *Chem. Phys. Lett.* **17**, 291 (1972).
15. D. M. Bishop, and A. Macias, *J. Chem. Phys.* **51**, 4997 (1969).
16. E. O. Kane, *Phys. Rev.* **125**, 1094 (1962).
17. R. M. Stevens, R. M. Pitzer, and W. N. Lipscomb, *J. Chem. Phys.* **38**, 550 (1963).
18. W. N. Lipscomb, *Advan. Magn. Resonance* **2**, 137 (1966).
19. J. Gerratt and I. M. Mills, *J. Chem. Phys.* **49**, 1719 (1968).
20. R. Moccia, *Chem. Phys. Lett.* **5**, 260 (1970).
21. G. Diercksen and R. McWeeny, *J. Chem. Phys.* **44**, 3554 (1966); **49**, 4852 (1969).
22. M. M. Mestechkin, *Theor. Exp. Chem.* (*USSR*) **4**, 98 (1968).
23. A. Dalgarno, *Proc. Roy. Soc. Ser. A* **251**, 282 (1959).
24. T. K. Rebane, *Opt. Spectrosc.* (*USSR*) **19**, 179 (1965).
25. P. W. Langhoff, J. D. Lyons, and R. P. Hurst, *Phys. Rev.* **148**, 18 (1966).
26. E. Hylleraas, *Z. Phys.* **65**, 209 (1930).
27. M. Cohen and A. Dalgarno, *Rev. Mod. Phys.* **35**, 506 (1963).
28. M. Cohen and A. Dalgarno, *Proc. Phys. Soc. London* **77**, 748 (1961).
29. M. Cohen and A. Dalgarno, *Proc. Roy. Soc. Ser. A* **275**, 492 (1963).
30. M. Cohen and A. Dalgarno, *J. Mol. Spectrosc.* **10**, 378 (1963), and references therein.
31. M. Cohen, *Proc. Roy. Soc. Ser. A* **293**, 365 (1966).
32. T. A. Weber, J. H. Weare, and R. G. Parr, *J. Chem. Phys.* **54**, 1865 (1971).
33. C. S. Sharma, *Proc. Roy. Soc. Ser. A* **304**, 513 (1968); also *Proc. Phys. Soc. London* (*At. Mol. Phys.*) **2**, 1010 (1969).
34. M. Cohen, *Proc. Phys. Soc. London* **82**, 778 (1963), and references therein; R. G. Wilson, *Proc. Phys. Soc. London* (*At. Mol. Phys.*) **4**, 228, 311 (1971).
35. A. Dalgarno, *Proc. Phys. Soc. London* **75**, 439 (1960).
36. J. Linderberg, *Phys. Rev.* **121**, 816 (1961).
37. R. L. Matcha and W. Byers Brown, *J. Chem. Phys.* **48**, 74 (1968), and references therein.

Chapter VII / The Hylleraas Variation Method

30. Perturbation Analysis of the Variation Principle and the Hylleraas Variation Method

As we pointed out at the end of Section 11, to derive results for eigenfunctions ψ' and eigenvalues E' from our results for $\hat{\varphi}$ and $\hat{E}$, it is sufficient to use as the set of trial functions the linear set $\psi = A$ with A completely arbitrary. Let us now do this for some of the formulas in Sections 24 and 26. From (24-10), (24-11), etc., we see, writing ψ' instead of $\hat{\varphi}$, that of the $\delta^m\psi'$, only $\delta^0\psi'$ is different from zero, and is given by

$$\delta^0\psi' = \delta A \equiv \delta\psi \tag{1}$$

where, as indicated, we will write $\delta\psi$ instead of δA. Further, (24-4)–(24-6), etc., simply reduce to the tautologies

$$\psi'^{(n)} = \psi'^{(n)} \tag{2}$$

so that the sequence of equations (24-14)-(24-15), etc., become (writing also E' instead of $\hat{E}$)

$$(\delta\psi, (H^{(0)} - E'^{(0)})\psi'^{(0)}) + \text{comp. conj.} = 0 \tag{3}$$

$$(\delta\psi, \{(H^{(0)} - E'^{(0)})\psi'^{(1)} + (H^{(1)} - E'^{(1)})\psi'^{(0)}\}) + \text{comp. conj.} = 0 \tag{4}$$

$$\begin{aligned}(\delta\psi, \{(H^{(0)} - E'^{(0)})\psi'^{(2)} &+ (H^{(1)} - E'^{(1)})\psi'^{(1)} \\ &+ (H^{(2)} - E'^{(2)})\psi'^{(0)})\}) + \text{comp. conj.} = 0\end{aligned} \tag{5}$$

etc.

Equations (3)–(5), etc., are to be true for all $\delta\psi$ therefore it follows that

$$(H^{(0)} - E'^{(0)})\psi'^{(0)} = 0 \tag{6}$$

$$(H^{(0)} - E'^{(0)})\psi'^{(1)} + (H^{(1)} - E'^{(1)})\psi'^{(0)} = 0 \tag{7}$$

$$(H^{(0)} - E'^{(0)})\psi'^{(2)} + (H^{(1)} - E'^{(1)})\psi'^{(1)} + (H^{(2)} - E'^{(2)})\psi'^{(0)} = 0 \tag{8}$$

etc., which will be recognized as the standard equations of Rayleigh–Schrödinger perturbation theory applied to H.

Turning to the energy formulas, (24-19) yields

$$(\psi'^{(0)}, (E'^{(1)} - H^{(1)})\psi'^{(0)}) = 0 \tag{9}$$

while (24-26) and (26-9) become, since $\bar{\psi}^{(2)}$ and $\bar{\psi}^{(3)}$ are identically zero,

$$\begin{aligned}(\psi'^{(0)}, (E'^{(2)} - H^{(2)})\psi'^{(0)}) = {} & (\psi'^{(1)}, (H^{(0)} - E'^{(0)})\psi'^{(1)}) \\ & + (\psi'^{(1)}, (H^{(1)} - E'^{(1)})\psi'^{(0)}) \\ & + (\psi'^{(0)}, (H^{(1)} - E'^{(1)})\psi'^{(1)})\end{aligned} \tag{10}$$

and

$$\begin{aligned}(\psi'^{(0)}, (E'^{(3)} - H^{(3)})\psi'^{(0)}) = {} & (\psi'^{(1)}, (H^{(2)} - E'^{(2)})\psi'^{(0)}) \\ & + (\psi'^{(0)}, (H^{(2)} - E'^{(2)})\psi'^{(1)}) \\ & + (\psi'^{(1)}, (H^{(1)} - E'^{(1)})\psi'^{(1)})\end{aligned} \tag{11}$$

Also, using (7), Eq. (10) can be rewritten alternatively as

$$(\psi'^{(0)}, (E'^{(2)} - H^{(2)})\psi'^{(0)}) = (\psi'^{(0)}, (H^{(1)} - E'^{(1)})\psi'^{(1)}) \tag{12}$$

or

$$(\psi'^{(0)}, (E'^{(2)} - H^{(2)})\psi'^{(0)}) = (\psi'^{(1)}, (H^{(1)} - E'^{(1)})\psi'^{(0)}) \tag{13}$$

and finally as

$$(\psi'^{(0)}, (E'^{(2)} - H^{(2)})\psi'^{(0)}) = -(\psi'^{(1)}, (H^{(0)} - E'^{(0)})\psi'^{(1)}) \tag{14}$$

All of these results again should be familiar from Rayleigh–Schrödinger perturbation theory, with (9) and (12) following in the usual way by taking the scalar product of (7) and (8) with $\psi'^{(0)}$ and using (6). However, although one can derive (11) by various manipulations involving the third-order equation, it is simplest to derive it, and analogous formulas for $E'^{(2n)}$ and $E'^{(2n+1)}$, as we in fact did derive (11), namely by extracting the terms of order ν^{2n} and ν^{2n+1} from

$$(\mathring{\psi}, (H - \mathring{E})\mathring{\psi}) = 0 \tag{15}$$

with

$$\mathring{\psi} = \psi'^{(0)} + \nu\psi'^{(1)} + \cdots + \nu^m\psi'^{(n)} \tag{16}$$

Assuming that (6) has been solved for $\psi'^{(0)}$ and $E'^{(0)}$, then (9) yields $E'^{(1)}$ and the next task is to solve (7) for $\psi'^{(1)}$, from which, using (11) and

(12), we can also calculate $E'^{(2)}$ and $E'^{(3)}$. Now in some cases of physical interest, for example, simple perturbations of harmonic oscillators[1] or simple perturbations of a hydrogen atom[2] (the latter are not only of direct physical interest but are also of interest in connection with interchange theorems in the $\mathscr{Z}^{-1}$ expansion, as we discussed in the previous chapter), one can solve (7) analytically. Also, if one can separate variables to some extent, then a direct numerical attack may be feasible, for example, this is the case for simple perturbations of central field problems generally.[3,4] However, one usually must resort to further approximations, notably use of variation methods as described in Section 25.

Since for the present case, (25-10) yields $\tilde{\psi}^{(2)} \equiv 0$, (25-8) becomes

$$J^{(2)} = (\psi^{(1)}, (H^{(0)} - E'^{(0)})\psi^{(1)}) + (\psi'^{(0)}, (H^{(1)} - E'^{(1)})\psi^{(1)}) \\ + (\psi^{(1)}, (H^{(1)} - E'^{(1)})\psi'^{(0)}) \equiv J_{\mathrm{H}}^{(2)} \tag{17}$$

with the set of $\psi^{(1)}$'s being some subset of the original A's which, since the latter encompassed all functions, means only that the $\psi^{(1)}$'s should be some set of functions, linear or not as the case may be. In the quantum mechanical context, the functional $J_{\mathrm{H}}^{(2)}$ was first introduced and used by Hylleraas [8], hence our introduction of the subscript H. Evidently, one can also write it down by inspection, thereby bypassing the discussion in Section 25, since obviously

$$\delta J_{\mathrm{H}}^{(2))} = 0, \qquad \text{all} \quad \delta\psi^{(1)} \tag{18}$$

is equivalent to Eq. (7).

Following the pattern of Section 25, one now proceeds to approximate the solution of (7) by requiring that

$$\delta \check{J}_{\mathrm{H}}^{(2)} = 0 \qquad \text{"all"} \quad \delta\check{\psi}^{(1)} \tag{19}$$

which we will call the Hylleraas variation method. The resultant $\check{\psi}^{(1)}$ can then be used to calculate an approximation $\check{E}^{(2)}$ to $E'^{(2)}$ according to

$$(\psi'^{(0)}, (\check{E}^{(2)} - H^{(2)})\psi'^{(0)}) = \check{J}_{\mathrm{H}}^{(2)} \tag{20}$$

and can also be used to calculate an approximation to $E'^{(3)}$, which we will

[1] See, for example, Shorb *et al.* [1]; also Ehlenberger and Mendelsohn [2].

[2] See, for example, Cohen and Dalgarno [3]; also Iafrate and Mendelsohn [4].

[3] See many papers by Sternheimer, a recent one published in 1973 [5]. See also Gupta and Sen [6].

[4] For more complex problems, see McKoy and Winter [7].

denote by $\dot{E}^{(3)}$, by imitating Eq. (11) and writing

$$(\psi'^{(0)}, (\dot{E}^{(3)} - H^{(3)})\psi'^{(0)}) = (\check{\psi}^{(1)}, (H^{(2)} - \check{E}^{(2)})\psi'^{(0)}) + (\psi'^{(0)}, (H^{(2)} - \check{E}^{(2)})\check{\psi}^{(1)}) + (\check{\psi}^{(1)}, (H^{(1)} - E^{(1)})\check{\psi}^{(1)}) \tag{21}$$

However, though by construction $\check{E}^{(2)}$ is a variational approximation to $E'^{(2)}$, $\dot{E}^{(3)}$ as we have just defined it is not in general a variational approximation to $E'^{(3)}$, hence our notation $\dot{E}^{(3)}$ rather than $\check{E}^{(3)}$.

If we write

$$\psi^{(1)} = \psi'^{(1)} + \Delta^{(1)} \tag{22}$$

then, using (7), one readily finds that

$$J_{\mathrm{H}}^{(2)} - J_{\mathrm{H}}'^{(2)} = (\Delta^{(1)}, (H^{(0)} - E'^{(0)})\,\Delta^{(1)}) \tag{23}$$

which is just (25-20) specialized to the present example. Thus $J_{\mathrm{H}}^{(2)} - J_{\mathrm{H}}'^{(2)}$, and hence in particular $\check{E}^{(2)} - E'^{(2)}$, will certainly be nonnegative if $\Delta^{(1)}$ is orthogonal to all eigenfunctions of $H^{(0)}$ whose eigenvalues are less than $E'^{(0)}$. Therefore it will be nonnegative if $E'^{(0)}$ is the lowest eigenvalue of $H^{(0)}$ or if all the low lying states are known to be orthogonal to $\psi'^{(1)}$ and to the $\psi^{(1)}$ for reasons of symmetry. In such cases, the Hylleraas variation is, by our standard arguments, very well founded indeed. Namely $J_{\mathrm{H}}^{(2)} - J_{\mathrm{H}}'^{(2)} \geq 0$ implies that $J_{\mathrm{H}}^{(2)}$ will have an absolute minimum which, ignoring as usual pathological cases, must be $\check{J}_{\mathrm{H}}^{(2)}$ (or the smallest $\check{J}_{\mathrm{H}}^{(2)}$ if (19) has more than one solution, which will often be the case if $\psi^{(1)}$ involves nonlinear parameters). Therefore the value of $\check{J}_{\mathrm{H}}^{(2)}$ can only improve (or at any rate not grow worse) as one enlarges the set of trial functions. In Section 33 we will discuss procedures which yield improvable upper bounds to the $E'^{(2)}$ of excited states.

The preceding discussion assumed that one could not solve (7) exactly. If, however, one can, but is then stumped by (8), one can instead apply variation techniques to the latter [or if one can also solve (8) but not the equation for $\psi'^{(3)}$, one can use variation techniques there, etc.]. To derive the relevant functional in the case of Eq. (8), one can follow the procedure suggested following (25-39), or, much more directly, one can simply write it down as

$$J_{\mathrm{H}}^{(4)} = (\psi^{(2)}, (H^{(0)} - E'^{(0)})\psi^{(2)}) + (\psi^{(2)}, (H^{(1)} - E'^{(1)})\psi'^{(1)}) + (\psi'^{(1)}, (H^{(1)} - E'^{(1)})\psi^{(2)}) + (\psi^{(2)}, (H^{(2)} - E'^{(2)})\psi'^{(0)}) + (\psi'^{(0)}, (H^{(2)} - E'^{(2)})\psi^{(2)}) \tag{24}$$

since $\delta J_{\mathrm{H}}^{(4)} = 0$, all $\delta\psi^{(2)}$, is clearly equivalent to (8). Further, one can verify, using (15) and (16) for $n = 2$, that

$$(\psi'^{(0)}, (E'^{(4)} - H^{(4)})\psi'^{(0)}) = (\psi'^{(1)}, (H^{(3)} - E'^{(3)})\psi'^{(0)}) + (\psi'^{(0)}, (H^{(3)} - E'^{(3)})\psi'^{(1)}) + (\psi'^{(1)}, (H^{(2)} - E'^{(2)})\psi'^{(1)}) + J'^{(4)}_{\mathrm{H}} \quad (25)$$

so that in the obvious way $\check{J}_{\mathrm{H}}^{(4)}$ will yield a variational approximation to $E'^{(4)}$. Also, if we write

$$\psi^{(2)} = \psi'^{(2)} + \Delta^{(2)} \quad (26)$$

then we find

$$J_{\mathrm{H}}^{(4)} - J'^{(4)}_{\mathrm{H}} = (\Delta^{(2)}, (H^{(0)} - E'^{(0)})\,\Delta^{(2)}) \quad (27)$$

so that one can make the by now standard remarks concerning the possibility of an improvable bound.

At the outset of Section 27, we made a rather cryptic comment which we would now like to explain. Consider the linear variation method with a set of trial functions which is independent of ν. Then one readily sees that $J^{(2)}, J^{(4)}, \ldots, J^{(2n)}$ of the linear variation method, if one were for some reason to write them down, would look just like (17), (24), etc., except that the $\psi'^{(m)}$ and $E'^{(m)}$, $m < n$, will be replaced by $\hat{\varphi}_k^{(m)}$ and $\hat{E}_k^{(m)}$. The point is now the following. Overlooking this necessary replacement, various authors[5] have implied that because of the similarity in the appearance of the functionals, perturbation theory within the linear variation method (which one can derive by varying the J's and letting the $\psi^{(n)}$ be arbitrary within the space, though this is not the easiest approach) provides variational approximations to each $E'^{(2n)}$ in turn, and also that the $\hat{\varphi}_k^{(1)}$, $\hat{\varphi}_k^{(2)}$, etc., are, in the sense of the Hylleraas variation method, approximate solutions of (7), (8), etc. However, clearly this is in general false. Thus $J^{(2)}$ of the linear variation method is not $J_{\mathrm{H}}^{(2)}$ unless $\hat{\varphi}_k^{(0)} = \psi'^{(0)}$. Now in fact this is often the case (for example, in the $\mathscr{Z}^{-1}$ expansion), and if it is, then $\hat{E}_k^{(2)}$ will be a variational approximation to $E'^{(2)}$, and even, as we have seen, a guaranteed and improvable bound under appropriate circumstances. Also, $\hat{\varphi}_k^{(1)}$ will be an approximation to the solution of (7) in the sense of the Hylleraas variation method. However, even granting this, it is unlikely that $\hat{\varphi}_k^{(1)} = \psi'^{(1)}$, in which case $J^{(4)}$ of the linear variation method is certainly not $J_{\mathrm{H}}^{(4)}$. Therefore $\hat{E}_k^{(4)}$ will not yield a variational approximation to $E'^{(4)}$

[5] See, however, Ahlrichs [9].

and $\hat{\psi}_k^{(2)}$ will not be an approximation to the solution of (8) in the sense of the Hylleraas variation method, etc.

Indeed this really goes back to the point which we made near the end of Section 23. Namely we know in general that

$$E' = \hat{E} + O(\psi' - \hat{\psi})^2 \tag{28}$$

and therefore

$$E'^{(0)} + \nu E'^{(1)} + \nu^2 E'^{(2)} + \cdots = \hat{E}^{(0)} + \nu \hat{E}^{(1)} + \nu^2 \hat{E}^{(2)} + \cdots + O(\psi' - \hat{\psi})^2 \tag{29}$$

However, only if $(\psi' - \hat{\psi}) = O(\nu)$ can we conclude that

$$E'^{(2)} = \hat{E}^{(2)} + O(\psi'^{(1)} - \hat{\psi}^{(1)})^2 \tag{30}$$

and only if $(\psi' - \hat{\psi}) = O(\nu^2)$ can we conclude that

$$E'^{(4)} = \hat{E}^{(4)} + O(\psi'^{(2)} - \hat{\psi}^{(2)})^2 \tag{31}$$

etc. As we have said before, in general $\hat{E}^{(n)}$ is simply the nth-order term in the expansion of $\hat{E}$. Thus even though with a sufficiently flexible set of trial functions it may very accurately approximate $E'^{(n)}$, still, based on the theory to the extent to which we have developed it, it is only in very special circumstances, as outlined above, that one can a priori claim that it has any special variational relationship to $E'^{(n)}$.

PROBLEMS

1. Use $\psi'^{(0)} + \nu\psi'^{(1)}$ as a trial function in the variation method and thereby derive an upper bound (if E' is the lowest state of a given type) to E'. Express your answer in terms of $E'^{(0)}, \ldots, E'^{(3)}$ and $(\psi'^{(1)}, \psi'^{(1)})$.

2. Improve on the upper bound of Problem 1 by use of the set of trial functions $\psi'^{(0)} + A\psi'^{(1)}$, where A is a variational parameter [10].

3. Show that $\dot{E}^{(3)} - E'^{(3)}$ is of first order in $\check{\psi}^{(1)} - \psi'^{(1)}$.

31. The Second-Order Hylleraas Variation Method: Details

We have seen in Eqs. (30-12)–(30-14) that $J_{\mathrm{H}}'^{(2)}$ can be written in various alternative forms, and so it is natural to ask whether or not the same is true of $\check{J}_{\mathrm{H}}^{(2)}$. Earlier, we exhibited one such general rewriting in (24-38) under the condition that the set of trial functions be independent of ν. Now the overall set of trial functions we are in effect using [recall (25-22), etc.] is

$$\psi'^{(0)} + \nu\psi^{(1)} + O(\nu^2) \tag{1}$$

where to the order to which we are working, the $O(\nu^2)$ terms are irrelevant. Therefore since ν is a real number, it follows that if the set of $\psi^{(1)}$ is invariant to multiplication by an arbitrary real number, then (24-38) will hold. Thus we will have, with appropriate changes in notation,

$$\breve{J}_{\mathrm{H}}^{(2)} = \tfrac{1}{2}[(\breve{\psi}^{(1)}, (H^{(1)} - E'^{(1)})\psi'^{(0)}) + (\psi'^{(0)}, (H^{(1)} - E'^{(1)})\breve{\psi}^{(1)})] \tag{2}$$

which is the analog of the average of (30-12) and (30-13). If we combine this with (30-17), then we have, alternatively,

$$\breve{J}_{\mathrm{H}}^{(2)} = -(\breve{\psi}^{(1)}, (H^{(0)} - E'^{(0)})\breve{\psi}^{(1)}) \tag{3}$$

which is the analog of (30-14), and also is simply (24-39) specialized to the present situation.

One can of course also derive (2) directly from $\delta\breve{J}_{\mathrm{H}}^{(2)} = 0$ by assuming that $\delta\breve{\psi}^{(1)} = \delta b\breve{\psi}^{(1)}$, with δb an arbitrary real number, is a possible variation of $\breve{\psi}^{(1)}$. However, unless all quantities are explicitly real, to derive the analog of (30-12) and (30-13), we need in addition to assume that δb can also be complex, that is, that there are no reality constraints on the overall scale of the $\psi^{(1)}$. If this is the case, then we can vary $\breve{\psi}^{(1)*}$ alone, whence $\delta\breve{J}_{\mathrm{H}}^{(2)} = 0$ yields, concelling a factor of δb,

$$(\breve{\psi}^{(1)}, (H^{(0)} - E'^{(0)})\breve{\psi}^{(1)}) + (\breve{\psi}^{(1)}, (H^{(1)} - E'^{(1)})\psi'^{(0)}) = 0 \tag{4}$$

This when combined with (30-17) then yields the desired result

$$\breve{J}_{\mathrm{H}}^{(2)} = (\psi'^{(0)}, (H^{(1)} - E'^{(1)})\breve{\psi}^{(1)}) \tag{5}$$

which in turn can be written

$$\breve{J}_{\mathrm{H}}^{(2)} = (\breve{\psi}^{(1)}, (H^{(1)} - E'^{(1)})\psi'^{(0)}) \tag{6}$$

since $\breve{J}_{\mathrm{H}}^{(2)}$ is certainly real.

Another property of the exact calculation is that though $\psi'^{(1)}$ is arbitrary to within an additive multiple of $\psi'^{(0)}$, still this has no effect on physical results. Thus since, with α an arbitrary number,

$$\psi'^{(0)} + \nu(\psi'^{(1)} + \alpha\psi'^{(0)}) + \cdots = (1 + \nu\alpha)(\psi'^{(0)} + \nu\psi'^{(1)} + \cdots)$$

this ambiguity has no effect on expectation values through first order in ν and also, as one can readily verify, it does not affect the values of $E'^{(2)}$ and $E'^{(3)}$ as calculated from, say, (30-12) and (30-11). Now, it is obvious that by the same argument, adding a multiple of $\psi'^{(0)}$ to $\check{\psi}^{(1)}$ will not affect approximate expectation values through first order ν. It is also of interest to note that it does not affect $\check{E}^{(2)}$, and if (4) is satisfied, it does not affect $\dot{E}^{(3)}$ either. First of all, from (30-6), (30-9), and (30-17), $J_{\mathrm{H}}^{(2)}$ is obviously invariant to the transformation $\psi^{(1)} \to \psi^{(1)} + \alpha\psi'^{(0)}$, which proves the point as far as $\check{E}^{(2)}$ is concerned. Turning to $\dot{E}^{(3)}$, under that same transformation, the right hand side of (30-21) changes by

$$
\begin{aligned}
&(\alpha + \alpha^*)(\psi'^{(0)}, (H^{(2)} - \check{E}^{(2)})\psi'^{(0)}) + \alpha^*(\psi'^{(0)}, (H^{(1)} - E'^{(1)})\check{\psi}^{(1)}) \\
&\qquad + \alpha(\check{\psi}^{(1)}, (H^{(1)} - E'^{(1)})\psi'^{(0)}) + \alpha^*\alpha(\psi'^{(0)}, (H^{(1)} - E'^{(1)})\psi'^{(0)}) \\
&\quad = \alpha[-\check{J}_{\mathrm{H}}^{(2)} + (\check{\psi}^{(1)}, (H^{(1)} - E'^{(1)})\psi'^{(0)}] \\
&\qquad + \alpha^*[-\check{J}_{\mathrm{H}}^{(2)} + (\psi'^{(0)}, (H^{(1)} - E'^{(1)})\check{\psi}^{(1)})] \\
&\qquad + \alpha^*\alpha(\psi'^{(0)}, (H^{(1)} - E'^{(1)})\psi'^{(0)})
\end{aligned}
$$

However, the third line on the right-hand side of this equation vanishes because of (30-9), while the first and second lines vanish because of (5) and (6), which proves the point.

When the set of trial functions $\psi^{(1)}$ has no constraints on its overall scale, then it is possible to replace $\check{J}_{\mathrm{H}}^{(2)}$ by another functional (sometimes called the Dirichlet functional) [11] which is scale invariant. We first replace the set $\psi^{(1)}$ by the then equivalent set

$$\phi^{(1)} = A\psi^{(1)}$$

where the new variational parameter A is an arbitrary complex number. With this replacement, $J_{\mathrm{H}}^{(2)}$ then becomes

$$
\begin{aligned}
J_{\mathrm{H}}^{(2)} = A^*A(\psi^{(1)}, (H^{(0)} - E'^{(0)})\psi^{(1)}) + A^*(\psi^{(1)}, (H^{(1)} - E'^{(1)})\psi'^{(0)}) \\
+ A(\psi'^{(0)}, (H^{(1)} - E'^{(1)})\psi^{(1)})
\end{aligned}
\tag{7}
$$

whence varying A^*, $\delta\check{J}_{\mathrm{H}}^{(2)} = 0$, yields

$$\check{A}(\check{\psi}^{(1)}, (H^{(0)} - E'^{(0)})\check{\psi}^{(1)}) + (\check{\psi}^{(1)}, (H^{(1)} - E'^{(1)})\psi'^{(0)}) = 0$$

which determines $\check{A}$ in terms of $\check{\psi}^{(1)}$:

$$\check{A} = -(\check{\psi}^{(1)}, (H^{(1)} - E'^{(1)})\psi'^{(0)})/(\check{\psi}^{(1)}, (H^{(0)} - E'^{(0)})\check{\psi}^{(1)}) \tag{8}$$

Using this result in (7), we have that $J_{\mathrm{H}}^{(2)}$ can be replaced by

$$\check{J}_{\mathrm{D}}^{(2)} \equiv -\frac{|(\check{\psi}^{(1)}, (H^{(1)} - E'^{(1)})\psi^{(0)})|^2}{(\check{\psi}^{(1)}, (H^{(0)} - E'^{(0)})\check{\psi}^{(1)})} \tag{9}$$

which one can then vary further so as to determine $\check{\psi}^{(1)}$.

Although we have derived (9) from the Hylleraas functional, it is in a certain sense a better functional.[6] Namely if contrary to our derivation we draw $\check{\psi}^{(1)}$ from a scale-fixed set of trial functions, then we expect (9) to yield better results than the use of (30-17) directly, and certainly no worse results, if $J_{\mathrm{H}}^{(2)}$ is a guaranteed upper bound to $J_{\mathrm{H}}'^{(2)}$, since clearly the use of $\check{J}_{\mathrm{D}}^{(2)}$ is then equivalent to using $J_{\mathrm{H}}^{(2)}$ but with the set of $\psi^{(1)}$ enlarged by allowing the overall scale to be arbitrary.

Throughout the discussion of the past two sections, we have assumed that $\psi'^{(0)}$ is known exactly, or at any rate with sufficient precision so that, for example, all relevant integrals involving $(H^{(0)} - E'^{(0)})\psi'^{(0)}$ vanish to some specified number of decimals. However, returning to a theme which we have explored somewhat before at the end of Section 25, often this is not the case. Rather one usually has some approximation $\chi^{(0)}$, often of dubious quality. Thus in the $\mathscr{Z}^{-1}$ expansion, $\psi'^{(0)}$ *is* known. On the other hand, if one is interested in calculating, say, the polarizability of an atom or molecule, then for $H^{(0)}$ we should take the Hamiltonian of the isolated system, while $H^{(1)}$ should be the interaction with the external electric field. However, only for very simple systems do we then know $\psi'^{(0)}$ with great accuracy.

In Section 25, we mentioned two responses which have been made to this situation. A third is to view $\chi^{(0)}$ as defining a "model system" in that everywhere in the previous discussion one replaces $H^{(0)}$ by a "model Hamiltonian" $H^{(00)}$ of which $\chi^{(0)}$ is an eigenfunction and proceeds to calculate the polarizability, say, of this model system. Subsequently, one might also supplement this calculation by treating $(H^{(0)} - H^{(00)})$ as a further perturbation in the hopes of improving the results (see Section 36). For example, the simplest uncoupled Hartree–Fock approximation, which we will describe in more detail in Section 36, can be thought of as a model calculation in which one uses for $\chi^{(0)}$ a UHF function, and where $H^{(00)}$ [recall the definition (10-91)] is $F^{(0)}$. (Although it is not often done in practice, one can envisage extension of this approach to modeling not only $H^{(0)}$ but also $H^{(1)}$, $H^{(2)}$, etc. For example, in the case of magnetic interactions, one might wish to do this in order to ensure that the model problem is gauge invariant.[7])

[6] See, for example, Lindner and Löwdin [12], and Goldstein [13].
[7] See, for example, Epstein [14]; also Starace [15].

Still another often used approach[8] is to imitate the Hylleraas variation method by using the functional

$$\dot{\mathscr{J}}^{(2)} = (\chi^{(1)}, (H^{(0)} - E^{(0)})\chi^{(1)}) + (\chi^{(1)}, (H^{(1)} - E^{(1)})\chi^{(1)}) + (\chi^{(0)}, (H^{(1)} - E^{(1)})\chi^{(1)}) \tag{10}$$

where

$$E^{(0)} = (\chi^{(0)}, H^{(0)}\chi^{(0)})/(\chi^{(0)}, \chi^{(0)}); \qquad E^{(1)} = (\chi^{(0)}, H^{(1)}\chi^{(0)})/(\chi^{(0)}, \chi^{(0)}) \tag{11}$$

to determine a $\hat{\chi}^{(1)}$. To see what is involved here, let us consider first a situation—for example, the calculation of the dipole polarizability of an atom—in which $\psi'^{(0)}$ and $\psi'^{(1)}$ are known to have different symmetries. If, then, as we will assume, $\chi^{(0)}$ at least has the same symmetry as $\psi'^{(0)}$, it is natural to restrict the $\chi^{(1)}$ to have the symmetry of $\psi'^{(1)}$. Since this means that the $\chi^{(1)}$ will be orthogonal to $\chi^{(0)}$, it therefore follows that in such a case we can replace $\dot{\mathscr{J}}^{(2)}$ by

$$\mathscr{J}^{(2)} = (\chi^{(1)}, (H^{(0)} - E^{(0)})\chi^{(1)}) + (\chi^{(1)}, H^{(1)}\chi^{(0)}) + (\chi^{(0)}, H^{(1)}\chi^{(1)}) \tag{12}$$

There are now several comments which can be made about the use of $\mathscr{J}^{(2)}$.

(i) Let us denote by $\chi'^{(1)}$ the optimal $\chi^{(1)}$ if the set $\chi^{(1)}$ were completely arbitrary (except for the requirement that the $\chi^{(1)}$ be orthogonal to $\chi^{(0)}$). Then, writing $\chi^{(1)} = \chi'^{(1)} + \Delta^{(1)}$, one readily finds that

$$\mathscr{J}^{(2)} - \mathscr{J}'^{(2)} = (\Delta^{(1)}, (H^{(0)} - E^{(0)})\,\Delta^{(1)}) \tag{13}$$

Therefore, by the usual argument, provided that $E^{(0)}$ is less than the first eigenvalue of $H^{(0)}$ whose associated eigenfunction has the same symmetry as the $\chi^{(1)}$, the smallest $\hat{\mathscr{J}}^{(2)}$ will be an upper bound to $\mathscr{J}'^{(2)}$, and its value will steadily decrease (or at least not increase) as one enlarges the set of trial functions. However, in this connection, it should be pointed out [19] that often even if $\chi^{(0)}$ is supposed to approximate the ground state, still $E^{(0)}$ may be well up among the eigenvalues of H, possibly even in the continuum. Indeed, it might be that under these circumstances, a more empirical choice for $E^{(0)}$ would be better.

(ii) If we write

$$\chi^{(1)} = \psi'^{(1)} + \Delta^{(1)}, \qquad \chi^{(0)} = \psi'^{(0)} + \Delta^{(0)}$$

[8] See Antanasoff [16], especially note added in proof. Also Buckingham [17]. For other calculations, see, for example, Kolos and Wolniewicz [18].

then we find since $E^{(0)} = E'^{(0)} + O(\Delta^{(0)2})$, that

$$\mathscr{J}^{(2)} - J_{\mathrm{H}}'^{(2)} = (\Delta^{(0)}, H^{(1)}\psi'^{(1)}) + (\psi'^{(1)}, H^{(1)}\Delta^{(0)}) + O(\Delta^{(0)2}, \Delta^{(1)2}, \Delta^{(0)}\Delta^{(1)})$$

so that the error is of first order in $\Delta^{(0)}$ with the first-order term having an unknown sign.

(iii) This will be a consistent variational calculation in the sense in which we used these words at the end of Section 25. Thus Eq. (25-51) or (25-55) will be satisfied for reasons of symmetry. Also, we can imagine that $\chi^{(2)}$ is to be as in (25-50), so that, as we showed there, its contribution to this order can be ignored, whence (25-45) becomes, in agreement with the present procedure,

$$(\chi^{(0)}, (E^{(2)} - H^{(2)})\chi^{(0)}) = \mathscr{J}^{(2)} \tag{14}$$

Turning now to situations in which symmetry plays no role, here, too, though there is no obvious compulsion to do so, it seems best to require that the $\chi^{(1)}$ be orthogonal to $\chi^{(0)}$. Namely suppose that instead of $\mathscr{J}^{(2)}$ one uses $\dot{\mathscr{J}}^{(2)}$. Then one can show [20] that at least in principle $\dot{\mathscr{J}}^{(2)\prime} - J_{\mathrm{H}}'^{(2)}$ may involve a zeroth-order (!) error. [See the discussion following Eq. (31) in Section 36.] Also, if $E^{(0)}$ is in the continuum of $H^{(0)}$, one may encounter singular equations because of vanishing "energy denominators" if one allows the $\chi^{(1)}$ to be too flexible. (These do not seem to occur in practice, probably because if $\chi^{(0)}$ is crude, one tends anyway to use relatively inflexible trial functions.)

If one requires that the $\chi^{(1)}$ be orthogonal to $\chi^{(0)}$, then comment (ii) remains unaltered. However, because symmetry is no help, (iii) will now be false. As for (i), its replacement becomes obvious when we note that use of $\mathscr{J}^{(2)}$ with the $\chi^{(1)}$ orthogonal to $\chi^{(0)}$ can be viewed as a Hylleraas variation calculation for a model problem.[9] Namely consider

$$H^{(00)} \equiv \pi H^{(0)}\pi + (1 - \pi)H^{(0)}(1 - \pi) \tag{15}$$

where π is the projector onto $\chi^{(0)}$. Then evidently $\chi^{(0)}$ is an eigenfunction of $H^{(00)}$, the eigenvalue being

$$(\chi^{(0)}, H^{(00)}\chi^{(0)})/(\chi^{(0)}, \chi^{(0)}) = (\chi^{(0)}, H^{(0)}\chi^{(0)})/(\chi^{(0)}, \chi^{(0)}) = E^{(0)} \tag{16}$$

Now suppose we were going to compute a variational approximation to the second-order energy within this model. Further, suppose we were to restrict

[9] See also Sadlej [21].

ourselves to trial functions orthogonal to $\chi^{(0)}$, this being no limitation since the "exact" first-order wave function is arbitrary to within additive multiples of $\chi^{(0)}$. Then the Hylleraas functional would be

$$(\chi^{(1)}, (\pi H^{(0)}\pi + (1-\pi)H^{(0)}(1-\pi))\chi^{(1)}) + (\chi^{(1)}, H^{(1)}\chi^{(0)}) + (\chi^{(0)}, H^{(1)}\chi^{(1)}) \tag{17}$$

But under our assumptions, $\pi\chi^{(1)} = 0$, $(1-\pi)\chi^{(1)} = \chi^{(1)}$, whence this functional is $\mathscr{J}^{(2)}$, which proves the point. Also, it enables us to say that in general $\tilde{\mathscr{J}}^{(2)}$ will be a guaranteed and improvable upper bound to $\mathscr{J}'^{(2)}$ if $E^{(0)}$ is the lowest eigenvalue of this $H^{(00)}$.[10]

Sometimes further approximations are made before the functionals are actually used, with results which can again be described in terms of a model calculation. For example, in many one- and two-electron problems where electron spin factors from the wave functions, $\chi^{(1)}$ in (14) is written as

$$\chi^{(1)} = f\chi^{(0)} \tag{18}$$

where f is a function of coordinates and variational parameters only. The point of this is that if $\chi^{(0)}$ were $\psi'^{(0)}$, one could write (we assume that $H^{(0)}$ involves no magnetic interactions)

$$(H^{(0)} - E^{(0)})\chi^{(1)} = (H^{(0)} - E^{(0)})f\chi^{(0)} = [H^{(0)}, f]\chi^{(0)} = [T, f]\chi^{(0)} \tag{19}$$

where T is the kinetic energy operator, thereby eliminating the complicated potential energy terms. Now with $\chi^{(0)} \neq \psi'^{(0)}$, (19) is not true. However, if one nevertheless makes the substitution in (14),[11] then it is easy to see that this again can be thought of as yielding a variational calculation for a model problem, this time with

$$H^{(00)} \equiv H_{\mathrm{S}}^{(00)} = T - \frac{T\chi^{(0)}}{\chi^{(0)}} + E^{(0)} \tag{20}$$

Namely this "Sternheimer Hamiltonian" does satisfy $(H_{\mathrm{S}}^{(00)} - E^{(0)})\chi^{(0)} = 0$, and further, to within an additive constant, it clearly is the unique $H^{(00)}$ of the form $T + V$ for which V is just a function, so that one does have

$$(H_{\mathrm{S}}^{(00)} - E^{(0)})f\chi^{(0)} = [T, f]\chi^{(0)}$$

as assumed. Also, from this observation, it follows that if $\chi^{(0)}$ has no

[10] For some relevant calculations, see Singh and Meath [22].

[11] For example, Chen and Dalgarno [23].

nodes so that $H_{\mathrm{S}}^{(00)}$ is nonsingular the resultant variational functional will be an improvable upper bound to its "exact" value, the latter being the optimal value when f can be completely arbitrary. Namely $\chi^{(0)}$ will then evidently be the ground state of $H_{\mathrm{S}}^{(00)}$ which proves the point.

This completes our discussion of the status of calculations based on the use of $\mathscr{J}^{(2)}$. One could now go on to consider the use of a $\mathscr{J}^{(4)}$ derived from $J_{\mathrm{H}}^{(4)}$ by replacing $\psi'^{(0)}$ *and* $\psi'^{(1)}$, etc., by approximations. However, just as with UHF, the situation obviously becomes even more obscure and we will not attempt to explore it except to mention a few points. First of all, if one is in fact doing perturbation theory within the linear variation method, then, as we pointed out at the end of Section 30, this is a perfectly consistent and correct thing to do. On the other hand, as we also pointed out, it is also an unnecessarily complicated approach to writing down the basic equations. Second, we would emphasize that if the approximate functions are sufficiently accurate so that various integrals, for example, those involving $(H^{(0)} - E^{(0)})\chi^{(0)}$, vanish to some appropriate number of decimals, then again this will be quite consistent. As a case which is presumably in point, we may mention the $\mathscr{Z}^{-1}$ expansion calculations of Scherr and collaborators,[12] who use $J_{\mathrm{H}}^{(4)}$, etc., with approximate functions determined in lower orders, these functions having been determined by sets of trial functions involving nonlinear as well as linear parameters. Finally, we have been concerned with questions of consistency, and a priori significance. If, however, one does not worry about such things, then one can simply view the use of $\mathscr{J}^{(2)}$, $\mathscr{J}^{(4)}$, etc., as devices for producing $\hat{\chi}^{(1)}$, $\hat{\chi}^{(2)}$, etc., and let it go at that.

PROBLEMS

1. Fill in the details of Epstein and Epstein [20] to show the possibility of a zeroth-order error.

32. The Linear Hylleraas Variation Method

If one uses trial functions of the form

$$\psi^{(1)} = \sum_{k=1}^{M} A_k^{(1)} \phi_k \tag{1}$$

[12] For a summary, see Scherr *et al.* [24].

where the ϕ_k are a given linearly independent basis set and the $A_k^{(1)}$ are arbitrary complex numbers, then $\delta \check{J}_{\mathrm{H}}^{(2)} = 0$ for all $\delta A_k^{(1)}$ yields the equations

$$\sum_{l=1}^{M} (\phi_k, (H^{(0)} - E'^{(0)})\phi_l)\check{A}_l^{(1)} + (\phi_k, (H^{(1)} - E'^{(1)})\psi'^{(0)}) = 0 \tag{2}$$

which, not surprisingly, have much the structure of the first-order equations of the linear variation method when the basis functions are independent of ν. Indeed they become identical aside from notation if $\psi'^{(0)}$ (or in the case of degeneracy, if all the eigenfunctions of $H^{(0)}$ degenerate with $\psi'^{(0)}$) is in the space of the ϕ_k, so let us consider that case first.

If $E'^{(0)}$ is nondegenerate, then we can simply take over the formulas of Section 27 with suitable changes in notation. Thus we can write the solution of (2) to within an additive arbitrary multiple of $A_l^{(0)}$, where

$$\psi'^{(0)} = \sum_{l=1}^{M} A_l^{(0)}\phi_l \tag{3}$$

as [8]

$$\check{A}_l^{(1)} = \sum_i{}' \frac{y_{(i)l}}{\lambda_i} \left\{\sum_j y_{(i)j}^*(\phi_j, (H^{(1)} - E'^{(1)})\psi'^{(0)})\right\} \tag{4}$$

where the $y_{(i)}$ are the orthonormal eigenvectors of the $M \times M$ Hermitian matrix whose elements are $(\phi_j, (H^{(0)} - E'^{(0)})\phi_k)$, and the λ_i are the associated eigenvalues. The prime on the sum means that we are to omit the eigenvector having eigenvalue zero, namely $A^{(0)}$.

If we introduce the functions

$$\theta_i \equiv \sum_{k=1}^{M} y_{(i)k}\phi_k \tag{5}$$

then, from (4) and (1), we can write $\check{\psi}^{(1)}$ as

$$\check{\psi}^{(1)} = -\sum_i{}' \theta_i \frac{(\theta_i, (H^{(1)} - E'^{(1)})\psi'^{(0)})}{\lambda_i} \tag{6}$$

whence (31-5) yields

$$(\psi'^{(0)}, (\check{E}^{(2)} - H^{(2)})\psi'^{(0)}) = -\sum_i{}' \frac{|(\theta_i, (H^{(1)} - E'^{(1)})\psi'^{(0)})|^2}{\lambda_i} \tag{7}$$

while (30-21) provides an analogous formula for $\dot{E}^{(3)}$. Note, however, that the functions θ_i are not in general orthogonal, since the fact that

$$\sum_{l=1}^{M} y_{(i)l}^* y_{(j)l} = \delta_{ij} \tag{8}$$

does not in general imply that

$$(\theta_i, \theta_j) = \sum_{k=1}^{M} \sum_{l=1}^{M} y_{(i)l}^{*}(\phi_l, \phi_k) y_{(j)k} \tag{9}$$

will equal δ_{ij}, unless of course $(\phi_l, \phi_k) = \delta_{lk}$.

As we discussed in Section 27, (6) is not the only way to represent the solution. Thus if

$$A_M^{(0)} \neq 0 \tag{10}$$

then we can equally well write

$$\breve{\psi}^{(1)} = -\sum_{i=1}^{M-1} \bar{\theta}_i \frac{(\bar{\theta}_i, (H^{(1)} - E'^{(1)})\psi'^{(0)})}{\bar{\lambda}_i} \tag{11}$$

where

$$\bar{\theta}_i \equiv \sum_{l=1}^{M-1} \bar{y}_{(i)l} \phi_l \tag{12}$$

the $\bar{y}_{(i)}$ and $\bar{\lambda}_i$ being the orthonormal eigenvectors and corresponding eigenvalues of the $(M-1) \times (M-1)$ Hermitian matrix whose elements are

$$(\phi_j, (H^{(0)} - E'^{(0)})\phi_k), \qquad j, k = 1, 2, \ldots, M-1 \tag{13}$$

[Strictly speaking, $\breve{\psi}^{(1)}$ in (11) will usually differ from $\breve{\psi}^{(1)}$ in (6) by a multiple of $\psi'^{(0)}$.] Further, using (31-5), we have

$$(\psi'^{(0)}, (\breve{E}^{(2)} - H^{(2)})\psi'^{(0)}) = -\sum_{i=1}^{M-1} \frac{|(\bar{\theta}_i, (H^{(1)} - E'^{(1)})\psi'^{(0)})|^2}{\bar{\lambda}_i}$$

with (30-21) again providing a formula for $\dot{E}^{(3)}$. Also, one can write $\breve{\psi}^{(1)}$ and $\breve{E}^{(2)}$ in compact matrix forms analogous to the formulas (27-23).

Finally, one can use the Rayleigh–Schrödinger approach. Thus introducing the $M-1$ orthonormal functions $\hat{\phi}_k$ which together with $\psi'^{(0)}$ diagonalize $H^{(0)}$ in the space of the ϕ_k (recall Result 2 of Section 6),

$$(\hat{\phi}_k, H^{(0)}\hat{\phi}_l) = \hat{E}_k^{(0)}\,\delta_{kl}, \qquad (\hat{\phi}_k, \hat{\phi}_l) = \delta_{kl}, \qquad (\hat{\phi}_k, \psi'^{(0)}) = 0 \tag{14}$$

one then readily finds by using $\psi'^{(0)}$ and the $\hat{\phi}_k$ as the basis set that

$$\breve{\psi}^{(1)} = -\sum_{i=1}^{M-1} \hat{\phi}_i \frac{(\hat{\phi}_i, H^{(1)}\psi'^{(0)})}{\hat{E}_i^{(0)} - E'^{(0)}} \tag{15}$$

$$(\psi'^{(0)}, (\breve{E}^{(2)} - H^{(2)})\psi'^{(0)}) = -\sum_{i=1}^{M-1} \frac{|(\hat{\phi}_i, H^{(1)}\psi'^{(0)})|^2}{\hat{E}_i^{(0)} - E'^{(0)}} \tag{16}$$

etc. [Again $\check{\psi}^{(1)}$ in (15) will usually differ from both (6) and (11) by a multiple of $\psi'^{(0)}$.]

Having determined $\check{\psi}^{(1)}$, one might be interested now in going on to higher orders, approximating $\psi'^{(2)}$, etc. However, it should be clear from our discussions at the end of Section 30 and in Section 31 that since $\check{\psi}^{(1)}$ is only an approximation to $\psi'^{(1)}$, there is no one correct thing to do at this point. Probably the most consistent procedure would be to simply do higher-order perturbation theory within the linear variation method with the set of trial functions

$$\psi = \sum_k A_k \phi_k \tag{17}$$

the ϕ_k being as in (1), perhaps augmented by basis functions of different symmetries if that is appropriate.

Turning to the degenerate case, the situation is straightforward only if the state one is interested in becomes nondegenerate in first order since only then will $\psi'^{(0)}$ be known from the outset through the usual requirement that in order for the first-order equation (2) to have a solution, the zeroth-order functions must diagonalize $H^{(1)}$. If this is the case, then, unambiguously, one readily finds

$$\check{\psi}^{(1)}_{\alpha\perp} = -\sum_{i=1}^{M-D} \frac{\hat{\phi}_i(\hat{\phi}_i, H^{(1)}\psi_\alpha'^{(0)})}{\hat{E}_i^{(0)} - E'^{(0)}} \tag{18}$$

and

$$(\psi_\alpha'^{(0)}, (\check{E}_\alpha^{(2)} - H^{(2)})\psi_\alpha'^{(0)}) = -\sum_{i=1}^{M-D} \frac{|(\hat{\phi}_i, H^{(1)}\psi_\alpha'^{(0)})|^2}{\hat{E}_i^{(0)} - E'^{(0)}} \tag{19}$$

where we have introduced a label α to denote the state of interest (the others of the degenerate set will be denoted by β), and where the $\hat{\phi}_i$ are now, if the degree of the degeneracy is D, the $M - D$ orthonormal functions which together with $\psi_\alpha'^{(0)}$ and the $\psi_\beta'^{(0)}$ diagonalize $H^{(0)}$ in the space of the ϕ_k. Also the subscript $\perp$ in (18) means that it yields only that part of $\psi_k^{(1)}$ which is orthogonal to the $\psi_\beta'^{(0)}$.

However, to calculate $\dot{E}^{(3)}$, and also to calculate most expectation values with some hope of first-order accuracy, one needs to make some assumption about the rest of $\check{\psi}_\alpha^{(1)}$. Here again there is no one correct thing to do, just as there was no one correct way to approximate $E'^{(3)}$ in the first place, the choice $\dot{E}^{(3)}$ simply being a plausible one. However, the obvious thing to do is simply to imitate the exact result

$$(\psi_\beta'^{(0)}, \psi_\alpha'^{(1)}) = -\frac{(\psi_\beta'^{(0)}, H^{(1)}\psi_{\alpha\perp}'^{(1)})}{E_\beta'^{(1)} - E_\alpha'^{(1)}} - \frac{(\psi_\beta'^{(0)}, H^{(2)}\psi_\alpha'^{(0)})}{E_\beta'^{(1)} - E_\alpha'^{(1)}} \tag{20}$$

and use

$$(\psi_\beta'^{(0)}, \check{\psi}_\alpha^{(1)}) = -\frac{(\psi_\beta'^{(0)}, H^{(1)}\check{\psi}_{\alpha\perp}^{(1)})}{E_\beta'^{(1)} - E_\alpha'^{(1)}} - \frac{(\psi_\beta'^{(0)}, H^{(2)}\psi_\alpha'^{(0)})}{E_\beta'^{(1)} - E_\alpha'^{(1)}} \tag{21}$$

which, incidently, as one easily sees, is also what results from the second-order equation if one were to follow the procedure based on (17).

If the state one is interested in does not become nondegenerate in first order, then again probably the most straightforward approach is to follow the procedure based on (17), now pursuing the perturbation theory at least far enough so that $\check{\psi}_\alpha$ is determined to whatever order one requires it. It should be kept in mind, however, that now, unless symmetry considerations intervene, even $\check{\psi}_\alpha^{(0)}$ will usually not be exact.

In our discussion so far, we have assumed that $\psi'^{(0)}$ (or $\psi_\alpha'^{(0)}$ and the $\psi_\beta'^{(0)}$) is in the space of the ϕ_k. However, this of course is by no means necessary (and in cases where symmetry considerations play a role, might even be quite unnatural). Let us therefore consider briefly the opposite case, in which $\psi'^{(0)}$ (or $\psi_\alpha'^{(0)}$ and the $\psi_\beta'^{(0)}$) is not in the space of the ϕ_k.

Then, since it sufficiently illustrates the points we want to make, confining attention to the Rayleigh–Schrödinger approach, one finds

$$\check{\psi}^{(1)} = -\sum_{k=1}^{M} \frac{\hat{\phi}_k(\hat{\phi}_k, (H^{(1)} - E'^{(1)})\psi'^{(0)})}{\hat{E}_i^{(0)} - E^{(0)}} \tag{22}$$

$$(\psi'^{(0)}, (\check{E}^{(2)} - H^{(2)})\psi'^{(0)}) = -\sum_{k=1}^{M} \frac{|(\hat{\phi}_k, (H^{(1)} - E'^{(1)})\check{\psi}^{(1)})|^2}{\hat{E}_k^{(0)} - E'^{(0)}} \tag{23}$$

etc., where the $\hat{\phi}_k$ are the M orthonormal functions which diagonalize $H^{(0)}$ in the space of the ϕ_k, and where we have assumed, as is presumably the case (and is certainly the case if $E'^{(0)}$ is the ground state), that

$$\hat{E}_k^{(0)} \neq E'^{(0)}, \qquad k = 1, 2, \ldots, M \tag{24}$$

For $M = 1$, (23) of course reduces to essentially (31-9).

Perhaps the most interesting remark is that (22) and (23) hold even if $E'^{(0)}$ is degenerate and can be applied to any $\psi'^{(0)}$ within the degenerate set. To put the matter another way, under these conditions, one could be quite unaware of any degeneracy since, from (24), the homogeneous part of Eq. (2) has no solutions so that no consistency conditions are imposed on $\psi'^{(0)}$, and there are no ambiguities in the solution of (2).

Turning to the nondegenerate case, the interesting remark here is that if we were to now augment the basis set by adding $\psi'^{(0)}$ as an $(M + 1)$st function, then there would be no change in the final results. (Incidentally, this is not true in the frequency-dependent situation mentioned at the end of Section 27 unless the $\hat{\phi}_k$ are anyway all orthogonal to $\psi'^{(0)}$.) Namely, as we pointed out before in the paragraph following Eq. (37-6), $J_{\mathrm{H}}^{(2)}$ is invariant to adding multiples of $\psi'^{(0)}$ to $\psi^{(1)}$. Indeed the step of going from (6) to (11) can be viewed as rather the inverse of this process in that we in effect removed $\psi'^{(0)}$ from the basis set, again without change in the final results.

Intermediate details are, however, different in the two procedures unless $\psi'^{(0)}$ is orthogonal to the rest of the set. To illustrate what is involved, suppose that to start with, there is only a single basis function ϕ_1, and to simplify the formulas, let us assume that it and $\psi'^{(0)}$ are normalized. Therefore we will have

$$\hat{\phi}_1 = \phi_1, \qquad \hat{E}_1^{(0)} = (\phi_1, H^{(0)}\phi_1) \tag{25}$$

whence (22) and (23) yield

$$\check{\psi}^{(1)} = -\,\phi_1 \frac{(\phi_1, (H^{(1)} - E'^{(1)})\psi'^{(0)})}{(\phi_1, H^{(0)}\phi_1) - E'^{(0)}} \tag{26}$$

$$(\psi'^{(0)}, (\check{E}^{(2)} - H^{(2)})\psi'^{(0)}) = -\frac{|(\phi_1, (H^{(1)} - E'^{(1)})\psi'^{(0)})|^2}{(\phi_1, H^{(0)}\phi_1) - E'^{(0)}} \tag{27}$$

Now suppose we introduce $\psi'^{(0)}$ as a second basis function. Then clearly we will now find

$$\hat{\phi}_1 = \frac{\phi_1 - \psi'^{(0)}(\psi'^{(0)}, \phi_1)}{(1 - |(\psi'^{(0)}, \phi_1)|^2)^{1/2}} \tag{28}$$

since this satisfies Eqs. (14), and hence

$$\hat{E}_1^{(0)} = (\hat{\phi}_1, H^{(0)}\hat{\phi}_1) = \frac{(\phi_1, H^{(0)}\phi_1) - E'^{(0)}\,|(\psi'^{(0)}, \phi_1)|^2}{1 - |(\psi'^{(0)}, \phi_1)|^2} \tag{29}$$

The last two results are obviously different from their counterparts in (25). However, since now

$$(\hat{\phi}^{(1)}, H^{(1)}\psi'^{(0)}) = \frac{(\phi_1, (H^{(1)} - E'^{(1)})\psi'^{(0)})}{(1 - |(\psi'^{(0)}, \phi_1)|^2)^{1/2}} \tag{30}$$

and

$$\hat{E}_1^{(0)} - E'^{(0)} = \frac{(\phi_1, H^{(0)}\phi_1) - E'^{(0)}}{(1 - |(\psi'^{(0)}, \phi_1)|^2)} \tag{31}$$

we see that (15) yields

$$\check{\psi}^{(1)} = -\frac{\phi_1(\phi_1, (H^{(1)} - E'^{(1)})\psi'^{(0)})}{(\phi_1, H^{(0)}\phi_1) - E'^{(0)}} + \psi'^{(0)} \frac{(\psi'^{(0)}, \phi_1)(\phi_1, (H^{(1)} - E'^{(1)})\psi'^{(0)})}{(\phi_1, H^{(0)}\phi_1) - E'^{(0)}} \tag{32}$$

which differs from (26) only by a multiple of $\psi'^{(0)}$. Further, it follows from (16) that

$$(\psi'^{(0)}, (\check{E}^{(2)} - H^{(2)})\psi'^{(0)}) \\ = -\frac{|(\phi_1, (H^{(1)} - E'^{(1)})\psi'^{(0)})|^2}{\{1 - |(\psi'^{(0)}, \phi_1)|^2\}\{(\phi_1, H^{(0)}\phi_1) - E'^{(0)}\}\{1 - |(\psi'^{(0)}, \phi_1)|^2\}^{-1}} \tag{33}$$

which is of course identical to (27) and hence, as we said, the final results are essentially identical in the two procedures.

PROBLEMS

1. Derive (2) by use of $\delta \check{J}_{\mathrm{D}}^{(2)} = 0$.

2. Write (22) and (23) compactly in terms of the inverse of a matrix.

3. Derive formulas analogous to (22) and (23) for the result of a linear variation calculation using $\mathscr{J}^{(2)}$ of Eq. (31-12).

4. The next few problems concern sum rules. Let $\psi'^{(0)}$ and $\psi_k'^{(0)}$, $k = 1, 2, \ldots$, be the orthonormal eigenfunctions of $H^{(0)}$ with eigenvalues $E'^{(0)}$ and $E_k'^{(0)}$, respectively. Show that

$$\sum_k |(\psi_k'^{(0)}, H^{(1)}\psi'^{(0)})|^2 = (\psi'^{(0)}, (H^{(1)} - E'^{(1)})^2\psi'^{(0)}) \equiv S_1'$$

and that

$$\sum_k (E_k'^{(0)} - E'^{(0)})|(\psi_k'^{(0)}, H^{(1)}\psi'^{(0)})|^2 = \tfrac{1}{2}(\psi'^{(0)}, [H^{(1)}, [H^{(0)}, H^{(1)}]]\psi'^{(0)}) \equiv S_0'$$

Let $H^{(1)}$ represent the interaction with a uniform electric field. Then show that if $H^{(0)}$ is of the form $T + V$, where V is a local potential, i.e., does not involve the momentum and/or integral operators (thus the $H^{(0)}$ of (1-1) is local and the potential in the Sternheimer Hamiltonian (31-20) is local, while the potential in the $F^{(0)}$ of UHF is not local because of the exchange term), then [Thomas–Reiche–Kuhn (TRK) sum rule]

$$S_0' = \tfrac{1}{2}N\mathscr{E}^2$$

independent of $\psi'^{(0)}$ and V.

5. Referring to (15), show that if $H^{(1)}\psi'^{(0)}$ is in the space of the ϕ_k, then

$$\sum_{i=1}^{M-1} |(\hat{\phi}_i, H^{(1)}\psi'^{(0)})|^2 = S_1',$$

$$\sum_{i=1}^{M-1} (\hat{E}_i^{(0)} - E'^{(0)})|(\hat{\phi}_i, H^{(1)}\psi'^{(0)})|^2 = S_0' \quad [25].$$

6. Referring to (22), show that if $(H^{(1)} - E'^{(1)})\psi'^{(0)}$ is in the space of the ϕ_k, then

$$\sum_{i=1}^{M} |(\hat{\phi}_i, (H^{(1)} - E'^{(1)})\psi'^{(0)})|^2 = S_1',$$

$$\sum_{i=1}^{M} (\hat{E}_i^{(0)} - E'^{(0)})|(\hat{\phi}_i, (H^{(1)} - E'^{(1)})\psi'^{(0)})|^2 = S_0'$$

7. Referring to Problem 3, consider the case in which $(\chi^{(0)}, H^{(1)}\chi^{(0)}) = 0$. Show, in what is hopefully obvious notation, that if $H^{(1)}\chi^{(0)}$ is in the basis set, then

$$\sum_{i=1}^{M} |(\hat{\phi}_i, H^{(1)}\chi^{(0)})|^2 = (\chi^{(0)}, H^{(1)^2}\chi^{(0)})$$

and

$$\begin{aligned}\sum_{i=1}^{M} (\hat{E}_i^{(0)} - E^{(0)})|(\hat{\phi}_i, H^{(1)}\chi^{(0)})|^2 = {} & \tfrac{1}{2}(\chi^{(0)}, [H^{(1)}, [H^{(0)}, H^{(1)}]]\chi^{(0)}) \\ & + \tfrac{1}{2}\{(H^{(1)^2}\chi^{(0)}, (H^{(0)} - E^{(0)})\chi^{(0)}) \\ & + \text{comp. conj.}\}\end{aligned}$$

Note that in the case of the TRK sum rule with a local V, the "error term" in the second and third lines is independent of the size of the basis set. Show that it will vanish if $\chi^{(0)}$ is a variational wave function and if $H^{(1)2}\chi^{(0)}$ is one of its possible variations. Try to imagine how one could arrange for this to be the case, if, for example, $\chi^{(0)}$ were the result of a linear variational calculation.

33. Improvable Upper Bounds to Second-Order Energies for Excited States

In this section, we will describe two sets of circumstances, really two ways of choosing the $\psi^{(1)}$, so that $\check{E}^{(2)}$ will be an improvable upper bound to $E'^{(2)}$ when $\psi'^{(0)}$ is an excited state. The discussion will closely parallel our earlier discussions in Sections 2 and 7 of conditions under which an $\hat{E}$

will be an upper bound to an excited state E'. The results also can be applied within the linear variation method, and also can be extended to higher orders if one knows the lower-order functions exactly.

In Section 2, we pointed out that we would have $\hat{E} \geq E'$ if $\hat{\psi}$ were orthogonal to all lower states, but we dismissed this as being of no practical importance (in the absence of symmetry considerations), since one would almost certainly not know the lower states exactly. However, in the present context, the analogous result becomes more interesting. The point is that one may know the exact *zero*th-order wave functions for the lower states and hence can enforce orthogonality through first order. In view of the ν^{2n+1} theorem, it is then not surprising that, as we will show in a moment, doing this will then imply

$$\check{E}^{(2)} \geq E'^{(2)} \tag{1}$$

Suppose then that there are T zeroth-order states with normalized eigenfunctions $\psi_t'^{(0)}$ and eigenvalues $E_t'^{(0)}$ such that $E_t'^{(0)} < E'^{(0)}$ (more precisely, if there are symmetry considerations involved, T is the number of states with $E_t'^{(0)} < E'^{(0)}$ and having the symmetry of $\psi'^{(1)}$). Suppose further that

$$\check{\psi} = \psi'^{(0)} + \nu\check{\psi}^{(1)} + \cdots \tag{2}$$

is orthogonal to all

$$\psi_t' = \psi_t'^{(0)} + \nu\psi_t'^{(1)} + \cdots, \qquad t = 1, 2, \ldots, T \tag{3}$$

through first order. Since they are already orthogonal in zeroth order, this means that

$$(\psi_t'^{(0)}, \check{\psi}^{(1)}) + (\psi_t'^{(1)}, \psi'^{(0)}) = 0 \tag{4}$$

However, from the familiar sum over states formula for $\psi_t'^{(1)}$, it follows that

$$(\psi_t'^{(1)}, \psi'^{(0)}) = -\frac{(\psi_t'^{(0)}, H^{(1)}\psi'^{(0)})}{E'^{(0)} - E_t'^{(0)}} \tag{5}$$

whence (4) yields

$$(\psi_t'^{(0)}, \check{\psi}^{(1)}) = -\frac{(\psi_t'^{(0)}, H^{(1)}\psi'^{(0)})}{E_t'^{(0)} - E'^{(0)}} \tag{6}$$

which, from the sum over states formula for $\psi'^{(1)}$, is precisely $(\psi_t'^{(0)}, \psi'^{(1)})$. Therefore if (4) and hence (6) is satisfied, we will have

$$(\psi_t'^{(0)}, (\check{\psi}^{(1)} - \psi'^{(1)})) = 0, \qquad t = 1, \ldots, T \tag{7}$$

so that $\check{\psi}^{(1)} - \psi'^{(1)}$ will be orthogonal to all $\psi_t'^{(0)}$ with $E_t'^{(0)} < E'^{(0)}$; from (30-23) we will then have

$$\check{J}_{\mathrm{H}}^{(2)} - J_{\mathrm{H}}'^{(2)} \geq 0$$

as desired.

One way [26] to ensure that (6) will be satisfied is simply to use trial functions of the form

$$\psi^{(1)} = -\sum_{t=1}^{T} \psi_t'^{(0)} \frac{(\psi_t'^{(0)}, H^{(1)}\psi'^{(0)})}{E_t'^{(0)} - E'^{(0)}} + X_{\perp} \tag{8}$$

where $X_{\perp}$ is orthogonal to the $\psi_t'^{(0)}$. Further, since with such a set, $\Delta^{(1)}$ of (30-22) is also obviously orthogonal to all the $\psi_t'^{(0)}$, it follows from (30-23) that $J_{\mathrm{H}}^{(2)} - J_{\mathrm{H}}'^{(2)} \geq 0$ and hence, as we discussed following (30-23), as we increase the flexibility of $X_{\perp}$, $\check{E}^{(2)}$ will steadily decrease (or at any rate, not increase) and hence improve.

Actually it is not necessary to be quite so rigid in specifying the form of $\psi^{(1)}$. Namely it is sufficient [27] to use trial functions of the form

$$\psi^{(1)} = \sum_{t=1}^{T} A_t \psi_t'^{(0)} + X \tag{9}$$

where the A_t are arbitrary complex numbers and where X can be anything so long as it does not involve the A_t explicitly. To see this, we first note that if we vary A_t^* and not A_t, then $\delta \check{J}_{\mathrm{H}}^{(2)} = 0$ yields

$$(\psi_t'^{(0)}, (H^{(0)} - E'^{(0)})\check{\psi}^{(1)}) + (\psi_t'^{(0)}, (H^{(1)} - E'^{(1)})\psi'^{(0)}) = 0$$

or

$$(E_t'^{(0)} - E'^{(0)})(\psi_t'^{(0)}, \check{\psi}^{(1)}) + (\psi_t'^{(0)}, H^{(1)}\psi'^{(0)}) = 0 \tag{10}$$

which is just (6). (See also the remarks following Eq. (35-60) in Section 35.) Moreover, the set (9) is completely equivalent to the set

$$\psi^{(1)} = \sum_{t=1}^{T} B_t \psi_t'^{(0)} + X_{\perp} \tag{11}$$

where now the B_t are arbitrary complex numbers, and therefore from (10) can in effect be replaced by

$$\psi^{(1)} = -\sum_{t=1}^{T} \psi_t'^{(0)} \frac{(\psi_t'^{(0)}, H^{(1)}\psi'^{(0)})}{E_t'^{(0)} - E'^{(0)}} + X_{\perp}$$

which proves the point.

In Section 7, we showed that even without knowing the lower states exactly, the linear variation method provided improvable upper bounds to the E' of excited states. We now turn to the perturbation analog of these results. In particular, we will show here that if one does a linear variation calculation of the type described in the preceding section, and if T of the $\hat{E}_i^{(0)}$ lie below $E'^{(0)}$ (there cannot be more, but there could be less), then $\check{J}^{(2)}$ will be an upper bound to $J'^{(2)}$. Further, in Appendix E, we will prove that this is an improvable bound.

Let us consider a trial function which is intermediate between $\psi^{(1)}$ of (32-1) and $\check{\psi}^{(1)}$ as given, say, in (32-22), namely

$$\psi_{\mathrm{I}}^{(1)} = \sum_{i=1}^{T} \alpha_i \hat{\phi}_i - \sum_{i=T+1}^{M} \hat{\phi}_i \frac{(\hat{\phi}_i, (H^{(1)} - E'^{(1)})\psi'^{(0)})}{\hat{E}_i^{(0)} - E'^{(0)}} \tag{12}$$

where the α_i are, for the moment, unspecified numbers. We now introduce $\bar{\Delta}_{\mathrm{I}}^{(1)}$, the difference between $\psi_{\mathrm{I}}^{(1)}$ and $\check{\psi}^{(1)}$, thus

$$\bar{\Delta}_{\mathrm{I}}^{(1)} \equiv \psi_{\mathrm{I}}^{(1)} - \check{\psi}^{(1)} = \sum_{i=1}^{T} \hat{\phi}_i\left(\alpha_i + \frac{(\hat{\phi}_i, (H^{(1)} - E'^{(1)})\psi'^{(0)})}{\hat{E}_i^{(0)} - E'^{(0)}}\right) \tag{13}$$

Then since $\bar{\Delta}_{\mathrm{I}}^{(1)}$ is a possible $\delta\check{\psi}^{(1)}$, it follows that $J_{\mathrm{HI}}^{(2)} - \check{J}_{\mathrm{H}}^{(2)}$ must be of second order in $\bar{\Delta}_{\mathrm{I}}^{(1)}$, so that we have

$$J_{\mathrm{HI}}^{(2)} - \check{J}_{\mathrm{H}}^{(2)} = (\bar{\Delta}_{\mathrm{I}}^{(1)}, (H^{(0)} - E'^{(0)})\, \bar{\Delta}_{\mathrm{I}}^{(1)})$$

Using (13) and the facts that the $\hat{\phi}_i$ are orthonormal and diagonalize $H^{(0)}$, we then find in detail that

$$J_{\mathrm{HI}}^{(2)} - \check{J}_{\mathrm{H}}^{(2)} = \sum_{i=1}^{T} \frac{|\, \alpha_i(\hat{E}_i^{(0)} - E'^{(0)}) + (\hat{\phi}_i, (H^{(1)} - E'^{(1)})\psi'^{(0)})\,|^2}{\hat{E}_i^{(0)} - E'^{(0)}} \tag{14}$$

But by assumption

$$\hat{E}_i^{(0)} - E'^{(0)} < 0, \qquad i = 1, 2, \ldots, T \tag{15}$$

therefore we conclude that whatever the values of the numbers α_i, we will have

$$J_{\mathrm{HI}}^{(2)} \le \check{J}_{\mathrm{H}}^{(2)} \tag{16}$$

However, we can certainly choose the T numbers α_i in such a way that

$$\Delta_{\mathrm{I}}^{(1)} \equiv \psi'^{(1)} - \psi_{\mathrm{I}}^{(1)}$$

is orthogonal to all the $\psi_t'^{(0)}$, whence it follows from (30-23) that for such a choice of the α_i, we will have

$$J_{\mathrm{H}}'^{(2)} \leq J_{\mathrm{HI}}^{(2)} \tag{17}$$

Thus, since (16) is true whatever the choice of the α_i, it follows, as announced, that under these conditions,

$$J_{\mathrm{H}}'^{(2)} \leq \check{J}_{\mathrm{H}}^{(2)} \tag{18}$$

If the $\psi_t'^{(0)}$, $t = 1, \ldots, T$, are in the space of the ϕ_k, then this second procedure is simply a special case of the first. Also from the discussion of Section 7, we can infer that this choice of the ϕ_k would give the best results. Nevertheless, as we have seen, it is not necessary to have this in order to guarantee a bound; all one really needs is that T of the $\hat{E}_i^{(0)}$ be less than $E'^{(0)}$. Also, one can include nonlinear factors in the ϕ_k, thereby making the second procedure more comparable to the first in flexibility. One would then of course have to be sure that at the optimal values of the nonlinear parameters, there were T of the $\hat{E}_i^{(0)}$ below $E'^{(0)}$.

The result (18) is closely related to an interesting and useful operator inequality [28, 29] which can be derived in a similar way. Let Ω be a Hermitian operator with T negative eigenvalues, possibly some positive eigenvalues, but with no zero eigenvalue. Also, let ζ_k be an orthonormal set of M functions which diagonalize Ω within their space, that is,

$$(\zeta_k, \Omega\zeta_l) = w_k\delta_{kl}, \qquad (\zeta_k, \zeta_l) = \delta_{kl} \tag{19}$$

We will now show that if T of the w_k are less than zero, then in Dirac bra-ket notation

$$\Omega^{-1} \geq \sum_{k=1}^{M} | \zeta_k \times \zeta_k |/w_k \tag{20}$$

To prove this, consider the inhomogeneous equation

$$\Omega\chi' + g = 0 \tag{21}$$

whose solution is

$$\chi' = -\Omega^{-1}g \tag{22}$$

With this equation, we can associate the Hylleraas-like functional

$$\mathscr{J}_{\mathrm{H}} = (\chi, \Omega\chi) + (\chi, g) + (g, \chi) \tag{23}$$

where the χ are trial functions. Also, from (21) and (22), we have that when $\chi = \chi'$, then

$$\mathscr{J}_{\mathrm{H}} = \mathscr{J}_{\mathrm{H}}' = (g, \chi') = -(g, \Omega^{-1}g) \tag{24}$$

Let us now use as the set of trial functions

$$\chi = \sum_{k=1}^{M} A_k \zeta_k \tag{25}$$

Then one readily finds that

$$\hat{\mathscr{J}}_{\mathrm{H}} = -\sum_{k=1}^{M} |(\zeta_k, g)|^2/w_k \tag{26}$$

However, by exactly the same argument which led to (18), one can prove that if T of the w_k are less than zero, then

$$\mathscr{J}_{\mathrm{H}}' < \hat{\mathscr{J}}_{\mathrm{H}} \tag{27}$$

Thus we have, from (24) and (26), that

$$(g, \Omega^{-1}g) \geq \sum_{k=1}^{M} \frac{(g, \zeta_k)(\zeta_k, g)}{w_k} \tag{28}$$

which, since g is arbitrary, implies (20). Also, it is easy to show that one can write (20) and (28) in the more general forms

$$\Omega^{-1} \geq \sum_{k=1}^{M} \sum_{l=1}^{M} |\eta_k\rangle(\mathbf{\Omega}^{-1})_{kl}\langle\eta_l| \tag{29}$$

and

$$(g, \Omega^{-1}g) \geq \sum_{k=1}^{M} \sum_{l=1}^{M} (g, \eta_k)(\mathbf{\Omega}^{-1})_{kl}(\eta_l, g) \tag{30}$$

respectively, where $\mathbf{\Omega}^{-1}$ is the inverse of the $M \times M$ Hermitian matrix whose elements are

$$(\eta_k, \Omega\eta_l) \tag{31}$$

and where the basis functions η_k are derived from the ζ_k by some linear transformation.

Obviously, there is a close connection between (27) and (18). In fact the latter is really a kind of limiting case of the former in the following sense. Suppose that we were to consider (30-7) to be the limit of

$$(H^{(0)} - E'^{(0)} - \sigma)\psi'^{(1)} + (H^{(1)} - E'^{(1)})\psi'^{(0)} = 0 \tag{32}$$

as the positive number σ tends to zero. Then since $(H^{(0)} - E'^{(0)} - \sigma)$ has the properties of Ω, we could use (27) and (26) to conclude that (since the analysis is more interesting if $\psi'^{(0)}$ is in the set, we consider that case)

$$\mathscr{J}_{\mathrm{H}}' \leq -\sum_{k=1}^{M-1} \frac{|(\hat{\phi}_k, (H^{(1)} - E'^{(1)})\psi'^{(0)})|^2}{\hat{E}_k^{(0)} - E'^{(0)} - \sigma} + \frac{|(\psi'^{(0)}, (H^{(1)} - E'^{(1)})\psi'^{(0)})|^2}{\sigma} \quad (33)$$

or, since $(\hat{\phi}_k, \psi'^{(0)}) = 0$ and $(\psi'^{(0)}, (H^{(1)} - E'^{(1)})\psi'^{(0)}) = 0$,

$$\mathscr{J}_{\mathrm{H}}' \leq -\sum_{k=1}^{M-1} \frac{|(\hat{\phi}_k, H^{(1)}\psi'^{(0)})|^2}{\hat{E}_k^{(0)} - E'^{(0)} - \sigma} \quad (34)$$

whence, letting $\sigma \to 0$, we have (18).

In case Ω has only positive eigenvalues, (28) can be thought of as a generalization of the Schwartz inequality since if there is only one function ζ_k, it becomes

$$(\zeta, \Omega\zeta)(g, \Omega^{-1}g) \geq (g, \zeta)(\zeta, g) \quad (35)$$

which can, if Ω has no negative eigenvalues, be written as

$$(\Omega^{1/2}\zeta, \Omega^{1/2}\zeta)(\Omega^{-1/2}g, \Omega^{-1/2}g) \geq (\Omega^{-1/2}g, \Omega^{1/2}\zeta)(\Omega^{1/2}\zeta, \Omega^{-1/2}g) \quad (36)$$

and this, of course, *is* the Schwartz inequality. Also, in general if Ω has only positive eigenvalues, we can write

$$\mathscr{J}_{\mathrm{H}} = (\Omega^{1/2}\chi + \Omega^{-1/2}g, \Omega^{1/2}\chi + \Omega^{-1/2}g) - (g, \Omega^{-1}g) \quad (37)$$

thereby making it obvious in this case that for any χ

$$\mathscr{J}_{\mathrm{H}} \geq -(g, \Omega^{-1}g) = \mathscr{J}_{\mathrm{H}}' \quad (38)$$

Finally, we will show that when Ω has only positive eigenvalues, (30) is equivalent to the statement that a certain Gram determinant (the determinant of a certain overlap matrix) is, as it must be, nonnegative. Thus for such Ω's, optimizing Hylleraas-like functionals using linear trial functions [30], and using Gram determinant techniques[13] are completely equivalent.[13]

To prove this, we start by writing (30) as

$$(f, f) - \sum_{k=1}^{M} \sum_{l=1}^{M} (f, \chi_k)(S^{-1})_{kl}(\chi_l, f) \geq 0 \quad (39)$$

[13] See Weinhold [29] and references there to many earlier papers.

where we have introduced the functions f and χ_k defined by

$$f \equiv \Omega^{-1/2} g, \qquad \chi_k \equiv \Omega^{1/2} \eta_k \tag{40}$$

and where S is the positive-definite overlap matrix

$$S_{kl} = (\chi_k, \chi_l) \tag{41}$$

If one now uses the standard formula that $(S^{-1})_{kl}$ is the cofactor of S_{lk} divided by the determinant of S, one readily sees that (39), when multiplied through by the determinant of S, is precisely the Gram inequality

$$\begin{vmatrix} (f,f) & (f,\chi_1) & \cdots & (f,\chi_M) \\ (\chi_1,f) & (\chi,\chi_1) & \cdots & (\chi_1,\chi_M) \\ \vdots & & & \\ (\chi_M,f) & (\chi_M,\chi_1) & \cdots & (\chi_M,\chi_M) \end{vmatrix} \geq 0 \tag{42}$$

where the left hand side of the latter has been expanded according to the elements of the first column [the first term of that expansion is evidently (f, f) det S].

PROBLEMS

1. Show directly that if one uses the set of trial functions (8), then $J^{(2)} \geq \check{J}_{\mathrm{H}}^{(2)}$, thereby providing a direct proof that one has an improvable bound.

2. Suppose that $T = 1$ and that $\hat{\phi}_1, \hat{\phi}_2, \ldots, \hat{\phi}_M$ yield $\hat{E}_1^{(0)} < E'^{(0)}$. Show in analogy to the results of Section 7 that use of $\psi_1'^{(0)}, \hat{\phi}_2, \ldots$ will yield better results.

34. The Second-Order Hylleraas Variation Method with Two Perturbations

If there are two perturbations, then

$$\begin{aligned} \nu H^{(1)} &\rightarrow \nu_1 H^{(10)} + \nu_2 H^{(01)} \\ \nu^2 H^{(2)} &\rightarrow \nu_1^2 H^{(20)} + \nu_1\nu_2 H^{(11)} + \nu_2^2 H^{(02)} \\ \nu\psi'^{(1)} &\rightarrow \nu_1\psi'^{(10)} + \nu_2\psi'^{(01)} \\ \nu E'^{(1)} &\rightarrow \nu_1 E'^{(10)} + \nu_2 E'^{(01)} \\ \nu^2 E'^{(2)} &\rightarrow \nu_1^2 E'^{(20)} + \nu_1\nu_2 E'^{(11)} + \nu_2^2 E'^{(02)} \end{aligned} \tag{1}$$

etc., where by the latter three replacements, we have assumed of course that there are no problems with degeneracy.

As far as $E'^{(20)}$ and $E'^{(02)}$ are concerned, there is nothing new to say. We can get improvable bounds to them by use of $J_{\mathrm{H}}^{(20)}$ and $J_{\mathrm{H}}^{(02)}$, respectively, where

$$J_{\mathrm{H}}^{(20)} = (\psi^{(10)}, (H^{(0)} - E'^{(0)})\psi^{(10)}) + (\psi^{(10)}, (H^{(10)} - E'^{(10)})\psi'^{(0)}) + (\psi'^{(0)}, (H^{(10)} - E'^{(10)})\psi^{(10)}) \tag{2}$$

and

$$(\psi'^{(0)}, (H^{(10)} - E'^{(10)})\psi'^{(0)}) = 0 \tag{3}$$

and similarly for $J_{\mathrm{H}}^{(02)}$. The interesting new quantity is $E'^{(11)}$ and here various possibilities present themselves. Thus from (26-37) or its interchange, we have, since in the present case $\bar{\varphi}^{(10)} = \bar{\varphi}^{(01)} = \bar{\varphi}^{(11)} = 0$, that

$$(\psi'^{(0)}, (E'^{(11)} - H^{(11)})\psi'^{(01)}) = (\psi'^{(01)}, (H^{(10)} - E'^{(10)})\psi'^{(0)}) + (\psi'^{(0)}, (H^{(10)} - E'^{(10)})\psi'^{(01)}) \tag{4}$$

and

$$(\psi'^{(0)}, (E'^{(11)} - H^{(11)})\psi'^{(0)}) = (\psi'^{(10)}, (H^{(01)} - E'^{(01)})\psi'^{(0)}) + (\psi'^{(0)}, (H^{(01)} - E'^{(01)})\psi'^{(10)}) \tag{5}$$

which since

$$(H^{(0)} - E'^{(0)})\psi'^{(10)} + (H^{(10)} - E'^{(10)})\psi'^{(0)} = 0 \tag{6}$$

we can also write as

$$(\psi'^{(0)}, (E'^{(11)} - H^{(11)})\psi'^{(0)}) = -(\psi'^{(01)}, (H^{(0)} - E'^{(0)})\psi'^{(10)}) - (\psi'^{(10)}, (H^{(0)} - E'^{(0)})\psi'^{(01)}) \tag{7}$$

and obviously there are other possibilities as well. Further, going on to third-order quantities, one has from (26-55) and the fact that in this case all the underlined functions vanish, that

$$\begin{aligned}(\psi'^{(0)}, (E'^{(12)} - H^{(12)})\psi'^{(0)}) = {} & \{(\psi'^{(01)}, (H^{(01)} - E'^{(01)})\psi'^{(10)}) \\ & + (\psi'^{(10)}, (H^{(02)} - E'^{(02)})\psi'^{(0)}) \\ & + (\psi'^{(01)}, (H^{(11)} - E'^{(11)})\psi'^{(0)}) \\ & + \text{comp. conj.}\} \\ & + (\psi'^{(01)}, (H^{(10)} - E'^{(10)})\psi'^{(01)})\end{aligned} \tag{8}$$

and a similar formula for $E'^{(21)}$.

In turn these formulas and others like them then suggest various approximations to $E'^{(11)}$, thus,

$$(\psi'^{(0)}, (\dot{E}^{(11)} - H^{(1)})\psi'^{(0)}) = (\check{\psi}^{(01)}, (H^{(10)} - E'^{(10)})\psi'^{(0)}) + (\psi'^{(0)}, (H^{(10)} - E'^{(10)})\check{\psi}^{(01)}) \tag{9}$$

$$(\psi'^{(0)}, (\ddot{E}'^{(11)} - H^{(11)})\psi'^{(0)}) = (\check{\psi}^{(10)}, (H^{(01)} - E'^{(01)})\psi'^{(0)}) + (\psi'^{(0)}, (H^{(01)} - E'^{(01)})\check{\psi}^{(10)}) \tag{10}$$

and

$$(\psi'^{(0)}, (\dddot{E}^{(11)} - H^{(11)})\psi'^{(0)}) = -(\check{\psi}^{(01)}, (H^{(0)} - E'^{(0)})\check{\psi}^{(10)}) - (\check{\psi}^{(10)}, (H^{(0)} - E'^{(0)})\check{\psi}^{(01)}) \tag{11}$$

where, as our notation implies, $\check{\psi}^{(10)}$ and $\check{\psi}^{(01)}$ will usually be determined by the use of $J_{\mathrm{H}}^{(20)}$ and $J_{\mathrm{H}}^{(02)}$, respectively. Also, (8) and its analog for $E'^{(21)}$ then obviously suggest corresponding approximations to $E'^{(12)}$ and $E'^{(21)}$ which we will not write out in detail.

In certain applications, notably in the calculation of spin–spin coupling constants,[14] these procedures are not quite straightforward because the conventional $H^{(01)}$ and $H^{(10)}$ are so singular that $E'^{(20)}$ and $E'^{(02)}$ are infinite, and therefore the $J_{\mathrm{H}}^{(20)}$ and $J_{\mathrm{H}}^{(02)}$ variational problems are really undefined. However, there are usually good physical reasons why such $H^{(10)}$ and $H^{(01)}$ are anyway only approximations and should be replaced by nonsingular operators, the exact details of the replacement usually being unimportant as far as $E'^{(11)}$ is concerned and therefore, hopefully, $\dot{E}^{(11)}$, $\ddot{E}^{(11)}$, and $\dddot{E}^{(11)}$ will also be insensitive to such replacements.

In general $\dot{E}^{(11)}$, $\ddot{E}^{(11)}$, and $\dddot{E}^{(11)}$ will provide different approximations to $E'^{(11)}$. An exception occurs if $\check{\psi}^{(01)}$ and $\check{\psi}^{(10)}$ are drawn from a common linear space. Then, as one easily shows, the common result, which we will denote by $\check{E}^{(11)}$, is given by

$$(\psi'^{(0)}, (\check{E}^{(11)} - H^{(11)})\psi'^{(0)}) = -\sum_{k=1}^{M} \frac{(\psi'^{(0)}, (H^{(10)} - E'^{(10)})\hat{\phi}_k)(\hat{\phi}_k, (H^{(01)} - E'^{(01)})\psi'^{(0)}) + \text{comp. conj.}}{\hat{E}_k^{(0)} - E'^{(0)}} \tag{12}$$

where to be specific we have assumed that $\psi'^{(0)}$ is not in the space. Further, usually $\dot{E}^{(11)}$, $\ddot{E}^{(11)}$, and $\dddot{E}^{(11)}$ all involve first-order errors since if we write

$$\check{\psi}^{(01)} = \psi'^{(01)} + \bar{\Delta}^{(01)}, \qquad \check{\psi}^{(10)} = \psi'^{(10)} + \bar{\Delta}^{(10)} \tag{13}$$

[14] See, for example, de Jeu [31], and Wrobel and Voitlander [32].

then one readily finds that

$$\begin{aligned} \dot{E}^{(11)} &= E'^{(11)} + O(\bar{\Delta}^{(01)}) \\ \ddot{E}^{(11)} &= E'^{(11)} + O(\bar{\Delta}^{(10)}) \\ \dddot{E}^{(11)} &= E'^{(11)} + O(\bar{\Delta}^{(10)}, \bar{\Delta}^{(01)}, \bar{\Delta}^{(10)}\bar{\Delta}^{(01)}) \end{aligned} \tag{14}$$

However, if $\check{\psi}^{(10)}$ and $\check{\psi}^{(01)}$ are drawn from a common linear space then, as we have said, the left hand sides in (14) are all equal to $\check{E}^{(11)}$, and therefore it must be that in this case

$$\check{E}^{(11)} = E'^{(11)} + O(\bar{\Delta}^{(10)}\bar{\Delta}^{(01)}) \tag{15}$$

since, from the first two of equations (14), $\check{E}^{(11)}$ must be exact if *either* $\bar{\Delta}^{(10)}$ or $\bar{\Delta}^{(01)}$ vanishes.

Moreover, it is easy to see how to ensure such a bilinear error in the general case provided that one is willing to calculate both $\check{\psi}^{(01)}$ and $\check{\psi}^{(10)}$. Namely if we make the replacements analogous to (1) in $\nu^2 \check{J}_{\mathrm{H}}^{(2)}$, then it must be that the $\nu_1\nu_2$ term, call it $\nu_1\nu_2(\psi'^{(0)}, (\check{E}^{(11)} - H^{(11)})\psi'^{(0)})$, will provide an approximation to $E'^{(11)}$ with the property (15) since we know that $\nu^2 \check{J}_{\mathrm{H}}^{(2)} - \nu^2 J_{\mathrm{H}}'^{(2)} = \nu^2 O(\psi'^{(1)} - \check{\psi}^{(1)})^2$. Therefore making this substitution, we find

$$\begin{aligned} (\psi'^{(0)}, (\check{E}^{(11)} - H^{(11)})\psi'^{(0)}) = {} & (\check{\psi}^{(10)}, (H^{(0)} - E'^{(0)})\check{\psi}^{(01)}) \\ & + (\check{\psi}^{(01)}, (H^{(0)} - E'^{(0)})\check{\psi}^{(10)}) \\ & + (\check{\psi}^{(10)}, (H^{(01)} - E'^{(01)})\psi'^{(0)}) \\ & + (\psi'^{(0)}, (H^{(01)} - E'^{(01)})\check{\psi}^{(10)}) \\ & + (\check{\psi}^{(01)}, (H^{(10)} - E'^{(10)})\psi'^{(0)}) \\ & + (\psi'^{(0)}, (H^{(10)} - E'^{(10)})\check{\psi}^{(01)}) \end{aligned} \tag{16}$$

the right hand side of (16) being −Eq. (11) + Eq. (9) + Eq. (10), and, as one easily sees, (15) *is* satisfied [see also Eq. (18) later], as expected.

As we have presented it, (16) is to be used with $\check{\psi}^{(10)}$ and $\check{\psi}^{(01)}$ derived, say, by use of $J_{\mathrm{H}}^{(20)}$ and $J_{\mathrm{H}}^{(02)}$, respectively. However, on occasion, it has been suggested[15] that one use

$$\delta \check{J}_{\mathrm{H}}^{(11)} = 0 \qquad \text{“all”} \quad \delta\check{\psi}^{(10)} \quad \text{and} \quad \delta\check{\psi}^{(01)} \tag{17}$$

$\check{J}_{\mathrm{H}}^{(11)}$ being simply the right hand side of (16), as an independent way of

[15] See, for example, de Jeu [31]. Also Mandan *et al.* [33] and Hambro [34].

determining an optimal $\psi^{(01)}$ and $\psi^{(10)}$. Note however, that although

$$J_{\mathrm{H}}^{(11)} - J_{\mathrm{H}}^{\prime(11)} = (\Delta^{(01)}, (H^{(0)} - E^{\prime(0)})\Delta^{(10)}) + \text{comp. conj.} \tag{18}$$

where

$$\Delta^{(01)} \equiv \psi^{(01)} - \psi^{(01)\prime} \qquad \text{and} \qquad \Delta^{(10)} = \psi^{(10)} - \psi^{(10)\prime} \tag{19}$$

so that, in accord with our expectations, the error is bilinear, this also means that $J_{\mathrm{H}}^{(11)} - J_{\mathrm{H}}^{\prime(11)}$, and hence in particular $\check{J}_{\mathrm{H}}^{(11)} - J_{\mathrm{H}}^{\prime(11)}$, is of indefinite sign. One practical consequence of this is that if one uses nonlinear parameters, then, since in general (17) will have more than one solution, there will be no obvious way[16] of choosing between them because a priori one does not know in what direction $J^{\prime(11)}$ lies. Another related feature of (17) is that unless one exercises some caution in choosing the trial functions, there may be no solution at all. Thus suppose one were to do a (doubly) linear variation calculation

$$\psi^{(01)} = \sum_{k=1}^{M} A_k^{(01)}\phi_k, \qquad \psi^{(10)} = \sum_{k=1}^{M'} A_k^{(10)}\chi_k \tag{20}$$

Then, using (17), one is led to

$$\sum_{l=1}^{M'} (\phi_k, (H^{(0)} - E^{\prime(0)})\chi_l)\check{A}_l^{(10)} + (\phi_k, (H^{(10)} - E^{\prime(10)})\psi^{\prime(0)}) = 0, \qquad k = 1, 2, \ldots, M \tag{21}$$

and

$$\sum_{l=1}^{M} (\chi_k, (H^{(0)} - E^{\prime(0)})\phi_l)\check{A}_l^{(01)} + (\chi_k, (H^{(01)} - E^{\prime(01)})\psi^{\prime(01)}) = 0, \qquad k = 1, 2, \ldots, M' \tag{22}$$

and the point is that Eq. (21) [Eq. (22)] will in general have no solution if $M > M'[M' > M]$.

As we saw from (19), $\check{E}^{(11)}$ is not a guaranteed bound on $E^{\prime(11)}$ and one easily sees that the same is true of $\dot{E}^{(11)}$, etc. Indeed this is really the basic difficulty with approximating $E^{\prime(11)}$ since whatever procedure one uses, one does not know what constitutes a best solution in the sense of being closest to $E^{\prime(11)}$, and one does not know in which direction improvement lies. Thus, for example, whatever procedure one uses, then, if one enlarges the

[16] See, for example, Mandan *et al.* [33]; also Hambro [34].

set of trial functions and the approximation changes, one does not know a priori (that is, without, say, appeal to experimental information) whether the direction of change is good or bad.

The situation would be different, however, if one had good lower bounding techniques for second-order energies. The point is the following: In general one has, however $\check{\psi}^{(01)}$ and $\check{\psi}^{(11)}$ have been determined, that

$$\begin{aligned}\nu_1{}^2(\check{J}_{\mathrm{H}}^{(20)} - J_{\mathrm{H}}^{\prime(20)}) + \nu_1\nu_2(\check{J}_{\mathrm{H}}^{(11)} - J_{\mathrm{H}}^{\prime(11)}) + \nu_2{}^2(\check{J}_{\mathrm{H}}^{(02)} - J_{\mathrm{H}}^{\prime(02)}) \\ = (\nu_1\bar{\Delta}^{(10)} + \nu_2\bar{\Delta}^{(01)}, (H^{(0)} - E^{\prime(0)})(\nu_1\bar{\Delta}^{(10)} + \nu_2\bar{\Delta}^{(01)}))\end{aligned} \tag{23}$$

and therefore under the obvious conditions one will have

$$\nu_1{}^2\check{E}^{(20)} + \nu_1\nu_2\check{E}^{(11)} + \nu_2{}^2\check{E}^{(02)} \geq \nu_1{}^2E^{\prime(20)} + \nu_1\nu_2E^{\prime(11)} + \nu_2{}^2E^{\prime(02)} \tag{24}$$

that is, under appropriate conditions, we do have a guaranteed upper bound to the *total* second-order energy. Suppose now that one would also calculate lower bounds to the total second-order energy. In particular, suppose one does this for $\nu_2 = \nu_1$ and for $\nu_2 = -\nu_1$, with the results $E_l(+)$ and $E_l(-)$, respectively, thus

$$E_l(+) \leq E^{\prime(20)} + E^{\prime(11)} + E^{\prime(02)} \tag{25}$$

$$E_l(-) \leq E^{\prime(0)} - E^{\prime(11)} + E^{\prime(02)} \tag{26}$$

These inequalities, when combined with (24) for $\nu_1 = \pm\nu_2$, namely

$$\check{E}^{(20)} + \check{E}^{(11)} + \check{E}^{(02)} \geq E^{\prime(20)} + E^{\prime(11)} + E^{\prime(02)} \tag{27}$$

and

$$\check{E}^{(20)} - \check{E}^{(11)} + \check{E}^{(02)} \geq E^{\prime(20)} - E^{\prime(11)} + E^{\prime(02)} \tag{28}$$

then yield

$$\tfrac{1}{2}[E_l(+) - \check{E}^{(20)} + \check{E}^{(11)} - \check{E}^{(02)}] \leq E^{\prime(11)} \leq \tfrac{1}{2}[\check{E}^{(20)} + \check{E}^{(11)} + \check{E}^{(02)} - E_l(-)]$$

whence one would have bounds on $E^{\prime(11)}$ and in particular at least know what direction of change constitutes improvement. However, although there are lower bound formulas for the second-order energy, they are just the perturbation analogs[17] of the lower bound formulas for E' mentioned in Section 3, and like the latter, have thus far been used effectively only for quite simple systems.[18]

[17] See, for example, Epstein [35].
[18] For a review, see Weinhold [29].

PROBLEMS

1. Derive (4) and (5) directly from the equation for $\psi'^{(11)}$ and the equations for $\psi'^{(10)}$ and $\psi'^{(01)}$.

2. Show that one can also write

$$(\psi'^{(0)}, (E'^{(11)} - H^{(11)})\psi'^{(0)}) = (\psi'^{(0)}, (H^{(10)} - E'^{(10)})\psi'^{(01)}) + (\psi'^{(0)}, (H^{(01)} - E'^{(01)})\psi'^{(10)})$$

Show that the approximation to $E'^{(11)}$ which this suggests also yields (12) under the appropriate conditions. Can you see a possible difficulty with this formula under more general circumstances?

3. The Hylleraas variational functional $J_{\mathrm{H}}^{(02)}$ will of course yield $\check{\psi}^{(01)} = \psi'^{(01)}$ if $\psi'^{(01)}$ is in the set of trial functions, and similarly for $J_{\mathrm{H}}^{(20)}$. As a check on (15), verify that if either $\psi'^{(10)}$ or $\psi'^{(01)}$ is in the space of the ϕ_k, then $\check{E}^{(11)}$, as given in (12) is exact, that is, equals $E'^{(11)}$.

4. Show as a corollary of the theorem of Problem 3 that if one does a linear variation calculation using $J_{\mathrm{H}}^{(02)}$ to determine $\check{\psi}^{(01)}$, then whether or not $\check{\psi}^{(01)}$ is exact, $\dot{E}^{(11)}$ as given in (9) will still be exact if $\psi'^{(10)}$ is in the space used for $\check{\psi}^{(01)}$.

5. Verify (18).

6. Show that use of (17) with $\psi^{(01)}$ and $\psi^{(10)}$ drawn from a common linear space again leads to (12).

7. Assuming that $M' = M$ and that the matrix whose elements are $(\phi_k, (H^{(0)} - E'^{(0)})\chi_l)$ has an inverse, write the results of the linear variation calculation based on (20) in compact form in terms of the inverse of that matrix.

References

1. A. M. Shorb, R. Schroeder, and E. R. Lippincott, *J. Chem. Phys.* **37**, 1043 (1962).
2. A. G. Ehlenberger and L. B. Mendelsohn, *J. Chem. Phys.* **56**, 586 (1972).
3. M. Cohen and A. Dalgarno, *Proc. Roy. Soc. Ser. A* **275**, 492 (1963).
4. G. J. Iafrate and L. B. Mendelsohn, *Phys. Rev. A* **2**, 561 (1970), and references therein.
5. R. M. Sternheimer, *Phys. Rev. A* **7**, 887 (1973).
6. R. P. Gupta and S. K. Sen, *Phys. Rev. A* **7**, 850 (1973), and references therein.
7. V. McKoy and N. W. Winter, *J. Chem. Phys.* **48**, 5514 (1968).

8. E. Hylleraas, *Z. Phys.* **65**, 209 (1930).
9. R. Ahlrichs, *Phys. Rev. A* **5**, 605 (1972).
10. A. Dalgarno and A. L. Stewart, *Proc. Phys. Soc. London* **77**, 467 (1961).
11. S. Prager and J. O. Hirschfelder, *J. Chem. Phys.* **39**, 3289 (1963).
12. P. Lindner and P.-O. Löwdin, *Int. J. Quantum Chem. Symp.* **2**, 161 (1968).
13. R. Goldstein, *J. Math. Phys.* (*N.Y.*) **8**, 473 (1967).
14. S. T. Epstein, *J. Chem. Phys.* **42**, 2897 (1965), and references therein.
15. A. F. Starace, *Phys. Rev. A* **3**, 1242 (1971).
16. J. V. Antanasoff, *Phys. Rev.* **36**, 1232 (1930).
17. R. A. Buckingham, *Proc. Roy. Soc. Ser. A* **160**, 94 (1937).
18. W. Kolos and L. Wolniewicz, *J. Chem. Phys.* **46**, 1426 (1967), and references therein.
19. T. J. Dougherty, T. Vladimiroff, and S. T. Epstein, *J. Chem. Phys.* **45**, 1803 (1966).
20. J. H. Epstein and S. T. Epstein, *J. Chem. Phys.* **42**, 3630 (1965).
21. A. J. Sadlej, *Chem. Phys. Lett.* **19**, 604 (1973), and references there to earlier work.
22. T. R. Singh and W. J. Meath, *J. Chem. Phys.* **54**, 1137 (1971).
23. J. C. Y. Chen and A. Dalgarno, *Proc. Phys. Soc. London* **85**, 399 (1965).
24. C. W. Scherr, F. C. Sanders, and R. E. Knight, *in* "Perturbation Theory and Its Applications in Quantum Mechanics" (C. H. Wilcox, ed.), Wiley, New York, 1966.
25. A. Dalgarno and S. T. Epstein, *J. Chem. Phys.* **50**, 2837 (1969).
26. O. Sinanoglu, *Phys. Rev.* **122**, 491 (1961); M. N. Adamov and V. A. Zubakov, *Sov. Phys. JETP* **13**, 169 (1963).
27. W. H. Miller, *J. Chem. Phys.* **44**, 2198 (1966); C. S. Sharma and I. Rebello, *Proc. Phys. Soc. London* (*Gen.*) **6**, 459 (1973), and references there to earlier work.
28. L. Rosenberg, L. Spruch, and T. F. O'Malley, *Phys. Rev.* **118**, 184 (1960); R. Sugar and R. Blankenbecler, *Phys. Rev. B* **136**, 472 (1964); P.-O. Löwdin, *Int. J. Quantum Chem. Symp.* **4**, 231 (1971).
29. F. Weinhold, *Advan. Quantum Chem.* **6**, 299 (1972).
30. P. S. C. Wang, *Int. J. Quantum Chem.* **3**, 57 (1969); *J. Chem. Phys.* **51**, 4767 (1969).
31. W. H. de Jeu, *Mol. Phys.* **20**, 573 (1971).
32. H. Wrobel and J. Voitlander, *Mol. Phys.* **25**, 323 (1973), and references therein.
33. R. N. Mandan, R. W. Haymaker, and R. Blankenbecler, *Phys. Rev.* **172**, 1788 (1968), and references therein.
34. L. Hambro, *Phys. Rev. A* [3], **5**, 2027 (1972).
35. S. T. Epstein, *J. Chem. Phys.* **48**, 1404 (1968).

Chapter VIII / Special Theorems Satisfied by Optimal First-Order Trial Functions

35. Derivation of Theorems

At the end of Section 26, we raised the question of whether or not when using the variation method within the variation method (VVM), the $\check{\psi}^{(1)}$ would satisfy theorems—hypervirial theorems, Hellmann–Feynman theorems, as the case may be—similar to those satisfied by $\hat{\psi}^{(1)}$. Similarly, when using the Hylleraas variation method (HVM), one may ask whether or not $\check{\psi}^{(1)}$ satisfies theorems analogous to those satisfied by $\psi'^{(1)}$. In this Section, we will derive sufficient conditions that the answers to these questions be yes. We will do this in the context of a single perturbation; however, the generalization to several perturbations should be obvious.

In VVM (HVM being a special case with $\hat{\psi}^{(0)} = \psi'^{(0)}$, $\hat{E}^{(0)} = E'^{(0)}$, $\hat{E}^{(1)} = E'^{(1)}$), we know from the discussion following (25-22) that, through second order in ν in the energy, what we are in effect doing is using the set of trial functions

$$\psi = \psi(\hat{A}^{(0)} + \nu A^{(1)}, \nu) \tag{1}$$

in the variation method. Here the set of $\hat{A}^{(0)} + \nu A^{(1)}$'s is a subset of the possible A's used in the variation method to determine $\hat{\psi}$, so that in turn, the set of trial functions (1) is a subset of the set of trial functions used there (and in what follows, we will consistently refer to that larger set of trial functions as *the* set of trial functions). Thus in VVM, one uses, for example, a subset of all Slater determinants in the case of UHF, or a subset of the functions in a linear space in the case of the linear variation method, etc.

For all the theorems which we have discussed in the context of *the* variation method (Hellmann–Feynman, hypervirial, and unitary invariance in particular), we gave as a sufficient condition for $\hat{\psi}$ to satisfy the theorem, that *the* set of trial functions be invariant to certain transformations, and this is the sort of condition that we now want to derive for VVM.

If *the* set of trial functions is invariant to a transformation τ, then this implies that under the transformation

$$\psi(\hat{A}^{(0)} + \nu A^{(1)}, \nu) \rightarrow \psi(\hat{A}^{(0)} + \nu A^{(1)} + \bar{A}_\tau, \nu) \tag{2}$$

where, since the transformations which we will consider are analytic in ν, we will assume, consistent with our earlier assumptions of this sort, that $\bar{A}_\tau$, which in general will depend on $\hat{A}^{(0)}$, $A^{(1)}$, ν, and, as indicated by the subscript, the particular transformation in question, can be expanded in powers of ν, thus

$$\bar{A}_\tau = \bar{A}_\tau^{(0)} + \nu \bar{A}_\tau^{(1)} + \cdots \tag{3}$$

In particular, if the transformation is independent of ν, then clearly $\bar{A}_\tau^{(0)}$ will be the change in $\hat{A}^{(0)}$, and $\bar{A}_\tau^{(1)}$ the change in $A^{(1)}$ produced by the transformation, or in symbols,

$$\bar{A}_\tau^{(0)} = \delta \hat{A}^{(0)}, \qquad \bar{A}_\tau^{(1)} = \delta A^{(1)} \tag{4}$$

On the other hand, if the transformation depends on ν and becomes the identity transformation for $\nu = 0$, then unless the transformation affects ν, a case which we will discuss separately later, one will have

$$\bar{A}_\tau^{(0)} = 0, \qquad \nu \bar{A}_\tau^{(1)} = \delta \hat{A}^{(0)} \tag{5}$$

and therefore in particular $\nu \bar{A}_\tau^{(1)}$ will be a definite known quantity independent of $A^{(1)}$.

In the case of unitary invariance, the basic observation in Section 13 was that the set of trial functions ψ used in the variation method for H would yield the same optimal energies as the set $U^+\psi$ in the variation method for

$$H^U = U^+HU \tag{6}$$

and that therefore if one uses the same set of trial functions with H^U as with H, and if that set is invariant to U, then one is guaranteed (unconstrained) unitary invariance.

Now in the present context, we will obviously *not* want to use the same set of trial functions for H and for H^U (unless $U = 1$ for $\nu = 0$) since for H^U one will want that for $\nu = 0$, $\hat{A}$ should reduce to $\hat{A}^{(0)U}$, where

$$\hat{\psi}^{(0)U} = \psi(\hat{A}^{(0)U}) = U^{(0)+}\psi(\hat{A}^{(0)}) \tag{7}$$

Therefore the question is rather: If one uses the set of trial functions (1) for H and the set of trial functions

$$\psi^U = \psi(\hat{A}^{(0)U} + \nu A^{(1)}, \nu) \tag{8}$$

(same $A^{(1)}$) for H^U, and if *the* set of trial functions is invariant to U, what conditions on the set $A^{(1)}$ are sufficient to guarantee invariance of the energy through second order?

To find the answer, we first note that if *the* set is invariant to U, then, in accord with (2), it must be that

$$U^+\psi(\hat{A}^{(0)} + \nu A^{(1)}, \nu) = \psi(\hat{A}^{(0)} + \nu A^{(1)} + \bar{A}_U, \nu) \tag{9}$$

for some $\bar{A}_U$. Then since obviously $\hat{A}^{(0)U}$ must be given by

$$\hat{A}^{(0)U} = \hat{A}^{(0)} + \bar{A}_U^{(0)} \tag{10}$$

it follows that we can write

$$U^+\psi(\hat{A}^{(0)} + \nu A^{(1)}, \nu) = \psi(\hat{A}^{(0)U} + \nu(A^{(1)} + \bar{A}_U^{(1)}) + \cdots, \nu) \tag{11}$$

Now it is very important to note that the nature of the dots in (11), which represent terms of order ν^2 and higher, is completely irrelevant to the determination of optimal first-order variational parameters $\check{A}^{(1)}$, and consequently to the determination of $\check{E}^{(2)}$, since we have seen in Section 25 that we can add $(\partial\hat{\varphi}^{(0)}/\partial\hat{A}^{(0)})A^{(2)}$ terms to $\tilde{\varphi}^{(2)}$ [and by the same argument can add $(\partial\hat{\varphi}^{(0)U}/\partial\hat{A}^{(0)U})A^{(2)}$ terms to $\tilde{\varphi}^{(2)U}$] without changing anything because these terms and their variations have the form of $\delta^0\hat{\varphi}$ [$\delta^0\hat{\varphi}^U$] and so contribute nothing to either $J^{(2)}$ or $\delta J^{(2)}$ [$J^{(2)U}$ or $\delta J^{(2)U}$]. Therefore effectively we can replace (9) by

$$U^+\psi(\hat{A}^{(0)} + \nu A^{(1)}, \nu) = \psi(\hat{A}^{(0)U} + \nu(A^{(1)} + \bar{A}_U^{(1)}), \nu) \tag{12}$$

whence it is clear that we will have invariance of the energy through second order and covariance of the wave function through first order, that is,

$$\begin{gathered} \hat{E}^{(0)} = \hat{E}^{(0)U}, \qquad \hat{E}^{(1)} = \hat{E}^{(1)U}, \qquad \check{E}^{(2)} = \check{E}^{(2)U}, \\ \hat{\varphi}^{(0)U} + \nu\check{\varphi}^{(1)U} = U^+(\hat{\varphi}^{(0)} + \nu\check{\varphi}^{(1)}) + \cdots \end{gathered} \tag{13}$$

if *the* set of trial functions is invariant to U, and if the set $A^{(1)}$ is invariant to addition of $\bar{A}_U^{(1)}$. Namely, from (8) and (12) we will have, as desired, that the set $U^+\psi$ is identical to the set ψ^U to the necessary order.

To be more specific, suppose that U is independent of ν, as for rotations and translations, and consider, for example, HVM. Then *the* set, which is the set of all functions, is invariant to any U. Also since $\psi = A$, we have that

$$\hat{A}^{(0)} + \nu A^{(1)} + \bar{A}_U = U^+\psi = U^+\psi'^{(0)} + \nu U^+\psi^{(1)} + \cdots \tag{14}$$

Therefore it follows that using the same set of $\psi^{(1)}$'s for H and for H^U, the energy will be invariant through second order in ν and the wave function will be covariant through first order in ν if the set of $\psi^{(1)}$ is invariant to U^+. As another example, consider VVM for UHF where *the* set is invariant to U's of the form

$$U = \prod_{s=1}^{N} u(s) \tag{15}$$

Then since

$$U^+ \mid \varphi_1, \varphi_2, \ldots, \varphi_N \mid = \mid u^+\varphi_1, u^+\varphi_2, \ldots, u^+\varphi_N \mid \tag{16}$$

it follows immediately that one will have unitary invariance in this VVM through second order in the energy, etc., if for each k the set of $\varphi_k^{(1)}$ is invariant to u^+.

An interesting example of a unitary transformation which depends on ν is provided by gauge transformations when the perturbation is the magnetic interaction. In this case, one can most simply identify ν with $1/c$ and U has the form

$$U = e^{i\nu\Lambda} = 1 + i\nu\Lambda + \cdots \tag{17}$$

where

$$\Lambda = \sum_{i=1}^{N} \lambda(\mathbf{r}_s)$$

Then, from (5), $\bar{A}_U^{(0)} = 0$ and $\bar{A}_U^{(1)}$ will be a definite known quantity so that, for example, use of trial parameters of the form

$$A^{(1)} = a\bar{A}_U^{(1)} + X \tag{18}$$

in both calculations, where a is an arbitrary number and where X does not involve a explicitly, will assure invariance since such a set is obviously invariant to $A^{(1)} \to A^{(1)} + \bar{A}_U^{(1)}$. In particular, in HVM,

$$U^+\psi = \psi'^{(0)} + \nu(\psi^{(1)} - i\Lambda\psi'^{(0)}) \tag{19}$$

and so in HVM we will have invariance to the gauge transformation Λ if the set of $\psi^{(1)}$ is invariant to the addition of $-i\Lambda\psi'^{(0)}$, something which is easily arranged by using trial functions of the form

$$\psi^{(1)} = ia\Lambda\psi'^{(0)} + X \tag{20}$$

where a is an arbitrary number, etc. Similarly in VVM for UHF one will have invariance if for each k, the set of $\varphi_k^{(1)}$ is invariant to addition of $-i\lambda\hat{\varphi}_k^{(0)}$, which in turn can be ensured by use of

$$\varphi_k^{(1)} = ia_k\hat{\varphi}^{(0)} + \chi_k \tag{21}$$

Of course in general one can also have constrained invariance simply by using

$$A^{(1)U} = \bar{A}_U^{(1)} + A^{(1)} \tag{22}$$

for H^U if one has used $A^{(1)}$ for H.

As we have just shown, use of (18) for both H and H^U guarantees that one will get the same energy through second order in both calculations. However, suppose that U is a member of a family of gauge transformations parameterized in some nonlinear way by a numerical parameter ξ. Then to get the same results for H and $H^{U(\xi_1)}$, one would use $a\bar{A}_{U(\xi_1)}^{(1)} + X$, say, while to get the same results for H and $H^{U(\xi_2)}$, one would use $a\bar{A}_{U(\xi_2)}^{(1)} + X$, and the point is that even with the same X, these will in general yield different answers for $\check{E}^{(2)}$ (and therefore also a different answer for $\check{E}^{(2)U(\xi_1)}$ than for $\check{E}^{(2)U(\xi_2)}$).

These considerations then obviously raise the question: Which is the best set to use as far as H is concerned? The obvious answer is, let the variation method decide. That is, one should let ξ be an additional variation parameter in the calculation for H. Indeed this is theoretically a particularly well-founded suggestion in those cases in which $\check{E}^{(2)}$ is a guaranteed and improvable upper bound to $\hat{E}^{(2)}$ since then introducing ξ as a variational parameter can only improve $\check{E}^{(2)}$. This last statement must of course then be qualified by the usual caveat that unless one knows that $\hat{E}^{(2)}$ is also an upper bound to $E'^{(2)}$, it may not be that it is the best approximation to $E'^{(2)}$ even though the result will be the closest to $\hat{E}^{(2)}$ among the $\check{E}^{(2)}(\xi)$.

The situation is much more straightforward when the Λ is linear in the gauge parameters, as in (13-23),

$$\Lambda = \sum_t \sum_{s=1}^{N} \alpha_t \bar{\lambda}_t(\mathbf{r}_s) \tag{23}$$

where the real numbers α_t are the gauge parameters and the $\bar{\lambda}_t$ are given functions, since then $\bar{A}_U^{(1)}$ will obviously have the form

$$\bar{A}_U^{(1)} = \sum_t \alpha_t \delta_t \tag{24}$$

where the δ_t are independent of the α_t and therefore use of

$$A^{(1)} = \sum_t a_t \delta_t + X \tag{25}$$

will then suffice to yield the same results for all α_t.

Turning now to hypervirial theorems, we showed in Section 16 that if *the* set is invariant to the family of unitary transformations $e^{i\alpha\mathscr{G}}$, where $\mathscr{G}$ is a Hermitian operator and where α is an arbitrary real number, then the hypervirial theorem for $\mathscr{G}$ would be satisfied. However, even assuming that *the* set is invariant to $e^{i\alpha\mathscr{G}}$, it is obvious that in general the set (1) will not be invariant since almost certainly $\bar{A}_{\mathscr{G}}^{(0)} \neq 0$. Thus there would seem to be a difficulty in trying to follow the pattern of Section 16. To overcome it, we now note that if *the* set is invariant to $e^{i\alpha\mathscr{G}}$, then it is also invariant to $e^{i\nu\alpha\mathscr{G}}$. Further, we note that this replacement of $\mathscr{G}$ by $\nu\mathscr{G}$ would change nothing in Section 16 except that if we were to make a perturbation expansion of (16-3) before canceling the factor of ν which would now appear there, the zeroth-order hypervirial theorem then would appear in first order, the first-order hypervirial theorem in second order, etc. Therefore we conclude that if *the* set of trial functions is invariant to $e^{i\alpha\mathscr{G}}$ and if the set (1) is essentially invariant to $e^{i\nu\alpha\mathscr{G}}$ through second order in ν, which, as we have discussed, requires only that $\hat{A}^{(0)} + \nu A^{(1)}$ should be invariant through first order, then the hypervirial theorem for $\mathscr{G}$ will be satisfied through first order. That is, we will have

$$(\hat{\psi}^{(0)}, [H^{(0)}, \mathscr{G}^{(0)}]\hat{\psi}^{(0)}) = 0 \tag{26}$$

and

$$(\hat{\psi}^{(0)}, [H^{(0)}, \mathscr{G}^{(1)}]\hat{\psi}^{(0)}) + (\hat{\psi}^{(0)}, [H^{(1)}, \mathscr{G}^{(0)}]\hat{\psi}^{(0)}) + (\check{\psi}^{(1)}, [H^{(0)}, \mathscr{G}^{(0)}]\hat{\psi}^{(0)}) + (\hat{\psi}^{(0)}, [H^{(0)}, \mathscr{G}^{(0)}]\check{\psi}^{(1)}) = 0 \tag{27}$$

and where often the first term on the left in (27) will vanish separately.

For the transformation $e^{i\nu\alpha\mathscr{G}}$, Eq. (5) applies and so

$$\hat{A}^{(0)} + \nu A^{(1)} + \bar{A}_{\mathscr{G}} = \hat{A}^{(0)} + \nu(A^{(1)} + \bar{A}_{\mathscr{G}}^{(1)}) + \cdots \tag{28}$$

where $\bar{A}_{\mathscr{G}}^{(1)}$ is a known quantity depending on $\hat{A}^{(0)}$ and $\mathscr{G}$. Therefore it is easy to arrange that the set $A^{(1)}$ be invariant to addition of $\bar{A}_{\mathscr{G}}^{(1)}$, and indeed, for example, one can do this for any number of $\mathscr{G}$'s just by use of $A^{(1)}$'s of the form

$$A^{(1)} = \sum_{\mathscr{G}} a_{\mathscr{G}} \bar{A}_{\mathscr{G}}^{(1)} + X \tag{29}$$

where the $a_{\mathscr{G}}$ are arbitrary numbers, etc. In particular, comparing (29) with (25), we see now that use of (25) will also ensure that the theorems (16-10) for the $(\partial\lambda/\partial\alpha_t)_{\alpha_t=0} = \bar{\lambda}_t$ are satisfied through first order.

For HVM, *the* set is invariant for any choice of $\mathscr{G}$ and

$$e^{i\nu\alpha\mathscr{G}}\psi = \psi'^{(0)} + \nu(\psi^{(1)} + i\alpha\mathscr{G}^{(0)}\psi'^{(0)}) + \cdots \tag{30}$$

so that trial functions of the form

$$\psi^{(1)} = \sum_{\mathscr{G}} ia_{\mathscr{G}}\mathscr{G}^{(0)}\psi'^{(0)} + X \tag{31}$$

will do the job. Similarly, for the VVM of UHF, where *the* set is invariant to any one particle $\mathscr{G}$,

$$\mathscr{G} = \sum_{s=1}^{N} g(s)$$

use of

$$\varphi_k^{(1)} = \sum_{g} ia_{kg}g^{(0)}\hat{\varphi}_k^{(0)} + \chi_k \tag{32}$$

will suffice. Also note that if, as implied, there are no reality conditions on the $a_{\mathscr{G}}$ or a_{kg}, then (16-7) will also be satisfied through first order.

Finally, let us consider the generalized Hellmann–Feynman theorem for a real parameter σ. We showed in Section 15 that a sufficient condition for it to be satisfied within the variation method is for the set of trial functions to be invariant to changes in σ. However, the situation within VVM is a little more complicated because three distinct kinds of situation can occur. Most simply, suppose first that $\partial H^{(0)}/\partial\sigma \neq 0$. Then from the fact that if *the* set is invariant to $\sigma \to \sigma + \delta\sigma$, it must also be invariant to $\sigma \to \sigma + \nu\,\delta\sigma$, it follows in the by now familiar way that if the set of $A^{(1)}$'s is invariant to the addition of the known quantity $\bar{A}_\sigma^{(1)}$, then the generalized Hellmann–Feynman theorem for σ will be satisfied through first order, that is, we will have

$$\left(\hat{\psi}^{(0)}, \left(\frac{\partial\hat{E}^{(0)}}{\partial\sigma} - \frac{\partial H^{(0)}}{\partial\sigma}\right)\hat{\psi}^{(0)}\right) = 0 \tag{33}$$

and

$$\left(\hat{\psi}^{(0)}, \left(\frac{\partial\hat{E}^{(1)}}{\partial\sigma} - \frac{\partial H^{(1)}}{\partial\sigma}\right)\hat{\psi}^{(0)}\right) + \left(\hat{\psi}^{(0)}, \left(\frac{\partial\hat{E}^{(0)}}{\partial\sigma} - \frac{\partial H^{(0)}}{\partial\sigma}\right)\check{\psi}^{(1)}\right)$$
$$+ \left(\check{\psi}^{(1)}, \left(\frac{\partial\hat{E}^{(0)}}{\partial\sigma} - \frac{\partial H^{(0)}}{\partial\sigma}\right)\hat{\psi}^{(0)}\right) = 0 \tag{34}$$

Further, since $\bar{A}_\sigma^{(1)}$ is a known quantity, use of

$$A^{(1)} = a\bar{A}_\sigma^{(1)} + X \tag{35}$$

with a an arbitrary number, etc., will ensure the theorem.

In particular, for HVM,

$$\psi(\sigma + \delta\sigma) = \psi'^{(0)} + \nu\left(\psi^{(1)} + \frac{\partial\psi'^{(0)}}{\partial\sigma}\,\delta\sigma\right) + \cdots \tag{36}$$

so one might use trial functions of the form

$$\psi^{(1)} = a(\partial\psi'^{(0)}/\partial\sigma) + X \tag{37}$$

where we have absorbed the $\delta\sigma$ into the a. Similarly, in VVM for UHF, one can use

$$\varphi_k^{(1)} = a_k(\partial\hat{\varphi}_k^{(0)}/\partial\sigma) + \chi_k \tag{38}$$

The preceding applies to situations with $\partial H^{(0)}/\partial\sigma \neq 0$. However, an obvious and important example of a σ for which $\partial H^{(0)}/\partial\sigma = 0$ is $\sigma = \nu$. Let us therefore consider the effect of $\nu \to \nu + \nu\,\delta\nu$. If *the* set is invariant to this transformation, then it must be that

$$\psi(\hat{A}^{(0)} + (\nu + \nu\,\delta\nu)A^{(1)}, \nu + \nu\,\delta\nu) = \psi(\hat{A}^{(2)} + \nu A^{(1)} + \bar{A}_\nu, \nu) \tag{39}$$

where $\bar{A}_\nu$ has the form

$$\bar{A}_\nu = \nu\bar{A}_\nu^{(1)} + \cdots$$

but where now, however, we cannot assume that $\bar{A}_\nu^{(1)}$ is a known quantity since it involves a contribution from $\nu\,\delta\nu A^{(1)}$. In fact, the latter is obviously the only contribution in cases where *the* individual trial functions have no explicit ν dependence so that $\psi = \psi(A)$ only. Therefore if *the* set of trial functions is invariant to changes in σ and if the set $A^{(1)}$ is invariant to addition of $\bar{A}_\nu^{(1)}$, then the Hellmann–Feynman theorem for $\sigma = \nu$ will be satisfied through first order in ν, the consequences being, of course, just (24-36) and (24-38), the latter with appropriate changes in notation. Namely if we write

$$\check{\psi} = \hat{\psi}^{(0)} + \nu\check{\psi}^{(1)} + \cdots \tag{40}$$

$$\check{E} = \hat{E}^{(0)} + \nu\hat{E}^{(1)} + \nu^2\check{E}^{(2)} + \cdots \tag{41}$$

and expand

$$\left(\check{\psi}, \left(\frac{\partial\check{E}}{\partial\nu} - \frac{\partial H}{\partial\nu}\right)\check{\psi}\right) = 0 \tag{42}$$

in powers of ν, then through first order, we just find

$$(\hat{\psi}^{(0)}, (\hat{E}^{(1)} - H^{(1)})\hat{\psi}^{(0)}) = 0 \tag{43}$$

and

$$2(\hat{\psi}^{(0)}, (\check{E}^{(2)} - H^{(2)})\hat{\psi}^{(0)}) + (\hat{\psi}^{(0)}, (\hat{E}^{(1)} - H^{(1)})\check{\psi}^{(1)}) + (\check{\psi}^{(1)}, (\hat{E}^{(1)} - H^{(1)})\hat{\psi}^{(0)}) = 0 \tag{44}$$

In particular, for HVM,

$$\psi(\nu + \delta\nu) = \psi'^{(0)} + \nu(1 + \delta\nu)\psi'^{(1)} + \cdots \tag{45}$$

and so, in agreement with what we found in Section 31, (44), which is equivalent to (31-2), will be satisfied if the set $\psi^{(1)}$ is invariant to multiplication by a real number. More generally, and as is also true, for example, in UHF, whenever *the* individual trial functions have no explicit ν dependence, then, as we have mentioned,

$$\bar{A}_\nu^{(1)} = \delta\nu\, A^{(1)}$$

and so in all such cases the theorem will be satisfied if the set $A^{(1)}$ is invariant to multiplication by a real number.

To remind ourselves of the virtues of (43) and (44), let us recall that if the perturbation is a uniform electric (magnetic) field, then (42) and (43) will ensure that the permanent electric (magnetic) dipole moment and the electric (magnetic) dipole polarizibility calculated from $\check{E}$ will agree with those calculated from the average dipole moment.

The third case we need to consider is where again $H^{(0)}$ does not involve σ but where in addition $\partial\nu/\partial\sigma = 0$. This means that $\partial H/\partial\sigma$ is of order ν to start with and therefore invariance of ψ, as given in (1), to $\sigma \to \sigma + \delta\sigma$ through second order in ν, something which is now possible since $\hat{A}^{(0)}$ does not involve σ and therefore $\delta\hat{A}^{(0)} = 0$, will imply that the generalized Hellmann–Feynman theorem for σ will be satisfied through *second* order in ν. (If $H^{(0)}$ involves σ, then of course to satisfy the theorem through second order in ν requires some approximation to $\hat{A}^{(2)}$ [1].) In detail, this means that we will have

$$\left(\hat{\psi}^{(0)}, \left(\frac{\partial\hat{E}^{(1)}}{\partial\sigma} - \frac{\partial H^{(1)}}{\partial\sigma}\right)\hat{\psi}^{(0)}\right) = 0 \tag{46}$$

and

$$\left(\hat{\psi}^{(0)}, \left(\frac{\partial\check{E}^{(2)}}{\partial\sigma} - \frac{\partial H^{(2)}}{\partial\sigma}\right)\hat{\psi}^{(0)}\right) + \left(\check{\psi}^{(1)}, \left(\frac{\partial\hat{E}^{(1)}}{\partial\sigma} - \frac{\partial H^{(1)}}{\partial\sigma}\right)\hat{\psi}^{(0)}\right) + \left(\hat{\psi}^{(0)}, \left(\frac{\partial\hat{E}^{(1)}}{\partial\sigma} - \frac{\partial H^{(1)}}{\partial\sigma}\right)\psi^{(1)}\right) = 0 \tag{47}$$

Moreover, they, like (43) and (44), should be more or less familiar. Namely if we think of σ as a second perturbation parameter, then (46) and (47) expanded in powers of σ yield standard looking formulas for $\hat{E}^{(1n)}$ and $\check{E}^{(2n)}$ in terms of the $\hat{\psi}^{(0m)}$ and $\check{\psi}^{(1m)}$ [2].

Since $H^{(0)}$ does not involve σ and *the* set of trial functions is invariant to $\sigma \to \sigma + \delta\sigma$, then

$$\psi(\hat{A}^{(0)} + \nu A^{(1)}(\sigma + \delta\sigma), \sigma + \delta\sigma) = \psi(\hat{A}^{(0)} + \nu A^{(1)}(\sigma) + \bar{A}_\sigma, \sigma) \tag{48}$$

where we have included possible explicit σ dependences in ψ and in $A^{(1)}$ and where now since $\hat{A}^{(0)}$ will also not involve σ,

$$\bar{A}_\sigma = \nu \bar{A}_\sigma^{(1)} + \cdots \tag{49}$$

Therefore in cases where $\partial H^{(0)}/\partial\sigma = \partial\nu/\partial\sigma = 0$, if *the* set of trial functions is invariant to changes in σ, and if the set $A^{(1)}$ is invariant to addition of $\bar{A}_\sigma^{(1)}$, then the generalized Hellmann–Feynman theorem for σ will be satisfied through second order in ν. In particular if as in HVM or UHF the individual trial functions have no explicit σ dependence, then $\bar{A}_\sigma^{(1)}$ will be simply

$$\bar{A}_\sigma^{(1)} = \nu(A^{(1)}(\sigma + \delta\sigma) - A^{(1)}(\sigma)) \tag{50}$$

Thus in such cases, the theorem will be satisfied if the set $A^{(1)}$ is independent of σ. In HVM, it would be sufficient to use a set of $\psi^{(1)}$ which is independent of σ, and similarly in VVM for UHF, sets of $\varphi_k^{(1)}$ which are independent of σ will suffice.

An interesting example in which $\partial H^{(0)}/\partial\sigma = \partial\nu/\partial\sigma = 0$ occurs in the calculation of the long-range forces between two atoms [3]. In carrying out this calculation, one usually refers the coordinates of one set of electrons to nucleus 1, the remainder to nucleus 2, and treats the Coulomb interaction between the electrons on different centers as a perturbation. In this case, $H^{(0)}$ does not involve the nuclear separation R explicitly, while $H^{(1)}$ does, and so, if the requisite conditions are satisfied, the generalized Hellmann–Feynman theorem for $\sigma = R$ in these coordinates will be satisfied through second order in the Coulomb interaction. However, as we mentioned following Eq. (47), this then simply provides standard formulas with which to compute the leading terms in the $1/R$ expansion of the energy and is completely equivalent to expanding $H^{(1)}$ in powers of $1/R$ and doing perturbation theory directly in $1/R$ through second order in ν. On the other hand, if one wants the generalized Hellmann–Feynman theorem for $\sigma = R$ to be *the* Hellmann–Feynman theorem, then one must refer the coordinates of all the electrons to a common origin and now $H^{(0)}$ will involve R. Therefore to satisfy *the* Hellmann–Feynman theorem to second order (or, less ambitiously, to calculate an approximation to the Hellmann–Feynman force to second order) will not surprisingly also require some approximation to $\hat{A}^{(2)}$ [3].

As we pointed out at the time, the theorems of Section 21 are not usually satisfied in variational calculations (the linear variation method being an exception in some cases—for example, linear calculations always yield orthogonality theorems) and hence in particular are not usually satisfied in zero order. However, in the latter respect, HVM is obviously an exception because $\psi'^{(0)}$ satisfies all zero-order theorems. ($1/\mathscr{Z}$ expansions when $\hat{\psi}^{(0)} = \psi'^{(0)}$ offer other exceptions. For example, recall the orthogonality theorems of Chapter 29.) Therefore it becomes of interest to note that by proper choice of trial functions in HVM, one can also ensure that the theorems will hold through first order. Namely, introducing an extra label as in Section 21, if one uses trial functions of the form

$$\psi_a^{(1)} = A^a \mathscr{G}^{(0)} \psi_b'^{(0)} + X_a \tag{51}$$

$$\psi_b^{(1)} = A^b \mathscr{G}^{(0)} \psi_a'^{(0)} + X_b \tag{52}$$

with the usual proviso that X_a and X_b not involve A^a and A^b explicitly, then the theorem (21-3) will be satisfied through first order. To see this, we note that varying A^{a*} (and not A^a), $\delta \check{J}_{Ha}^{(2)} = 0$ yields

$$(\mathscr{G}^{(0)}\psi_b'^{(0)}, (H_a^{(0)} - E_a'^{(0)})\check{\psi}_a^{(1)}) + (\mathscr{G}^{(0)}\psi_b'^{(0)}, (H_a^{(1)} - E_a'^{(1)})\psi_a'^{(0)}) = 0 \tag{53}$$

while varying A^b (and not A^{b*}), $\delta \check{J}_{Hb}^{(2)} = 0$ yields

$$(\check{\psi}_b^{(1)}, (H_b^{(0)} - E_b'^{(0)})\mathscr{G}^{(0)}\psi_a'^{(0)}) + (\psi_b'^{(0)}, (H_b^{(1)} - E_b'^{(1)})\mathscr{G}^{(0)}\psi_a'^{(0)}) = 0 \tag{54}$$

We now write (53) as

$$\begin{aligned}(\psi_b'^{(0)}, \mathscr{G}^{(0)}H_a^{(0)}\check{\psi}_a^{(1)}) &+ (\psi_b'^{(0)}, \mathscr{G}^{(0)}H_a^{(1)}\psi_a'^{(0)}) \\ &= E_a'^{(0)}(\psi_b'^{(0)}, \mathscr{G}^{(0)}\check{\psi}_a^{(1)}) + E_a'^{(1)}(\psi_b'^{(0)}, \mathscr{G}^{(0)}\psi_a'^{(0)})\end{aligned} \tag{55}$$

If then we add to (55) the identities

$$(\check{\psi}_b^{(1)}, \mathscr{G}^{(0)}H_a^{(0)}\psi_a'^{(0)}) = E_a'^{(0)}(\check{\psi}_b^{(1)}, \mathscr{G}^{(0)}\psi_a'^{(0)}) \tag{56}$$

and

$$(\psi_b'^{(0)}, \mathscr{G}^{(1)}H_a^{(0)}\psi_a'^{(0)}) = E_a'^{(0)}(\psi_b'^{(0)}, \mathscr{G}^{(1)}\psi_a'^{(0)})$$

then we obviously have the first-order part of (21-1). Similarly, (54) yields the first-order part of (21-2) and therefore, as we claimed, (21-3) with appropriate changes in notation will be satisfied through first order.

To satisfy several theorems at once, one can of course simply use

$$\psi_a^{(1)} = \sum_{\mathscr{G}} A_{\mathscr{G}}^{a} \mathscr{G}^{(0)} \psi_b'^{(0)} + X_a \tag{57}$$

etc. In particular, we see that use of

$$\begin{aligned} \psi_a^{(1)} &= A^a \psi_b'^{(0)} + X_a \\ \psi_b^{(1)} &= A^b \psi_b'^{(0)} + X_b \end{aligned} \tag{58}$$

when $H_a = H_b = H$ will guarantee orthogonality of $\check{\psi}_a$ and $\check{\psi}_b$ through first order, that is,

$$(\psi_b'^{(0)}, \check{\psi}^{(1)}) + (\check{\psi}_b^{(1)}, \psi_a'^{(0)}) = 0 \tag{59}$$

since of course to begin with we have

$$(\psi_a'^{(0)}, \psi_b'^{(0)}) = 0 \tag{60}$$

(Earlier, in Section 33, we saw that the use of (58) would guarantee that $\check{\psi}_a$ and $\check{\psi}_b$ would also be orthogonal to ψ_b' and ψ_a', respectively, through first order.)

PROBLEMS

1. Prove directly that $E'^{(1)U} = E'^{(1)}$.

2. Prove directly that $E'^{(2)U} = E'^{(2)}$.

3. In this problem we consider connections between unitary invariance and hypervirial theorems when U is independent of ν. In the text we briefly discussed a case, that of gauge transformations, in which U depends on ν, becoming 1 when $\nu = 0$. Here we consider the opposite extreme in which $U(\beta) = \exp i\beta\mathscr{G}$ where β is a real number and $\mathscr{G} = \mathscr{G}^{(0)}$ thus, for example, rotations and translations. Show that unitary invariance plus the vanishing of $(\check{\psi}, [H, \mathscr{G}]\check{\psi})$ through first order implies the vanishing of $(\check{\psi}^U, [H^U, \mathscr{G}]\check{\psi}^U)$ through first order. Using HVM as an example, show that if in accord with (31) one would use the *different* sets $\psi^{(1)} = ia\mathscr{G}\psi'^{(0)} + X$ and $\psi^{(1)U} = iaU^{+}(\beta)\psi'^{(0)} + X$

(same set X) in order to ensure the hypervirial theorems, that one would also have what one might call partially constrained unitary invariance if the set X is invariant to $U^{+}(\beta)$. Show that formally use of the single set

$$\psi^{(1)} = \int_{-\infty}^{\infty} ia(\gamma)U^{+}(\gamma)\psi'^{(0)}\, d\gamma + X$$

will ensure both of the hypervirial theorems and also (unconstrained) unitary invariance if the set X is invariant to $U^{+}(\beta)$.

References

1. L. Salem and E. B. Wilson, Jr., *J. Chem. Phys.* **36**, 3421 (1962).
2. R. Yaris, *J. Chem. Phys.* **39**, 863 (1963).
3. J. O. Hirschfelder and M. A. Eliason, *J. Chem. Phys.* **47**, 1164 (1967).

Chapter IX / Corrections to Approximate Calculations

36. Mostly First-Order Corrections

Returning to a theme, aspects of which we have already discussed more than once, let us suppose that from somewhere, though *not* from a consistent variation calculation, we have produced an approximation χ

$$\chi = \chi^{(0)} + \nu\chi^{(1)} + \cdots \tag{1}$$

to an eigenfunction of H

$$H = H^{(0)} + \nu H^{(1)} + \nu^2 H^{(2)} + \cdots \tag{2}$$

(We will consider only a single perturbation. However, the discussion is readily generalized.) We now wish to correct this approximation in some systematic way. As far as the energy is concerned, it is natural as a first step simply to calculate

$$(\chi, H\chi)/(\chi, \chi) \equiv \Xi \tag{3}$$

since from the variation principle, this quantity will differ from an eigenvalue by second-order terms and therefore would appear to automatically include first-order corrections.

Indeed this procedure, in one form or another, is the one which has usually been used to calculate first-order corrections.[1] However, it should be kept in mind that the statement that Ξ involves only a second-order error can be misleading. First of all, χ is generally in error because $\chi^{(0)}$ is in error. Let $\gamma^{(0)}$ denote some measure of the latter error. Further, $\chi^{(1)}$ will also be in error by $\gamma^{(1)}$ say; $\chi^{(2)}$ by $\gamma^{(2)}$; etc.; so that all together,

$$\chi - \psi' = O(\gamma^{(0)}, \nu\gamma^{(1)}, \nu^2\gamma^{(2)}, \ldots)$$

[1] For example, Tuan *et al.* [1], Epstein and Johnson [2], Tuan and Wu [3], Riemenschneider *et al.* [4], Riemenschneider and Kestner [5], and Sadlej and Jaszuński [6].

whence

$$\Xi - E' = O(\gamma^{(0)2}, \nu\gamma^{(0)}\gamma^{(1)}, \nu^2\gamma^{(0)}\gamma^{(2)}, \nu^2\gamma^{(1)2}, \ldots) \tag{4}$$

so that, except when $\nu = 0$, the error cannot in general be said to be of second order. To ensure a true second-order error, what one needs is that χ be exact when $\gamma^{(0)} = 0$, since this will imply that $\gamma^{(1)}$, $\gamma^{(2)}$, ... are all of order $\gamma^{(0)}$,

$$\gamma^{(n)} \sim \gamma^{(0)} \tag{5}$$

whence (4) will become

$$\Xi - E' = O(\gamma^{(0)2}, \nu\gamma^{(0)2}, \ldots) \tag{6}$$

If we expand (3) in powers of ν, then we find

$$\Xi = \Xi^{(0)} + \nu\Xi^{(1)} + \nu^2\Xi^{(2)} + \cdots \tag{7}$$

where

$$(\chi^{(0)}, (\Xi^{(0)} - H^{(0)})\chi^{(0)}) = 0 \tag{8}$$

$$\begin{aligned}(\chi^{(0)}, (H^{(1)} - \Xi^{(1)})\chi^{(0)}) &+ (\chi^{(1)}, (H^{(0)} - \Xi^{(0)})\chi^{(0)}) \\ &+ (\chi^{(0)}, (H^{(0)} - \Xi^{(0)})\chi^{(1)}) = 0\end{aligned} \tag{9}$$

$$\begin{aligned}(\chi^{(0)}, (H^{(2)} - \Xi^{(2)})\chi^{(0)}) &+ (\chi^{(1)}, (H^{(1)} - \Xi^{(1)})\chi^{(0)}) \\ &+ (\chi^{(0)}, (H^{(1)} - \Xi^{(1)})\chi^{(1)}) + (\chi^{(2)}, (H^{(0)} - \Xi^{(0)})\chi^{(0)}) \\ &+ (\chi^{(1)}, (H^{(0)} - \Xi^{(0)})\chi^{(1)}) + (\chi^{(0)}, (H^{(0)} - \Xi^{(0)})\chi^{(2)}) = 0\end{aligned} \tag{10}$$

etc.

In particular, if we write (9) as

$$\begin{aligned}\Xi^{(1)} = &\frac{(\chi^{(0)}, H^{(1)}\chi^{(0)})}{(\chi^{(0)}, \chi^{(0)})} \\ &+ \frac{(\chi^{(1)}, (H^{(0)} - \Xi^{(0)})\chi^{(0)}) + (\chi^{(0)}, (H^{(0)} - \Xi^{(0)})\chi^{(1)})}{(\chi^{(0)}, \chi^{(0)})}\end{aligned} \tag{11}$$

then evidently the second term on the right provides a correction to the zeroth-order approximation to the average value of $H^{(1)}$ provided by $(\chi^{(0)}, H^{(1)}\chi^{(0)})/(\chi^{(0)}, \chi^{(1)})$. Here of course $H^{(1)}$ could represent the effect of an external field, the correction then being a correction to the approximate value of a permanent moment, say, but also $H^{(1)}$ could be a "fictitious perturbation" like W of Eq. (26-41) et seq. We will discuss the formula for $\Xi^{(2)}$ later.

Thus far, we have not specified where $\chi^{(0)}$ and $\chi^{(1)}$ have come from. Usually $\chi^{(0)}$ will be the result of some variational calculation. As for $\chi^{(1)}$, $\chi^{(2)}, \ldots$, they are often derived from a model problem as described in Section 31. Namely given a model Hamiltonian $H^{(00)}$ such that

$$(H^{(00)} - \Xi^{(0)})\chi^{(0)} = 0 \tag{12}$$

$\chi^{(1)}$ and $\chi^{(2)}$ are then to be determined by solving (hopefully exactly, but perhaps only approximately, usually by variational techniques)

$$(H^{(00)} - \Xi^{(0)})\chi^{(1)} + (H^{(1)} - E'^{(10)})\chi^{(0)} = 0 \tag{13}$$

and

$$(H^{(00)} - \Xi^{(0)})\chi^{(2)} + (H^{(1)} - E'^{(10)})\chi^{(1)} + (H^{(2)} - E'^{(20)})\chi^{(0)} = 0 \tag{14}$$

where

$$E'^{(10)} = (\chi^{(0)}, H^{(1)}\chi^{(0)})/(\chi^{(0)}, \chi^{(0)}) \tag{15}$$

and

$$E'^{(20)} = [(\chi^{(0)}, (H^{(1)} - E'^{(10)})\chi^{(1)}) + (\chi^{(0)}, H^{(2)}\chi^{(0)})]/(\chi^{(0)}, \chi^{(0)}) \tag{16}$$

etc. Moreover, if one does this (exactly), then the error in Ξ *will* be of second order in $\gamma^{(0)}$ since obviously if $H^{(00)}$ were $H^{(0)}$, that is, if $\gamma^{(0)}$ were equal to zero, then $\chi^{(1)}$, $\chi^{(2)}$, etc., would be exact. However, this argument is, as it stands, only a formal one since, as we have mentioned before, given $\chi^{(0)}$, there are infinitely many $H^{(00)}$'s which satisfy (10) and the size of the correction terms in (11), for example, will in general depend on the choice of $H^{(00)}$.

Ignoring the latter point for the moment, if we use this approach, then, recalling the results of Section 26, evidently what one is doing is using straightforward Rayleigh–Schrödinger perturbation theory in interchanged form, with $H^{(00)}$ as the zero-order Hamiltonian, to calculate the first order in $(H^{(0)} - H^{(00)})$ corrections. However, it is perhaps useful to see this in detail. Considering $(H^{(0)} - H^{(00)})$ as the second perturbation with ν_2 simply an order parameter whose numerical value is one, we have from (26-49) with suitable changes in notation

$$E'^{(11)}(\chi^{(0)}, \chi^{(0)}) = (\chi^{(1)}, (H^{(0)} - H^{(00)})\chi^{(0)}) + (\chi^{(0)}, (H^{(0)} - H^{(00)})\chi^{(1)}) \tag{17}$$

where we have used the fact that in this case

$$E'^{(01)} = (\chi^{(0)}, (H^{(0)} - H^{(00)})\chi^{(0)})/(\chi^{(0)}, \chi^{(0)}) = 0 \tag{18}$$

Then using (12), we can write (17) as

$$E'^{(11)} = [(\chi^{(1)}, (H^{(0)} - \Xi^{(0)})\chi^{(0)}) + (\chi^{(0)}, (H^{(0)} - \Xi^{(0)})\chi^{(1)})]/(\chi^{(0)}, \chi^{(0)}) \quad (19)$$

which, as expected, agrees with the second term on the right in (11).

Similarly from (26-38) with the roles of ν_1 and ν_2 interchanged, we have

$$\begin{aligned} E'^{(21)}(\chi^{(0)}, \chi^{(0)}) = (\chi^{(2)}, (H^{(0)} - H^{(00)})\chi^{(0)}) &+ (\chi^{(1)}, (H^{(0)} - H^{(00)})\chi^{(1)}) \\ &+ (\chi^{(0)}, (H^{(0)} - H^{(00)})\chi^{(2)}) \\ &- E'^{(11)}[(\chi^{(1)}, \chi^{(0)}) + (\chi^{(0)}, \chi^{(1)})] \end{aligned} \quad (20)$$

which from (12) can be written as

$$\begin{aligned} E'^{(21)} = \{&(\chi^{(2)}, (H^{(0)} - \Xi^{(0)})\chi^{(0)}) + (\chi^{(1)}, (H^{(0)} - H^{(00)})\chi^{(1)}) \\ &+ (\chi^{(0)}, (H^{(0)} - \Xi^{(0)})\chi^{(2)}) \\ &- E'^{(11)}[(\chi^{(1)}, \chi^{(0)}) + (\chi^{(0)}, \chi^{(1)})]\}/(\chi^{(0)}, \chi^{(0)}) \end{aligned} \quad (21)$$

To compare with (10), we write the latter as

$$\begin{aligned} \Xi^{(2)} = &\left\{\frac{(\chi^{(0)}, H^{(2)}\chi^{(0)})}{(\chi^{(0)}, \chi^{(0)})} - \frac{(\chi^{(1)}, (H^{(00)} - \Xi^{(0)})\chi^{(1)})}{(\chi^{(0)}, \chi^{(0)})}\right\} \\ &+ \frac{\begin{aligned}(\chi^{(2)}, (H^{(0)} - \Xi^{(0)})\chi^{(0)}) + (\chi^{(1)}, (H^{(0)} - H^{(00)})\chi^{(1)}) \\ + (\chi^{(0)}, (H^{(0)} - \Xi^{(0)})\chi^{(2)})\end{aligned}}{(\chi^{(0)}, \chi^{(0)})} \\ &+ \frac{\begin{aligned}2(\chi^{(1)}, (H^{(00)} - \Xi^{(0)})\chi^{(1)}) + (\chi^{(1)}, (H^{(1)} - \Xi^{(1)})\chi^{(0)}) \\ + (\chi^{(0)}, (H^{(1)} - \Xi^{(1)})\chi^{(1)})\end{aligned}}{(\chi^{(0)}, \chi^{(0)})} \end{aligned} \quad (22)$$

the first line being the second-order energy of the model problem [recall (30-14) also note that it equals $E'^{(20)}$], and yielding, for example, an approximate polarizability, so that the second and third lines are therefore corrections to this zeroth-order result. However, from (15), (11), and (19)

$$E'^{(10)} = \Xi^{(1)} - E'^{(11)} \quad (23)$$

and therefore from (13)

$$\begin{aligned} 2(\chi^{(1)}, (H^{(00)} - \Xi^{(0)})\chi^{(1)}) &+ (\chi^{(1)}, [H^{(1)} - (\Xi^{(1)} - E'^{(11)})]\chi^{(0)}) \\ &+ (\chi^{(0)}, [H^{(1)} - (\Xi^{(1)} - E'^{(11)})]\chi^{(1)}) = 0 \end{aligned}$$

which, when used in the third line of (22), shows, as expected, that the correction is precisely $E'^{(21)}$. Further, having recognized what is involved in the use of (13) and (14), we also have an obvious prescription for calculating higher-order corrections, namely apply Rayleigh–Schrödinger perturbation theory, with the perturbations $\nu H^{(1)} + \nu^2 H^{(2)} + \cdots$ and $(H^{(0)} - H^{(00)})$, to higher order in $(H^{(0)} - H^{(00)})$.

If $\chi^{(0)}$ is a variational approximation with, therefore from (8), $\Xi^{(0)}$ the corresponding energy, then sometimes Brillouin's theorem can be helpful in simplifying (9) and (10) since the quantity $(H^{(0)} - \Xi^{(0)})\chi^{(0)}$ figures rather prominently in these formulas. As a prime example,[1] consider any one of the many versions of uncoupled Hartree–Fock theory [7] with the $H^{(1)}$, $H^{(2)}$, etc., one-electron operators, and with $\chi^{(0)}$ a UHF approximation to an eigenfunction of $H^{(0)}$. The simplest [8] of these uncoupled approximations, as we have mentioned before, and as is discussed in more detail in the problems, is a model theory in the sense of (12)–(14) with $(H^{(00)} - \Xi^{(0)}) = (F^{(0)} - \sum \varepsilon_i^{(0)})$. However, probably most of the others cannot be so categorized.

In any case, in these theories, χ is a single determinant, the spin orbitals being of the form

$$\varphi_i = \hat{\varphi}_i^{(0)} + \nu\varphi_i^{(1)} + \nu^2\varphi_i^{(2)} + \cdots \tag{24}$$

Therefore $\chi^{(1)}$ is a sum of a term proportional to $\chi^{(0)}$ and of one-electron excitations of $\chi^{(0)}$, while $\chi^{(2)}$ consists of zero, one, and two excitations of $\chi^{(0)}$, the single excitations involving the $\varphi_i^{(2)}$, and the double excitations involving two $\varphi_i^{(1)}$'s. If now we examine (7), then we see that since from Brillouin's theorem

$$(\chi^{(1)}, (H^{(0)} - \Xi^{(0)})\chi^{(0)}) = 0 \tag{25}$$

$\Xi^{(1)}$ will simply be given by

$$\Xi^{(1)} = (\chi^{(0)}, H^{(1)}\chi^{(0)})/(\chi^{(0)}, \chi^{(0)}) \tag{26}$$

that is, there are no first-order corrections. (For the simple uncoupled theory, this is of course precisely the same result we found in Section 10, namely that there are no first-order corrections to UHF expectation values of one-electron operators.)

Turning to $\chi^{(2)}$, we see similarly that the zero- and one-electron excitation parts of $\chi^{(2)}$ will make no contribution to $(\chi^{(2)}, (H^{(0)} - \Xi^{(0)})\chi^{(0)})$ and its complex conjugate, and so one has the simplification that one does not need to make any assumptions about, or more to the point, even calculate, the $\varphi_i^{(2)}$ in order to calculate $\Xi^{(2)}$ (of course in any case with χ a determinant, and whether or not $\chi^{(0)}$ is UHF, one could choose[2] to put $\varphi_i^{(2)} \equiv 0$, though physically there is every reason not to do so).

Further, $\Xi^{(2)}$ will be an upper bound to the $\hat{E}^{(2)}$ of UHF if $\hat{E}$ is a minimum. The point is simply that χ is a Slater determinant and therefore a member of *the* set of trial functions for UHF. Therefore since we have seen that Ξ and $\hat{E}$ first begin to differ in second order in ν, it follows that if $\hat{E}$ is a minimum, then

$$\Xi^{(2)} \geq \hat{E}^{(2)} \tag{27}$$

Moreover, it has been found [9–11] that the approximation to $\hat{E}^{(2)}$ (the latter often being quite a good approximation to $E'^{(2)}$ and indeed, empirically at least, often systematically an upper bound)[3] can be markedly improved by writing

$$\Xi^{(2)} = \Xi^{(20)} + \Xi^{(21)} \tag{28}$$

where $\Xi^{(20)}$ is the uncoupled value and $\Xi^{(21)}$ is the correction, and then using the so-called "geometric approximation" (first Padé approximant)

$$\dot{\Xi}^{(2)} = \frac{\Xi^{(20)}}{1 - (\Xi^{(21)}/\Xi^{(20)})} \tag{29}$$

instead of $\Xi^{(2)}$ to approximate $\hat{E}^{(2)}$.

Although the model Hamiltonian approach has much to recommend it since formally $\Xi^{(1)}$ and $\Xi^{(2)}$ involve only second-order errors, and since, at least in principle, one sees how to proceed to make higher-order corrections, it does suffer from the defect that the choice of $H^{(00)}$ is, formally at least, very arbitrary, which in turn means that the numerical values of the corrections of first, second, etc., order (though not that of their sum) are rather arbitrary. In this connection, we would now like to mention some methods in the literature which at first sight seem to avoid this ambiguity, and show that in fact they also fall in the model Hamiltonian category, the ambiguity (really the symbol $H^{(00)}$) having disappeared simply because, in effect, a particular choice of $H^{(00)}$ has been made.

For example, Delves[4] has suggested that optimally $\chi^{(1)}$ should be determined by solving

$$(H^{(0)} - \Xi^{(0)})\chi^{(1)} + (H^{(1)} - \zeta)\chi^{(0)} = 0 \tag{30}$$

[2] See, for example, Riemenschneider *et al.* [5].

[3] See, for example, Epstein and Johnson [2].

[4] Delves [12]. See also Aranoff and Percus [13] and Deal *et al.* [14].

where the number ζ is to be chosen so that

$$(\chi^{(1)}, \chi^{(0)}) = 0 \tag{31}$$

The point is simply that if $\chi^{(0)}$ were exact, that is, equal to $\psi'^{(0)}$, then $\chi^{(1)}$ would also be exact and so, as discussed earlier, $\Xi^{(1)}$ would involve no first-order error. Although it is not immediately apparent, the requirement (31) actually plays an essential role here. To see this, we take the scalar product of (30) with $\psi'^{(0)}$, to find

$$(\psi'^{(0)}, \chi^{(1)}) = -\frac{(\psi'^{(0)}, (H^{(1)} - \zeta)\chi^{(0)})}{E'^{(0)} - \Xi^{(0)}} \tag{32}$$

Now the denominator in (32) is of order $\gamma^{(0)^2}$, while if, say, one chooses $\zeta = E'^{(10)}$ as given in (15), then the numerator is of order $\gamma^{(0)}$ and so the limit $\gamma^{(0)} \to 0$ will be singular. [This also accounts for the "zero-order error" phenomenon mentioned following Eq. (31-14).] On the other hand, requiring (31) obviously suffices to avoid this catastrophe since it will ensure that $(\psi'^{(0)}, \chi^{(1)})$ becomes zero as $\gamma^{(0)} \to 0$.

To see that, in spite of appearances, this really does involve an $H^{(00)}$,[5] we note that (28) can be replaced (and in practice has to be replaced) by the variational functional

$$(\chi^{(1)}, (H^{(0)} - \Xi^{(0)})\chi^{(1)}) + (\chi^{(1)}, (H^{(1)} - \zeta)\chi^{(0)}) + (\chi^{(0)}, (H^{(1)} - \zeta)\chi^{(1)}) \tag{33}$$

where the Lagrange multiplier ζ is to be chosen so that $(\check{\chi}^{(1)}, \chi^{(0)}) = 0$. However, equivalently, then, we may use the functional

$$(\chi^{(1)}, (H^{(0)} - \Xi^{(0)})\chi^{(1)}) + (\chi^{(1)}, H^{(1)}\chi^{(0)}) + (\chi^{(0)}, H^{(1)}\chi^{(1)}) \tag{34}$$

with the requirement that the trial functions all be orthogonal to $\chi^{(0)}$. We now observe that the latter functional in just $\mathscr{J}^{(2)}$ of (31-12) and we showed then that its use implies that one is in fact doing a model calculation with

$$H^{(00)} = \pi H^{(0)}\pi + (1 - \pi)H^{(0)}(1 - \pi) \tag{35}$$

π being the projection onto $\chi^{(0)}$.

Similarly the functional

$$\begin{aligned}(f\chi^{(0)}, [T, f]\chi^{(0)}) + (f\chi^{(0)}, (H^{(1)} - E'^{(10)})\chi^{(0)}) \\ + (\chi^{(0)}, (H^{(1)} - E'^{(10)})f\chi^{(0)})\end{aligned} \tag{36}$$

[5] See also Aranoff and Percus [13].

has often been used[6] in calculations for helium to produce a $\chi^{(1)}$ according to

$$\chi^{(1)} = f\chi^{(0)} \tag{37}$$

and again, though the symbol $H^{(00)}$ does not explicitly appear, this is equivalent, as we have shown in Section 31, to doing a model calculation using as $H^{(00)}$ the Sternheimer Hamiltonian (31-20).

Similarly, the Schwartz [16] method, which also seems to avoid any mention of an $H^{(00)}$, is, in practice at least [9–11, 17], also a model calculation with $H^{(00)}$ again the Sternheimer Hamiltonian, that is, in practice it is equivalent to the procedure based on (36).

So far, we have assumed that $\chi^{(0)}$ and $\chi^{(1)}$ were to be gotten from somewhere and then simply inserted into the formulas for $\Xi^{(1)}$ and $\Xi^{(2)}$. However, we have seen that the error in $\Xi^{(1)}$ is of order $\gamma^{(0)}\gamma^{(1)}$ and therefore it should come as no surprise that taking the first variation of Eq. (9) and requiring

$$\delta\Xi^{(0)} = \delta\Xi^{(1)} = 0 \qquad \text{all} \quad \delta\chi^{(0)}, \quad \text{all} \quad \delta\chi^{(1)} \tag{38}$$

just yields

$$(H^{(0)} - \Xi^{(0)})\chi^{(0)} = 0 \tag{39}$$

$$(H^{(0)} - \Xi^{(0)})\chi^{(1)} + (H^{(1)} - \Xi^{(1)})\chi^{(0)} = 0 \tag{40}$$

that is,

$$\chi^{(0)} = \psi'^{(0)}, \qquad \chi^{(1)} = \psi'^{(1)} \tag{41}$$

Thus, just as with $J_{\mathrm{H}}^{(11)}$ of Section 34, it has been suggested [18] that

$$\delta\Xi^{(0)} = \delta\Xi^{(1)} = 0 \qquad \text{``all''} \quad \delta\chi^{(0)}, \quad \text{``all''} \quad \delta\chi^{(1)} \tag{42}$$

perhaps supplemented by the requirement

$$(\chi^{(0)}, \chi^{(1)}) = 0 \tag{43}$$

be used as the source of $\chi^{(0)}$ and $\chi^{(1)}$. Moreover, although this approach suffers from many of the same ills as the corresponding use of $J_{\mathrm{H}}^{(11)}$, recently [19] it has been applied with some success. However, we refer the reader to the literature for details.

[6] For example, Chen and Dalgarno [15].

PROBLEMS

1. Whatever the nature of $\chi^{(0)}$, if, for some reason, one chooses $\chi^{(1)} = 0$, then (11) yields no first-order correction. How can one understand this result? [Hint: What then is the order of magnitude of $\gamma^{(1)}$?]

2. In Section 28, we mentioned that the so-called uncoupled SCF theory results if one omits the G terms from Eq. (28-8). Similarly the simplest uncoupled HF equations result when one omits the $G^{(1)}$ terms from (28-39) and their analogs in the higher-order equations. Thus the uncoupled HF spin orbitals satisfy

$$(\hat{f}^{(0)} + \nu h^{(1)} + \nu^2 h^{(2)} + \cdots)\varphi_i = \varepsilon_i \varphi_i \tag{*}$$

Show that, as stated in the text, the Slater determinant χ formed from the φ_i satisfies

$$(F^{(0)} + \nu H^{(1)} + \nu^2 H^{(2)} + \cdots)\chi = \mathfrak{z}\chi; \qquad \mathfrak{z} = \sum \varepsilon_i \tag{**}$$

where $\mathfrak{z} = \sum_n E'^{(n0)}$ (recall also Problem 3 of Section 10).

3. Referring to Problem 2, show, however, as one would expect from our discussion of UHF, that (**) plus the requirement that the φ_i are to be orthonormal implies more generally that

$$(\hat{f}^{(0)} + \nu h^{(1)} + \nu^2 h^{(2)} + \cdots)\varphi_i = \sum \varphi_j \varepsilon_{ji},$$
$$\varepsilon_{ji} = \varepsilon_{ij}^*, \qquad \mathfrak{z} = \sum \varepsilon_{ii} \tag{***}$$

What is the relation between the solutions of (***) and the solutions of (**)? On occasion, various authors, for one reason or another, have used the more general form (***). However, show explicitly, from the relation which you have just established, that the solutions of (*) and (***) are really quite equivalent, yielding essentially the same χ and exactly the same $\mathfrak{z}$.

4. In the variation method within the variation method for UHF, use $\varphi_i^{(1)} = \check{A}\bar{\varphi}_i^{(1)}$, where the $\bar{\varphi}_i^{(1)}$ are the solutions of the simple uncoupled equations, and where A is a variational parameter. Show that the resultant $\check{E}^{(2)}$ is the geometric approximation (Tuan [20]; also Burrows [21]). What does this imply about the relationship between the geometric approximation and $\hat{E}^{(2)}$ of UHF?

5. Although in introducing the discussion of uncoupled Hartree–Fock theories we said that $H^{(1)}$ and $H^{(2)}$ should be one-electron operators, this in fact seems to play no role in the subsequent formal discussion. Explain.

6. Show that if one would use (30), but with $\zeta = E'^{(10)}$, then (11) would yield no first-order correction.

References

1. D. F.-T. Tuan, S. T. Epstein, and J. O. Hirschfelder, *J. Chem. Phys.* **44**, 431 (1966).
2. S. T. Epstein and R. E. Johnson, *J. Chem. Phys.* **47**, 2275 (1967).
3. D. F.-T. Tuan and K. K. Wu, *J. Chem. Phys.* **53**, 620 (1970).
4. B. R. Riemenschneider, N. R. Kestner, and B. K. Rao, *Chem. Phys. Lett.* **12**, 396 (1971).
5. B. R. Riemenschneider and N. R. Kestner, *Chem. Phys. Lett.* **5**, 381 (1970).
6. A. J. Sadlej and N. Jaszuński, *Mol. Phys.* **22**, 761 (1971).
7. P. W. Langhoff, M. Karplus, and R. P. Hurst, *J. Chem. Phys.* **44**, 505 (1966); A. J. Sadlej, *Mol. Phys.* **22**, 705 (1971), and references therein.
8. A. Dalgarno, *Proc. Roy. Soc. Ser. A* **251**, 282 (1959).
9. J. M. Schulman and J. I. Musher, *J. Chem. Phys.* **49**, 4845 (1968).
10. A. T. Amos and H. G. Ff. Roberts, *J. Chem. Phys.* **50**, 2375 (1969).
11. D. F.-T. Tuan, *Chem. Phys. Lett.* **7**, 115 (1970).
12. L. M. Delves, *Proc. Phys. Soc. London* **92**, 55 (1967).
13. S. Aranoff and J. Percus, *Phys. Rev.* **166**, 1255 (1968).
14. W. J. Deal, Jr., R. H. Young, and N. R. Kestner, *J. Chem. Phys.* **49**, 3395 (1968).
15. J. C. Y. Chen and A. Dalgarno, *Proc. Phys. Soc. London* **85**, 399 (1965).
16. C. Schwartz, *Ann. Phys. (New York)* **2**, 170 (1959).
17. R. J. Weiss, *Proc. Phys. Soc. London* **81**, 439 (1963).
18. L. M. Delves, *Nucl. Phys.* **41**, 497 (1963).
19. L. J. Shustek and J. B. Kreiger, *Phys. Rev. A* **3**, 1253 (1971); B. Kirtman, *J. Chem. Phys.* **55**, 1457 (1971).
20. D. F. T. Tuan, *Chem. Phys. Lett.* **7**, 115 (1970).
21. B. L. Burrows, *Int. J. Quantum Chem.* **7**, 345 (1973).

Appendix A / The Max-Min Theorem

From Result 6 of Section 2, we can characterize E_k', the kth smallest eigenvalue of H, by

$$E_k' = \underset{\psi}{\mathrm{Min}}(\psi, H\psi)/(\psi, \psi), \qquad (\psi_i', \psi) = 0, \qquad i = 1, 2, \ldots, k-1 \tag{1}$$

where the ψ_i' are the eigenfunctions of H associated with the lower eigenvalues. That is, one minimizes E subject to the constraint that the ψ be orthogonal to the lower eigenfunctions. We now want to point out that there exists another variation approach, the so-called "Max-Min theorem,"[1] which does not require explicit information about lower states. Namely

$$E_k' = \underset{\omega_i}{\mathrm{Max}}\ \underset{\psi}{\mathrm{Min}}(\psi, H\psi)/(\psi, \psi), \qquad (\omega_i, \psi) = 0, \qquad i = 1, 2, \ldots, k-1 \tag{2}$$

where the ω_i are $k - 1$ arbitrary functions. In words, one first fixes the ω_i and determines the minimum of E subject to the constraint that ψ be orthogonal to the ω_i. This minimum is then a functional of the ω_i. To find E_k', one then maximizes with respect to the ω_i. We will now give a brief proof that these two definitions of E_k' are equivalent.

We first note that whatever functions one chooses for the ω_i, they span a space which is at most $(k - 1)$ dimensional, and that therefore there is at least one linear combination of the $\psi_1', \ldots, \psi_k'$ which is orthogonal to all the ω_i, and which therefore can be used as ψ in a minimization stage of (2). However, for this ψ we have, writing it as $\psi = \sum_{l=1}^{k} C_l \psi_l'$, that

$$E = \frac{(\psi, H\psi)}{(\psi, \psi)} = \frac{\sum_{l=1}^{k} | C_l |^2 E_l'}{\sum_{l=1}^{k} | C_l |^2} \leq E_k'$$

Therefore none of the minima in (2) can exceed E_k' and hence in particular the maximum of the minima cannot exceed E_k'. On the other hand, if we

[1] See, for example, Gould [1].

choose the ω_i to be equal to the ψ_i', $i = 1, \ldots, k-1$, then from (1) it follows that the minimum for this choice of the ω_i is precisely E_k', hence (2) follows.

We will now use (2) to give an elegant derivation of some of the results in Section 7. First we will derive the separation theorem (7-10). Referring to the earlier discussion, let $\bar{\bar{H}}$ be the projection of H onto the $(M+1)$-dimensional space spanned by the ϕ_k and ϕ. Then evidently

$$\hat{E}_k(M+1) = \underset{\omega_i}{\text{Max}}\ \underset{\psi}{\text{Min}}(\psi, \bar{\bar{H}}\psi)/(\psi, \psi), \quad (\omega_i, \psi) = 0, \quad i = 1, 2, \ldots, k-1$$

$$\hat{E}_k(M) = \underset{\omega_i}{\text{Max}}\ \underset{\psi}{\text{Min}}(\psi, \bar{\bar{H}}\psi)/(\psi, \psi), \quad (\omega_i, \psi) = 0, \quad i = 1, 2, \ldots, k-1$$
$$(\phi, \psi) = 0,$$

$$\hat{E}_{k-1}(M) = \underset{\omega_i}{\text{Max}}\ \underset{\psi}{\text{Min}}(\psi, \bar{\bar{H}}\psi)/(\psi, \psi), \quad (\omega_i, \psi) = 0, \quad i = 1, 2, \ldots, k-2$$
$$(\phi, \psi) = 0$$

where the ψ are selected from the $(M+1)$-dimensional space.

Comparing $\hat{E}_{k-1}(M)$ and $\hat{E}_k(M+1)$, we see that the prescriptions are similar except that for $\hat{E}_k(M+1)$, ω_{k-1} is permitted to vary, while for $\hat{E}_{k-1}(M)$, it is in effect fixed at ϕ. Thus the Max in the latter case cannot be higher than in the former case and we have

$$\hat{E}_{k-1}(M) \leq \hat{E}_k(M+1)$$

Now let us compare $\hat{E}_k(M+1)$ and $\hat{E}_k(M)$. As far as the ω_i are concerned, the prescriptions are the same. However, in the latter case, ψ is more restricted so that the Min cannot be lower and we have

$$\hat{E}_k(M+1) \leq \hat{E}_k(M)$$

which completes the derivation of (7-10).

As an additional application, we will now derive the analog of (7-12) for an arbitrary excited state [2]. Referring to the earlier discussion, if $\bar{H}$ is the projection of H onto the $(M+1)$-dimensional space spanned by the ϕ_k, then the two procedures can be characterized by

$$\hat{E}_k(M+1) = \underset{\omega_i}{\text{Max}}\ \underset{\psi}{\text{Min}}(\psi, \bar{H}\psi)/(\psi, \psi), \quad (\omega_i, \psi) = 0, \quad i = 1, 2, \ldots, k-1$$

and

$$\hat{E}_1 = \text{Min}(\psi, \bar{H}\psi)/(\psi, \psi), \qquad (\psi_i', \psi) = 0, \qquad i = 1, \ldots, k-1$$

respectively. Thus whether or not the ψ_i' are eigenfunctions of H, we have

$$\hat{E}_1 \leq \hat{E}_k(M+1)$$

However, the ψ_i' *are* the eigenfunctions of H associated with the lower eigenvalues, so we also know that

$$E_k' \leq \hat{E}_1$$

and therefore

$$E_k' \leq \hat{E}_1 \leq \hat{E}_k(M+1)$$

which is the desired generalization of (7-12).

References

1. S. H. Gould, "Variational Methods for Eigenvalue Problems," 2nd ed. Section 6, Chapter 2. Univ. of Toronto Press, Toronto, 1966.
2. J. F. Perkins, *J. Chem. Phys.* **45**, 2156 (1966).

Appendix B / *Lagrange Multipliers*

We wish to find the consequences of

$$\delta \hat{E} = 0$$

in a situation in which the A's satisfy certain equations of constraint

$$\mathscr{C}_\alpha = 0, \qquad \alpha = 1, 2, \ldots, C \tag{1}$$

where the $\mathscr{C}_\alpha$ are numbers. To be specific, and since it is the most easily visualized case, suppose that the A's are M real numbers and that the $\mathscr{C}_\alpha$ are also real. The direct approach is first to use Eqs. (1) to extract an independent set of parameters, say $A_1, A_2, \ldots, A_R$, where $R = M - C$, in terms of which all the others may be expressed. Next one writes $\hat{E}$ in terms of these independent parameters, and, denoting the result by $\hat{\mathscr{E}}$, calculates $\delta \hat{E}$ from

$$\delta \hat{E} = \sum_{i=1}^{R} \frac{\partial \hat{\mathscr{E}}}{\partial \hat{A}_i} \delta A_i = 0 \tag{2}$$

Then since the δA_i, $i = 1, \ldots, R$, are arbitrary, $\delta \hat{E} = 0$ yields

$$\partial \hat{\mathscr{E}} / \partial \hat{A}_i = 0, \qquad i = 1, 2, \ldots, R \tag{3}$$

as the equations to be solved.

Another approach is the method of Lagrange multipliers.[1] Here one introduces numbers l_α, the Lagrange multipliers, and the prescription is to require (1) and

$$\delta \hat{E} - \sum_{\alpha=1}^{C} l_\alpha \, \delta \hat{\mathscr{C}}_\alpha = 0 \tag{4}$$

or in more detail

$$\sum_{j=1}^{M} \left(\frac{\partial \hat{E}}{\partial \hat{A}_j} + \sum_{\alpha=1}^{C} l_\alpha \frac{\partial \hat{\mathscr{C}}_\alpha}{\partial \hat{A}_j} \right) \delta A_j = 0 \tag{5}$$

[1] For references and generalizations, see Einhorn and Blankenbecler [1].

without regard to the lack of independence of the A_j. That is, one is to solve the M equations

$$\frac{\partial \hat{E}}{\partial \hat{A}_j} + \sum_{\alpha=1}^{C} l_\alpha \frac{\partial \hat{\mathscr{C}}_\alpha}{\partial \hat{A}_j} = 0 \tag{6}$$

together with the C equations (1) for the $M + C$ unknowns A_i and l_α. [Note that since $\hat{E}$ is real and since we have assumed the $\hat{\mathscr{C}}_\alpha$ to be real, (4) implies that the l_α *are* real.]

We will now show that the two procedures are equivalent. The point is simply that if Eqs. (1) are satisfied when $A = \hat{A}$ so that we can use them to determine $\hat{A}_{R+1}, \ldots, \hat{A}_M$ in terms of $\hat{A}_1, \ldots, \hat{A}_R$, then Eqs. (1) will also imply that

$$\sum_{j=1}^{M} \frac{\partial \hat{\mathscr{C}}_\alpha}{\partial \hat{A}_j} \frac{\partial \hat{A}_j}{\partial \hat{A}_i} = 0, \qquad i = 1, 2, \ldots, R \tag{7}$$

Therefore, multiplying (6) by $\partial \hat{A}_j / \partial \hat{A}_i$ and summing over j, we find

$$\sum_{j=1}^{M} \frac{\partial \hat{E}}{\partial \hat{A}_j} \frac{\partial \hat{A}_j}{\partial \hat{A}_E} = 0, \qquad i = 1, 2, \ldots, R$$

which clearly is the same as (3) since the left hand side is $\partial \hat{E} / \partial \hat{A}_i$ but with the understanding that we have used the constraints to express $\hat{A}_{R+1}, \ldots, A_M$ in terms of $\hat{A}_1, \ldots, \hat{A}_R$, i.e., it is just $\partial \hat{\mathscr{E}} / \partial \hat{A}_i$.

Often in practice the constraints are most naturally expressed in a complex form—for example, that certain scalar products should vanish. Since a complex constraint involves two real constraints, it should be clear that it will require two Lagrange multipliers, one for the real part and one for the imaginary part or, equivalently, one for it and one for its complex conjugate. Further, by the same argument which led to the conditions that the l_α are real, it will follow that the Lagrange multipliers associated with a constraint and its complex conjugate should themselves be complex conjugates of one another.

The preceding discussion has assumed that in fact the constraints can be satisfied by some subset of the A's. If this is not the case, then of course neither procedure works, since the original problem as stated has no solution. To put the matter another way—Lagrange multipliers cannot work miracles.

PROBLEM

1. Use the Lagrange multiplier technique to minimize $x^2 + 2xy + 3y^2$ first under the constraint that $x = y + 7$ and then under the constraint that $x^2 + y^2 = 7$.

Reference

1. M. B. Einhorn and R. Blankenbecler, *Ann. Phys.* (*N.Y.*) **67**, 480 (1971).

Appendix C / Theorems Satisfied by Optimal Time-Dependent Variational Wave Functions

Variational calculations in the time-dependent case[1] are usually based on the Frenkel variation method [2], which is summarized by

$$\left(\delta\hat{\varphi}, \left(H - i\frac{\partial}{\partial t}\right)\hat{\varphi}\right) + \left(\left(H - i\frac{\partial}{\partial t}\right)\hat{\varphi}, \delta\hat{\varphi}\right) = 0 \qquad \text{“all”} \quad \delta\hat{\varphi} \tag{1}$$

We will now show that just as in the time-independent case, invariance of the set of trial functions to various transformations will ensure that $\hat{\varphi}$ satisfies certain theorems. Since many of the details simply parallel those in the body of the text, we will, for the most part, simply sketch proofs and summarize results, leaving the filling in of details as a problem.

Suppose first of all that the set of trial functions is invariant to a phase change, that is, to multiplication by $e^{i\alpha}$, where α is any real number. This will then imply that

$$i\hat{\varphi} \tag{2}$$

is a possible $\delta\hat{\varphi}$ and hence, from (1), that

$$-i(\hat{\varphi}, H\hat{\varphi}) - \left(\hat{\varphi}, \frac{\partial\hat{\varphi}}{\partial t}\right) + i(H\hat{\varphi}, \hat{\varphi}) - \left(\frac{\partial\hat{\varphi}}{\partial t}, \hat{\varphi}\right) = 0 \tag{3}$$

or (since H is Hermitian)

$$\frac{\partial}{\partial t}(\hat{\varphi}, \hat{\varphi}) = 0 \tag{4}$$

Thus under these conditions, the norm of $\hat{\varphi}$ will be constant in time.

If the set of trial functions is invariant to multiplication by an arbitrary real number, then

$$\hat{\varphi} \tag{5}$$

[1] For a recent review see Langhoff *et al.* [1].

will be a possible variation of $\hat{\psi}$, and hence from (1) we will have

$$\left(\hat{\psi}, \left(H - i\frac{\partial}{\partial t}\right)\hat{\psi}\right) + \left(\left(H - i\frac{\partial}{\partial t}\right)\hat{\psi}, \hat{\psi}\right) = 0 \tag{6}$$

If then (4) is also satisfied, it further follows that

$$\left(\hat{\psi}, \left(H - i\frac{\partial}{\partial t}\right)\hat{\psi}\right) = 0 \tag{7}$$

since (6) and (4) are the real and imaginary parts, respectively, of (7). The result (7) is of course analogous to $(\hat{\psi}, (H - \hat{E})\hat{\psi}) = 0$ in time-independent theory, with the difference that the latter is simply the definition of $\hat{E}$.

The preceding discussion implicitly assumed that the variational problem (1) has a solution. That this is not just an idle remark is shown by the following simple but interesting examples. First of all, consider the set of trial functions $e^{iA}\phi$, where A is an arbitrary real number and ϕ is a given function. Then, varying $\hat{A}$, (1) yields

$$\frac{\partial}{\partial t}(\phi, \phi) = 0$$

which of course need not be true. In particular, if (4) is not already satisfied, then this shows that simply enlarging the set of trial functions by multiplying by e^{iA} will not do the job. Similarly consider the set $A\phi$, where again A is an arbitrary real number and ϕ is a given function. Then varying $\hat{A}$ yields

$$\left(\phi, \left(H - i\frac{\partial}{\partial t}\right)\phi\right) + \left(\left(H - i\frac{\partial}{\partial t}\right)\phi, \phi\right) = 0$$

which again need not be true, with the corollary that if (6) is not already satisfied, then simply enlarging the set of trial functions by multiplying by A will not do the job.

If, as is often the case, the individual trial functions are normalized to one, then $\hat{\psi}$ will certainly not be a possible variation of $\hat{\psi}$ and thus (6) will not be satisfied automatically. However, in this case it is easy to arrange that it be satisfied. First of all, under these conditions, (4) will be trivially true. Second, as we will show in a moment, if $\hat{\psi}'$ is a solution of the variational problem, then so is $\hat{\psi} = e^{iF(t)}\hat{\psi}'$ with $F(t)$ any real function of t alone. Third, by proper choice of $F(t)$ we can then satisfy (6), namely, as one easily sees, F should satisfy

$$\frac{\partial F}{\partial t} + \left(\hat{\psi}', \left(H - i\frac{\partial}{\partial t}\right)\hat{\psi}'\right) = 0 \tag{8}$$

[Note that $(\partial/\partial t)(\hat{\psi}', \hat{\psi}') = 0$ means that $(\hat{\psi}', (H - i\,\partial/\partial t)\hat{\psi}')$ *is* real.] Finally, $\hat{\psi}$ is physically equivalent to $\hat{\psi}'$ in that it yields the same expectation values (as long as the operator does not involve $\partial/\partial t$).

Now it may be that $\hat{\psi}$, like $\hat{\psi}'$, is a member of the set of trial functions (time-dependent UHF theory in which the set of trial functions consists of all normalized Slater determinants is a case in point), in which case we are simply exploiting an ambiguity in the solutions of (1) in order to satisfy (6). However, even if $\hat{\psi}$ is not a member of the original set ψ, it is a member of the enlarged set $e^{iA(t)}\psi$ where $A(t)$ is an arbitrary real function of t alone, the only effect of the enlargement evidently being to make the solutions of (1) arbitrary to the desired extent.

Of course[2] given any $\hat{\psi}'$ derived from any set of trial functions whatsoever, one can always find a complex function of time alone, $\phi(t)$ say, such that the physically equivalent (as far as expectation values are concerned) wave function χ defined by

$$\chi = e^{i\phi}\hat{\psi}' \tag{9}$$

satisfies $(\chi, (H - i\,\partial/\partial t)\chi) = 0$; namely ϕ should satisfy

$$\frac{\partial\phi}{\partial t} + \left(\hat{\psi}', \left(H - i\frac{\partial}{\partial t}\right)\hat{\psi}'\right) = 0 \tag{10}$$

However, the point of the remarks of the preceding paragraph was to show that if the individual trial functions are normalized, then χ itself will be an optimal variational wave function within either the original set of trial functions, or within a trivial extension thereof.

To complete the discussion, we must now show that, as claimed, $e^{iF}\hat{\psi}'$ is a solution of the variational equations. Thus we examine

$$\begin{aligned}
&\left(\delta(e^{iF}\hat{\psi}'), \left(H - i\frac{\partial}{\partial t}\right)e^{iF}\hat{\psi}'\right) + \left(\left(H - i\frac{\partial}{\partial t}\right)e^{iF}\hat{\psi}', \delta(e^{iF}\hat{\psi}')\right) \\
&\quad = \left\{\left(\delta\hat{\psi}', \left(H - i\frac{\partial}{\partial t} + \frac{\partial F}{\partial t}\right)\hat{\psi}'\right) + \left(\left(H - i\frac{\partial}{\partial t} + \frac{\partial F}{\partial t}\right)\hat{\psi}', \delta\hat{\psi}'\right)\right\} \\
&\qquad -i\,\delta F\left\{\left(\hat{\psi}', \left(H - i\frac{\partial}{\partial t} + \frac{\partial F}{\partial t}\right)\hat{\psi}'\right) - \left(\left(H - i\frac{\partial}{\partial t} + \frac{\partial F}{\partial t}\right)\hat{\psi}', \hat{\psi}'\right)\right\} \qquad (11) \\
&\quad = \left\{\left(\delta\hat{\psi}', \left(H - i\frac{\partial}{\partial t}\right)\hat{\psi}'\right) + \left(\left(H - i\frac{\partial}{\partial t}\right)\hat{\psi}', \delta\hat{\psi}'\right)\right\} \\
&\qquad + \frac{\partial F}{\partial t}\{(\delta\hat{\psi}', \hat{\psi}') + (\hat{\psi}', \delta\hat{\psi}')\} - \delta F\frac{\partial}{\partial t}(\hat{\psi}', \hat{\psi}') \qquad (12)
\end{aligned}$$

[2] See, for example, Löwdin and Mukherjee [3].

Then we note that the first line on the right in (12) vanishes because $\hat{\psi}'$ is assumed to be a solution of the variational problem, that the second term of the second line vanishes because $(\hat{\psi}', \hat{\psi}')$ is normalized, while the first term of the second line vanishes because if the individual trial functions are normalized, then we will have

$$(\delta\hat{\psi}', \hat{\psi}') + (\hat{\psi}', \delta\hat{\psi}') = 0$$

which completes the proof.

Let us now suppose that the set of trial functions is invariant to some unitary transformation U and that one uses that same set of trial functions with H and with H^U, where now

$$H^U = U^+HU - iU^+(\partial U/\partial t) \tag{13}$$

Then since

$$\left(\delta\hat{\psi}, \left(H - i\frac{\partial}{\partial t}\right)\hat{\psi}\right) + \left(\left(H - i\frac{\partial}{\partial t}\right)\hat{\psi}, \delta\hat{\psi}\right) = 0 \tag{14}$$

can be written

$$\left(\delta U^+\hat{\psi}, \left(H^U - i\frac{\partial}{\partial t}\right)U^+\hat{\psi}\right) + \left(\left(H^U - i\frac{\partial}{\partial t}\right)U^+\hat{\psi}, \delta U^+\hat{\psi}\right) = 0 \tag{15}$$

we see that if $\hat{\psi}$ is a solution for H, then $U^+\hat{\psi}$ (which by assumption is a member of the set) will be a solution for H^U. Thus under these conditions the optimal trial functions transform in the expected manner under U.

If the set of trial functions is invariant to $\sigma \to \sigma + \delta\sigma$, where σ is a real parameter in H, then this implies that

$$\partial\hat{\psi}/\partial\sigma \tag{16}$$

is a possible variation of $\hat{\psi}$, whence (1) yields

$$\left(\frac{\partial\hat{\psi}}{\partial\sigma}, \left(H - i\frac{\partial}{\partial t}\right)\hat{\psi}\right) + \left(\left(H - i\frac{\partial}{\partial t}\right)\hat{\psi}, \frac{\partial\hat{\psi}}{\partial\sigma}\right) = 0 \tag{17}$$

This can also be written as

$$\frac{\partial}{\partial\sigma}\left\{\left(\hat{\psi}, \left(H - i\frac{\partial}{\partial t}\right)\hat{\psi}\right)\right\} - \left(\hat{\psi}, \frac{\partial H}{\partial\sigma}\hat{\psi}\right) + i\frac{\partial}{\partial t}\left(\hat{\psi}, \frac{\partial\hat{\psi}}{\partial\sigma}\right) = 0 \tag{18}$$

from which it follows that if (7) is also satisfied, then $\hat{\psi}$ will satisfy the "time-dependent Hellmann–Feynman Theorem" [4]

$$\left(\hat{\varphi}, \frac{\partial H}{\partial \sigma} \hat{\varphi}\right) = i \frac{\partial}{\partial t} \left(\hat{\varphi}, \frac{\partial \hat{\varphi}}{\partial \sigma}\right) \tag{19}$$

If the set of trial functions is invariant to the unitary operator

$$U = e^{i\alpha\mathscr{G}} \tag{20}$$

where $\mathscr{G}$ is a Hermitian operator and α an arbitrary real number, then this implies that

$$i\mathscr{G}\hat{\varphi} \tag{21}$$

is a possible $\delta\hat{\varphi}$, whence (1) yields

$$\left(i\mathscr{G}\hat{\varphi}, \left(H - i\frac{\partial}{\partial t}\right)\hat{\varphi}\right) + \left(\left(H - i\frac{\partial}{\partial t}\right)\hat{\varphi}, i\mathscr{G}\hat{\varphi}\right) = 0 \tag{22}$$

which can also be written as

$$\frac{\partial}{\partial t}(\hat{\varphi}, \mathscr{G}\hat{\varphi}) = \left(\hat{\varphi}, \frac{\partial \mathscr{G}}{\partial t} \hat{\varphi}\right) + i(\hat{\varphi}, [H, \mathscr{G}]\hat{\varphi}) \tag{23}$$

Thus if the set of trial functions is invariant to (20), then (23), which is variously known as the time-dependent hypervirial theorem for $\mathscr{G}$, or the generalized Ehrenfest theorem for $\mathscr{G}$, will be satisfied. If, further, (4) is satisfied [that is, if (23) is satisfied for $\mathscr{G} = 1$], then we can write (23) as

$$(\partial/\partial t)\langle\mathscr{G}\rangle = \langle\partial\mathscr{G}/\partial t\rangle + i\langle[H, \mathscr{G}]\rangle \tag{24}$$

the angle brackets as usual denoting average values.

In time-dependent UHF, the set of trial functions consists of all normalized Slater determinants and therefore in UHF theory, (23) is satisfied for all one-electron $\mathscr{G}$'s. Similarly, in time-dependent SUHF, (23) is satisfied for all spin-independent one-electron $\mathscr{G}$'s.

In particular, if $\mathscr{G}$ corresponds to a gauge transformation

$$\mathscr{G} = \sum_s g(\mathbf{r}_s, t) \tag{25}$$

then one finds after an integration by parts that (23) becomes

$$\int d\mathbf{r}\, [\boldsymbol{\nabla} \cdot \mathbf{j} + (\partial \varrho/\partial t)]g = 0 \tag{26}$$

where $\mathbf{j}$ is the one-particle current and ϱ is the one-particle density. If then one has invariance for all g's as in UHF and SUHF, then (26) implies that

$$\nabla \cdot \mathbf{j} + (\partial \varrho / \partial t) = 0 \tag{27}$$

that is, charge-current conservation.

The choice $\mathscr{G} = H$ in (23) yields what we will call the "work-energy theorem":

$$\frac{\partial}{\partial t}(\hat{\psi}, H\hat{\psi}) = \left(\hat{\psi}, \frac{\partial H}{\partial t}\hat{\psi}\right)$$

However if H contains two particle operators $iH\hat{\psi}$ will *not* be a possible $\delta\hat{\psi}$ within time dependent UHF, and therefore from the above one would perhaps not expect the work-energy theorem to be satisfied in that approximation. Nevertheless it *is* satisfied there, and, more generally, it is satisfied by any $\hat{\psi}$ which is derived from a set of trial functions which is invariant to time translations. Thus, for example, it is satisfied in the time dependent linear variational method if the basis functions are time independent. The proof is as follows: under the given conditions $\partial\hat{\psi}/\partial t$ *will* be a possible $\delta\hat{\psi}$. Therefore from (1) we will have

$$\left(\frac{\partial \hat{\psi}}{\partial t}, H\hat{\psi}\right) + \left(H\hat{\psi}, \frac{\partial \hat{\psi}}{\partial t}\right) = 0$$

or

$$\left(\frac{\partial \hat{\psi}}{\partial t}, H\hat{\psi}\right) + \left(\hat{\psi}, H\frac{\partial \hat{\psi}}{\partial t}\right) = 0 \tag{28}$$

However

$$\frac{\partial}{\partial t}(\hat{\psi}, H\hat{\psi}) = \left(\hat{\psi}, \frac{\partial H}{\partial t}\hat{\psi}\right) + \left(\frac{\partial \hat{\psi}}{\partial t}, H\hat{\psi}\right) + \left(\hat{\psi}, H\frac{\partial \hat{\psi}}{\partial t}\right)$$

which, when combined with (28), proves the point.

In the notation of Sections 17 and 18, if the theorem (24) is satisfied for the components of **D** and the components of **P**, then one finds for a molecule in a spatially uniform electric field that

$$(\partial^2/\partial t^2)\langle \mathbf{D} \rangle = \langle \mathbf{F}_{eN} \rangle - N\mathscr{E}(t) \tag{29}$$

whence, generalizing (18-12), we have that the average total force on the nuclei is

$$\langle \mathscr{F} \rangle = q\mathscr{E}(t) - (\partial^2/\partial t^2)\langle \mathbf{D} \rangle \tag{30}$$

If the field is simple harmonic, one then has to first order in the field

$$\langle \mathbf{D} \rangle = \langle \mathbf{D} \rangle^{(0)} - \boldsymbol{\alpha}(\omega) \cdot \mathscr{E}(t) + \cdots \tag{31}$$

where $\langle \mathbf{D} \rangle^{(0)}$ is the static dipole moment and where $\boldsymbol{\alpha}(\omega)$ is the dipole polarizability tensor. Then since for a simple harmonic field

$$\partial^2 \mathscr{E} / \partial t^2 = -\omega^2 \mathscr{E}$$

it follows that we will have, through first order in the field,

$$\langle \mathscr{F} \rangle = \{q\mathbf{I} - \omega^2 \boldsymbol{\alpha}(\omega)\} \cdot \mathscr{E}(t) \tag{32}$$

a relation which, for example, should be true in UHF and SUHF, and which has proven useful for testing approximate wave functions [5].

When H is a periodic function of the time, then in the steady-state the solutions of the time-dependent Schrödinger equation have the form

$$\psi' = [\exp(-i\eta' t)]\chi' \tag{33}$$

where η' is a real constant and χ' is a periodic function with (χ', χ') independent of t [6]. Since $(\psi', (H - i\,\partial/\partial t)\psi') = 0$, η' and χ' are related by $(\chi', (H - i\,\partial/\partial t - \eta')\chi') = 0$ or equivalently by

$$\left(\chi', \left(H - i\frac{\partial}{\partial t} - \eta'\right)\chi'\right)_T = 0 \tag{34}$$

where T denotes the average over a period.

Suppose now that one uses trial functions of the form

$$\psi = e^{-i\eta t}\chi \tag{35}$$

where η is a real constant and where χ is periodic with (χ, χ) independent of t. Suppose further that η and χ are related by

$$\left(\chi, \left(H - i\frac{\partial}{\partial t} - \eta\right)\chi\right)_T = 0 \tag{36}$$

Then one can show [1] that (1) then implies that

$$\left[\left(\delta\hat{\chi}, \left(H - i\frac{\partial}{\partial t} - \hat{\eta}\right)\hat{\chi}\right) + \left(\hat{\chi}, \left(H - i\frac{\partial}{\partial t} - \hat{\eta}\right)\delta\hat{\chi}\right)\right]_T = 0 \tag{37}$$

or, in short, comparing with (34),

$$\delta\hat{\eta} = 0 \tag{38}$$

Therefore, comparing (36) and (34), we see that under these conditions, the Frenkel variation method yields a variational approximation, the Heinrichs [7] variational approximation, to η'.

Now let us write

$$H = H^{(0)} + \nu H^{(1)} + \nu^2 H^{(2)} + \cdots \tag{39}$$

where $H^{(0)}$ is independent of time and where $H^{(1)}$, $H^{(2)}$, etc., are periodic. Then we can derive a ν^{2n+1} theorem for $\hat{\eta}$ in much the same way that we derived the theorem for $\hat{E}$ in Section 26. Namely suppose that we know the variation parameters in $\hat{\chi}$ through nth order in ν and define $\mathring{A}$ by

$$\mathring{A} = \hat{A}^{(0)} + \nu\hat{A}^{(1)} + \nu^2\hat{A}^{(2)} + \cdots + \nu^n\hat{A}^{(n)} \tag{40}$$

Further, define the function $\mathring{\chi}$ and the constant $\mathring{\eta}$ by

$$\mathring{\chi} = \chi(\hat{A}^{(0)} + \nu\hat{A}^{(1)} + \cdots + \nu^n\hat{A}^{(n)}, \nu) \tag{41}$$

and

$$\left[\left(\mathring{\chi}, \left(H - i\frac{\partial}{\partial t} - \mathring{\eta}\right)\mathring{\chi}\right)\right]_T = 0 \tag{42}$$

Then, writing

$$\mathring{\chi} = \hat{\chi} + \mathring{\Delta} \tag{43}$$

one readily finds that

$$\mathring{\eta} - \hat{\eta} = \left[\left(\mathring{\Delta}, \left(H - i\frac{\partial}{\partial t} - \hat{\eta}\right)\hat{\chi}\right) + \left(\left(H - i\frac{\partial}{\partial t} - \hat{\eta}\right)\hat{\chi}, \mathring{\Delta}\right)\right.$$
$$\left. + \left(\mathring{\Delta}, \left(H - i\frac{\partial}{\partial t} - \hat{\eta}\right)\mathring{\Delta}\right)\right]_T \Big/ (\mathring{\chi}, \mathring{\chi}) \tag{44}$$

where we have also used the fact that since $\mathring{A}$ is a possible A, $(\mathring{\chi}, \mathring{\chi})$ will be independent of time [which incidently also shows that $\mathring{\eta}$ defined by (42) is real], and $\mathring{\chi}$, $\hat{\chi}$, and $\mathring{\Delta}$ will all be periodic. However,

$$\mathring{\Delta} = (\partial\hat{\chi}/\partial\hat{A})(\mathring{A} - \hat{A}) + O(\nu^{2n+2}) \tag{45}$$

and therefore since $(\partial\hat{\chi}/\partial A)(\mathring{A} - \hat{A})$ is certainly a possible $\delta\hat{\chi}$, it follows that the contribution of $(\partial\hat{\chi}/\partial\hat{A})(\mathring{A} - \hat{A})$ to the sum of first two terms on the right in (44) vanishes, whence we have that, as desired,

$$\mathring{\eta} = \hat{\eta} + O(\nu^{2n+2}) \tag{46}$$

If the time-dependent Hellmann–Feynman theorem (19) for $\sigma = \nu$ is satisfied, then $\hat{\eta}$ has a simple interpretation (in any case, $\hat{\eta}^{(0)}$ is of course $\hat{E}^{(0)}$ of the time-independent variation method). Thus we then have

$$(\hat{\varphi}, (H^{(1)} + 2\nu H^{(2)} + \cdots)\hat{\varphi}) = i\frac{\partial}{\partial t}\left(\hat{\varphi}, \frac{\partial \hat{\varphi}}{\partial \nu}\right) \tag{47}$$

or

$$(\hat{\chi}, (H^{(1)} + 2\nu H^{(2)} + \cdots)\hat{\chi}) = \frac{\partial \hat{\eta}}{\partial \nu}(\hat{\chi}, \hat{\chi}) + i\frac{\partial}{\partial t}\left(\hat{\chi}, \frac{\partial \hat{\chi}}{\partial \nu}\right) \tag{48}$$

If now we average over a period, then since $\hat{\chi}$ and $\partial\hat{\chi}/\partial\nu$ are periodic, it follows that

$$\frac{\partial \hat{\eta}}{\partial \nu} = \frac{[(\hat{\chi}, (H^{(1)} + 2\nu H^{(2)} + \cdots)\hat{\chi})]_T}{(\hat{\chi}, \hat{\chi})}$$

In particular, if the system is interacting with a spatially uniform electric field, then $H^{(1)} = -\mathscr{D}\cdot\mathscr{E}(t)$, where $\mathscr{D}$ is the dipole moment operator, while $H^{(2)} = H^{(3)} = \cdots$, etc., all equal zero. Therefore in this case $\partial\hat{\eta}/\partial\nu$ (for which, from the preceding, we have a ν^{2n} theorem) yields the average over one period of $\langle -\mathscr{D}\rangle \cdot \mathscr{E}$. If, further, the electric field is simple harmonic, then this means that $\partial\hat{\eta}/\partial\nu$ yields that part of the average dipole moment which is in phase with the electric field.

PROBLEMS

1. Consider an atom in a static uniform electric field and do not neglect nuclear motion, i.e., do not clamp the nucleus. By introducing the center of mass coordinates and the coordinates of the electrons with respect to the nucleus as new variables, show that one can separate center of mass motion and nuclear motion. Then consider a wave function of the form

   (wave packet for center of mass)(eigenstate of internal energy)

   Show from the hypervirial theorem for the total internal momentum that

   $$\text{average total force on electrons due to nucleus} - N\left(\frac{M + \mathscr{Z}}{M + N}\right)\mathscr{E} = 0$$

where M is the nuclear mass. Then show, using various time-dependent hypervirial theorems, that

$$M \frac{d^2\langle \mathbf{R} \rangle}{dt^2} = \mathscr{X}\left(1 - \frac{N}{\mathscr{X}} \frac{M + \mathscr{X}}{M + N}\right)\mathscr{E} \qquad (*)$$

where $\mathbf{R}$ is the nuclear coordinate [8]. In which, if any, of the various Born–Oppenheimer approximations [9] will a result analogous to (*) hold?

References

1. P. W. Langhoff, S. T. Epstein, and M. Karplus, *Rev. Mod. Phys.* **44**, 602 (1972).
2. J. Frenkel, "Wave Mechanics, Advanced General Theory," p. 436. Oxford Univ. Press (Clarendon), London and New York, 1934.
3. P.-O. Löwdin and P. K. Mukherjee, *Chem. Phys. Lett.* **14**, 1 (1972).
4. E. F. Hayes and R. G. Parr, *J. Chem. Phys.* **43**, 1831 (1965).
5. S. T. Epstein and R. E. Johnson, *J. Chem. Phys.* **51**, 188 (1969); V. G. Kaveeshwar, A. Dalgarno, and R. P. Hurst, *Proc. Phys. Soc. London* (*At. Mol. Phys.*) **2**, 984 (1969).
6. R. H. Young and W. J. Deal, Jr., *J. Math. Phys.* (*N.Y.*) **11**, 3298 (1970).
7. J. Heinrichs, *Phys. Rev.* **172**, 1315 (1968); **176**, 2167 (1968).
8. R. M. Sternheimer, *Phys. Rev.* **96**, 951, footnote 25 (1954).
9. J. O. Hirschfelder and W. J. Meath, *in* "Intermolecular Forces" (J. O. Hirschfelder, ed.). Academic Press, New York, 1967.

Appendix D / Various Hypervirial Theorems in the Presence of Magnetic Fields

Here again we will simply summarize results, leaving the filling in of details as a problem.[1] Also, we will note at the outset that all theorems are satisfied by UHF and SUHF because the $\mathscr{G}$'s are all spinless one-electron operators. We write the Hamiltonian (1-1) as

$$H = \sum_{s=1}^{N} \left\{ \frac{1}{2} \boldsymbol{v}_s^2 - \sum_a \frac{\mathscr{Z}_a}{|\mathbf{R}_a - \mathbf{r}_s|} - \Phi(r_s) \right\} + \sum_{s>t}\sum \frac{1}{|\mathbf{r}_s - \mathbf{r}_t|}$$

$$+ \sum_{a>b}\sum \frac{\mathscr{Z}_a \mathscr{Z}_b}{|\mathbf{R}_a - \mathbf{R}_b|} + \sum_a \mathscr{Z}_a \Phi(\mathbf{R}_a) \tag{1}$$

where $\boldsymbol{v}_s$ is the velocity operator for the sth electron

$$\boldsymbol{v}_s \equiv \mathbf{p}_s + (1/c)\mathscr{A}(\mathbf{r}_s) \tag{2}$$

and where $\mathscr{A}(\mathbf{r})$ and $\Phi(\mathbf{r})$ are the vector and scalar potentials, respectively, of the external field

$$\mathscr{B}(\mathbf{r}) = \nabla \times \mathscr{A}, \qquad \mathscr{E}(\mathbf{r}) = -\nabla \Phi \tag{3}$$

With

$$\mathbf{D} = \sum_{s=1}^{N} \mathbf{r}_s \tag{4}$$

then

$$i[H, \mathbf{D}] = \sum_{s=1}^{N} \boldsymbol{v}_s \tag{5}$$

and therefore the hypervirial theorems for the components of $\mathbf{D}$ now become velocity theorems, that is,

$$\langle \mathbf{V} \rangle = 0 \tag{6}$$

[1] See also Young [1] and Carr [2].

where

$$\mathbf{V} = \sum_{s=1}^{N} \boldsymbol{v}_s \tag{7}$$

is the operator for the total velocity of the electrons. These theorems then will be satisfied if the set of trial functions is invariant to the transformation $\exp(i\boldsymbol{\alpha}\cdot\mathbf{D})$ and possibly also for reasons of symmetry.

Although we could now go on and look at the hypervirial theorems for $\mathbf{P}$ (and these will be satisfied if the set of trial functions is invariant to translation), it is clearly more interesting to look at the hypervirial theorems for the components of $\mathbf{V}$. One then readily finds that $i[H, \mathbf{V}]$ is the operator for the total force acting on the electrons, that is,

$$\begin{aligned} i[H, \mathbf{V}] &= \sum_{s=1}^{N} \left\{ \sum_{a} \frac{\mathscr{Z}_a(\mathbf{R}_a - \mathbf{r}_s)}{|\,\mathbf{R}_a - \mathbf{r}_s\,|^3} - \mathscr{E}(\mathbf{r}_s) - \frac{1}{2c}(\boldsymbol{v}_s \times \mathscr{B}(\mathbf{r}_s) + \mathscr{B}(\mathbf{r}_s) \times \boldsymbol{v}_s) \right\} \\ &= \sum_{s=1}^{N} \mathbf{F}_s \end{aligned} \tag{8}$$

so that the hypervirial theorems for the components of $\mathbf{V}$ are the force theorems. If the magnetic field is a uniform one, then the average total magnetic force on the electrons is evidently

$$-(\langle \mathbf{V} \rangle / c) \times \mathscr{B} \tag{9}$$

Thus in this case, if the velocity theorems are satisfied in directions perpendicular to $\mathscr{B}$, the magnetic part of the average force on the electrons will vanish separately and hence, in particular, the various considerations of Section 18 concerning the forces on the (fixed) nuclei will remain unchanged.

Also, in the case of a uniform magnetic field, it is easy to generalize (18-3) so as to ensure that the force theorems will be satisfied. Taking the direction of the magnetic field as the z direction, and using the gauge (13-22), the components of $\mathscr{A}$ are then

$$\mathscr{A} = (-y\mathscr{B}/2, x\mathscr{B}/2, 0) \tag{10}$$

whence one readily finds that

$$\begin{aligned} \psi = &\left\{ \exp\left[i(\mathscr{B}/2c) \sum_{s=1}^{N} (-y_s A_x + x_s A_y) + i(N\mathscr{B}/2) A_x A_y \right] \right\} \\ &\times \theta(\mathbf{r}_1 + \mathbf{A}, \mathbf{r}_2 + \mathbf{A}, \ldots, \mathbf{r}_N + \mathbf{A}) \end{aligned} \tag{11}$$

with A_x, A_y, and A_z arbitrary real numbers, will do the job (the factor of $e^{i(N\mathscr{B}/2c)A_xA_y}$ of course being irrelevant). The point is simply that, as one can easily verify, then to within irrelevant additive multiples of ψ

$$\partial\psi/\partial\mathbf{A} = i\mathbf{V}\psi \tag{12}$$

and so, as desired (recall Section 10), $i\mathbf{V}\hat{\psi}$ will, in effect, be a possible variation of $\hat{\psi}$.

Since in this gauge, $V_z = P_z$, the hypervirial theorem for P_z will obviously also be satisfied by use of this set. To satisfy in addition the hypervirial theorems for P_x and P_y, one can, in accord with (18-3), introduce two more arbitrary real numbers C_x and C_y and use

$$\psi = \left\{\exp\left[i(\mathscr{B}/2c)\sum_{s=1}^{N}\{(-y_s - C_y)A_x + (x_s + C_x)A_y\} + i(N/2)A_xA_y\right]\right\}$$
$$\times\, \theta(\mathbf{r}_1 + \mathbf{A} + \mathbf{C}, \mathbf{r}_2 + \mathbf{A} + \mathbf{C}, \ldots, \mathbf{r}_N + \mathbf{A} + \mathbf{C}) \tag{13}$$

or, omitting irrelevant factors,

$$\psi = \left\{\exp\left[i(\mathscr{B}/2c)\sum_{s=1}^{N}(-y_sA_x + x_sA_y)\right]\right\}$$
$$\times\, \theta(\mathbf{r}_1 + \mathbf{A} + \mathbf{C}, \mathbf{r}_2 + \mathbf{A} + \mathbf{C}, \ldots, \mathbf{r}_N + \mathbf{A} + \mathbf{C}) \tag{14}$$

In the presence of a magnetic field, the operator for the total mechanical orbital angular momentum about the coordinate origin is

$$\mathbf{L} = \sum_{s=1}^{N}\mathbf{r}_s\times\boldsymbol{v}_s \tag{15}$$

and one readily finds that the hypervirial theorems for the components of $\mathbf{L}$ are just the torque theorems, the torque operator being

$$\boldsymbol{\tau} = \sum_{s=1}^{N}\tfrac{1}{2}(\mathbf{r}_s\times\mathbf{F}_s - \mathbf{F}_s\times\mathbf{r}_s) \tag{16}$$

Turning to the virial theorem, with $\mathscr{V}$ now generalized to

$$\mathscr{V}^{\mathscr{B}} = \tfrac{1}{2}\sum_{s=1}^{N}(\mathbf{r}_s\cdot\boldsymbol{v}_s + \boldsymbol{v}_s\cdot\mathbf{r}_s) \tag{17}$$

then

$$i[H, \mathscr{V}^{\mathscr{B}}] = \sum_{s=1}^{N}\{\boldsymbol{v}_s^{\,2} + \tfrac{1}{2}\mathbf{r}_s\cdot\mathbf{F}_s + \tfrac{1}{2}\mathbf{F}_s\cdot\mathbf{r}_s\} \tag{18}$$

Thus when the hypervirial theorem for $\mathscr{V}^{\mathscr{B}}$ is satisfied, we will have

$$\sum_{s=1}^{N} \langle \boldsymbol{v}_s^2 + \tfrac{1}{2}\mathbf{r}_s\cdot\mathbf{F}_s + \tfrac{1}{2}\mathbf{F}_s\cdot\mathbf{r}_s\rangle = 0 \tag{19}$$

In particular, in the case of a uniform magnetic field and using the gauge (10), it is easy to see how to satisfy this theorem, since then

$$\mathbf{r}_s\cdot\mathscr{A}(\mathbf{r}_s) = 0 \tag{20}$$

whence

$$\mathscr{V}^{\mathscr{B}} = \tfrac{1}{2}\sum(\mathbf{r}_s\cdot\mathbf{p}_s + \mathbf{p}_s\cdot\mathbf{r}_s) = \mathscr{V} \tag{21}$$

Therefore in this case, the theorem will be satisfied if the set of trial functions is invariant to positive scaling. (This will of course always ensure that the theorem for $\mathscr{V}$ is satisfied whatever $\mathscr{B}$ and the gauge may be.)

PROBLEMS

1. Show that $\langle\mathbf{V}\rangle = 0$ is equivalent to $\int \mathbf{j}\, d^3r = 0$, where $\mathbf{j}$ is the one-electron current density.

2. Derive the various theorems which would be satisfied if one had invariance to what might be called the Landau gauge transformation

$$\lambda = \xi xy$$

the uniform magnetic field being in the z direction and the starting gauge being that of (13-22).

3. Show explicitly that the set (11) is essentially invariant to the action of $\exp(i\boldsymbol{\alpha}\cdot\mathbf{V})$, where the components of α are any three real numbers. (Hint: Recall or become acquainted with the useful theorem in Messiah [3].)

4. Write Eq. (19) in a form more like that of Eq. (20-11).

5. Show that the sets of trial functions (11) and (13) are, possibly to within irrelevant overall scale factors, further examples of the sort discussed in Problem 1 of Section 21 with however the difference that the $\mathscr{G}_i$, instead of forming a Lie algebra, have instead the property that their commutators are c-numbers. Show in general that in such a case use of such trial functions will again ensure the hypervirial theorems for the $\mathscr{G}_i$. [Hint: Use [3] again.]

References

1. R. H. Young, *Mol. Phys.* **16**, 509 (1969).
2. W. J. Carr, Jr., *Phys. Rev.* **106**, 414 (1957).
3. A. Messiah, "Quantum Mechanics," Vol. 1, p. 442. Wiley, New York, 1966.

Appendix E / Proof That (33-18) Is an Improvable Bound

Just as the discussion in the text was patterned after that in the first part of Section 7, so our discussion here will be analogous to that of Eq. (17-3), etc. We start with the set

$$\psi^{(1)} = \sum_{l=1}^{M} A_l \phi_l \tag{1}$$

and add one more function; thus we replace (1) by

$$\psi^{(1)} = \sum_{l=1}^{M} A_l \phi_l + B\phi \tag{2}$$

where clearly there is no loss in generality in assuming that ϕ is orthogonal to all the ϕ_k:

$$(\phi, \phi_k) = 0, \qquad k = 1, 2, \ldots, M \tag{3}$$

It is now convenient to switch from the ϕ_k to the $\hat{\phi}_k$ as a basis and write instead of (2) and (3)

$$\psi^{(1)} = \sum_{l=1}^{M} A_l \hat{\phi}_l + B\phi, \qquad (\phi, \hat{\phi}_k) = 0 \tag{4}$$

Also, for definiteness, we will assume that $\psi'^{(0)}$ is not in the space. Then by use of (32-14), one readily finds that $\delta \check{J}_{\mathrm{H}}^{(2)} = 0$ yields

$$(\hat{E}_j^{(0)} - E'^{(0)})\check{A}_j + (\hat{\phi}_j, H^{(0)}\phi)\check{B} + (\hat{\phi}_j, (H^{(1)} - E'^{(1)})\psi'^{(0)}) = 0 \tag{5}$$

and

$$\sum_k (\phi, H^{(0)}\phi_k)\check{A}_k + (\phi, (H^{(0)} - E'^{(0)})\phi)\check{B} + (\phi, (H^{(1)} - E'^{(1)})\psi'^{(0)}) = 0 \tag{6}$$

Equation (5) then yields

$$\check{A}_j = -\frac{(\hat{\phi}_j, (H^{(1)} - E'^{(1)})\psi'^{(0)})}{\hat{E}_j^{(0)} - E'^{(0)}} - \frac{(\hat{\phi}_j, H^{(0)}\phi)\check{B}}{\hat{E}_j^{(0)} - E'^{(0)}} \tag{7}$$

or, in more explicit notation like that of Section 7,

$$\check{A}_j(M+1) = \check{A}_j(M) - \frac{(\hat{\phi}_j, H^{(0)}\phi)\check{B}}{\hat{E}_j^{(0)} - E'^{(0)}} \tag{8}$$

Therefore from (4) we have that

$$\check{\psi}^{(1)}(M) = \check{\psi}(M+1) + \Delta B \tag{9}$$

where

$$\Delta = -\phi + \sum_{j=1}^{M} \hat{\phi}_j \frac{(\hat{\phi}_j, H^{(0)}\phi)}{\hat{E}_j^{(0)} - E'^{(0)}} \tag{10}$$

We now note that $\hat{\psi}^{(1)}(M)$ is a member of the set (4), whence it follows (or one can check by detailed calculations) that

$$\hat{J}_{\mathrm{H}}^{(2)}(M) - \hat{J}_{\mathrm{H}}^{(2)}(M+1) = |B|^2(\Delta, (H^{(0)} - E'^{(0)})\Delta) \tag{11}$$

so that the sign of $\hat{J}_{\mathrm{H}}^{(2)}(M) - \hat{J}_{\mathrm{H}}^{(2)}(M+1)$ will be the sign of $(\Delta, (H^{(0)} - E'^{(0)})\Delta)$ and hence what we want to prove is that if there are T of the $\hat{E}_i^{(0)}$ below $E'^{(0)}$, then this quantity is positive, so that on enlarging the basis set, $\hat{J}_{\mathrm{H}}^{(2)}$ will steadily decrease (or in any case not increase).

If one uses (10) and (13-14), one finds that

$$(\Delta, (H^{(0)} - E'^{(0)})\Delta) = (\phi, (H^{(0)} - E'^{(0)})\phi) - \sum_{j=1}^{M} \frac{|(\phi, H^{(0)}\hat{\phi}_j)|^2}{\hat{E}_j^{(0)} - E'^{(0)}} \equiv D(E'^{(0)}) \tag{12}$$

We now note, comparing (12) and (7-9), that the solutions of

$$D(\varepsilon) = 0 \tag{13}$$

are just the $\hat{E}_j^{(0)}(M+1)$ which would arise from diagonalizing $H^{(0)}$ with the set of functions (4) and so we know, from (7-11), that the kth root of this equation, call it $\hat{E}_k^{(0)}(M+1)$, will satisfy

$$\hat{E}_k^{(0)}(M+1) \leq \hat{E}_k^{(0)}, \qquad k = 1, 2, \ldots, M \tag{14}$$

Also, by assumption,

$$\hat{E}_T^{(0)} \leq E'^{(0)} \leq \hat{E}_{T+1}^{(0)} \tag{15}$$

and therefore, since there are only T exact eigenvalues below $E'^{(0)}$, it follows from (14) and (15) that

$$\hat{E}_T^{(0)} \leq E'^{(0)} \leq \hat{E}_{T+1}^{(0)}(M+1) \leq \hat{E}_{T+1}^{(0)} \tag{16}$$

The fact that under these conditions, $D(E'^{(0)})$ is positive then follows most simply by using a graph. Thus define

$$F(\varepsilon) \equiv (\phi, (H^{(0)} - \varepsilon)\phi), \qquad G(\varepsilon) \equiv \sum_{J=1}^{M} \frac{|(\phi, H^{(0)}\hat{\phi}_j)|^2}{\hat{E}_j^{(0)} - \varepsilon}$$

Then in order to ensure (16), we must have the graph shown in Fig. E-1, and so $D(E'^{(0)}) = F(E'^{(0)}) - G(E'^{(0)})$ is obviously ≥ 0, as desired.

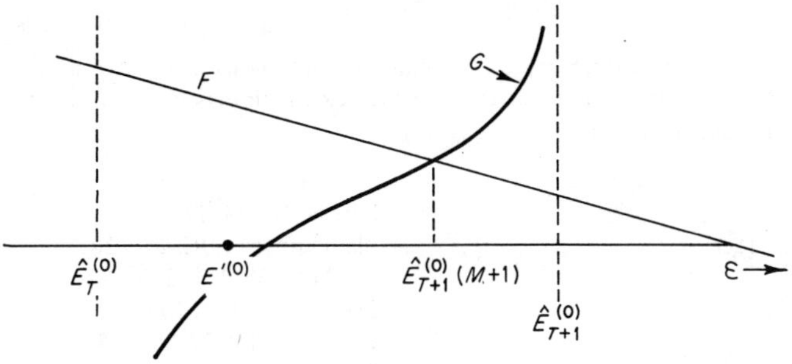

Fig. E-1.

Author Index

Numbers in parentheses are reference numbers and indicate that an author's work is referred to although his name is not cited in the text. Numbers in italics show the page on which the complete reference is listed.

A

Aashamar, K., 123, *154,* 156, 157(9), *185*
Åberg, T., 46, *56*
Adam, G., 27, *55,* 97(34), *118*
Adamov, M. N., 208(26), *220*
Adams, W. H., 66(25), *68*
Ahlrichs, R., 191, *220*
Albat, R., 54, *56*
Alder, B. J., 11(24), *25*
Amos, A. T., 239(10), 241(10), *243*
Amos, T., 86(18), 91, *118*
Antanasoff, J. V., 196, *220*
Aranoff, S., 239, 240, *243*

B

Bagus, P. S., 54, *56,* 69, 115, *118, 119*
Bangudu, E. A., 21, *25*
Barr, T. L., 123(19), *154*
Barraclough, C. G., 30, *55*
Bartlett, R. J., 123(19), *154*
Bauche, J., 60, *67*
Bell, R. P., 112(49), 115(49), *119*
Bender, C. F., 64, *68*
Benston, M. L., 115, *119*
Berthier, G., 54, *56,* 60, *67*
Bertoncini, B. J., 11, *25*
Beveridge, D. L., 54, *56*
Bhabha, H. J., 20, *25*
Biedenharn, L., 121, *154*
Billingsley II, F. P., 123, *154*
Bingel, W. A., 110(47), *119*
Birkhoff, G., 2, *24*
Bishop, D. M., 10, *25,* 123, *154,* 165(15), *186*
Bjorna, N., 93, *118*
Blankenbecler, R., 210(28), 216(33), 217(33), *220,* 247, *249*
Blatt, J. M., 121, *154*
Blinder, S. M., 23(35), *25*
Bloomer, W. L., 14, 25
Bonačić-Koutecký, V., 14, *25*
Born, M., 1, *24*
Boys, S. F., 23(34), *25*
Brändas, E. J., 72(4), 97(34), *118,* 123(19), *154,* 161, *185*
Bratoz, S., 162(13), *186*
Brillouin, L., 58, *67*
Brooks, H., 121, *154*
Brown, R. D., 54, *56*
Brown, R. T., 110(47), *119*
Brueckner, K. A., 46, *56*
Bruner, B. L., 14, *25*
Buckingham, R. A., 196, *220*
Burden, F. R., 9, *25*
Burrows, B. L., 242, *243*
Byers Brown, W., 82, *118,* 124(28), 146(37), 147(28), *154,* 185, *186*

C

Carbo, R., 162(13), *186*
Carr, Jr., W. J., 260, *264*
Caspers, W. J., 157(7), *185*
Catana, F., 54, *56*
Chang, T. C., 59, *67*
Chen, J. C. Y., 198, *220,* 241, *243*
Chong, D. P., 115, *119*
Chung, K. T., 123(13), *154*
Čižek, J., 48, *56*

Claveire, P., 123(20), *154,* 156(3), 161(3), *185*
Claxton, T. A., 86, *118*
Clementi, E., 54, *56*
Clinton, W., 105, *118*
Cohen, H. D., 123, *154*
Cohen, M., 44(23), *55,* 61(16), 65(24), *68,* 115, *119,* 123, *154,* 179(27), 180(28, 29), 181, 183, 185, *186,* 189, *219*
Coulson, C. A., 72(4), 96(32), 112(49), 113(50), 115(49), *118, 119,* 146(38), *154*
Courant, R., 2, 8(11), *24,* 31(7), *55*
Csizmadia, I. G., 80, *118*

D

Dalgarno, A., 61(16), 65(24), *68,* 115, *119,* 120, 147, *154, 155,* 161, 167(23), 179(27), 180(28, 29), 181, 183, 185, *185, 186,* 189, 192(10), 198, 206(25), *219, 220,* 238(8), 241, *243,* 256(5), *259*
Das, G., 11, *25*
Davidson, E. R., 47(29), *56,* 63, 64(23), *68,* 133(31a), *154*
Davies, D. W., 37(16), *55*
Deal, Jr., W. J., 239, *243,* 256(6), *259*
Dehn, J. T., 52, *56*
de Jeu, W. H., 215, 216, *220*
Delos, J. B., 23(35), *25*
Delves, L. M., 239, 241(18), *243*
Diercksen, G., 165, *186*
Diner, S., 123, *154,* 156, 161, *185*
Dion, D. A., 21(32), *25*
Ditchfield, R., 79, *118*
DiToro, M., 54(45), *56*
Domany, E., 117(62), *119*
Dougherty, T. J., 196(19), *220*
Doyle, H., 123, *154,* 156, *185*
Drake, G. W. F., 123(25), *154,* 156(2), 161, *185*
Dunning, Jr., J. H., 54(49), *56*
Dupont-Bourdelet, F., 120, *154*

E

Eger, M., 115, *119*
Ehlenberger, A. G., 189, *219*
Einhorn, M. B., 247, *249*
Eliason, M. A., 230(3), 231(3), *233*
England, W., 44, *55*
Epstein, I. R., 80, *118*
Epstein, J. H., 197(20), 199, *220*
Epstein, S. T., 24(37), *25,* 59, *67,* 79(7), 87(21), 96(33), 98(33), 115(33), 117(33, 57), *118, 119,* 124(28), 147(28), *154,* 195, 196(19), 197(20), 199, 206(25), 218, *220,* 234(1), 239, *243,* 250(1), 256(1, 5), *259*

F

Farmer, C. M., 117(61), *119*
Feinberg, M. J., 1(6), *24*
Feynman, R. P., 1, *24*
Finlayson, B. A., 2, 21, *24,* 31(8), *55*
Foldy, L. L., 39(17), *55*
Foley, H. M., 147, *155*
Folland, N.-O., 51, *56*
Fraga, S., 64, *68*
Freeman, A. J., 123, *154*
Frenkel, J., 250(2), *259*
Froese Fischer, C., 51, *56*
Frost, A. A., 81(10), *118*

G

Geratt, J., 64, *68,* 165, *186*
Gershgorn, Z., 123, *154,* 156, 161, *185*
Gianinetti, E., 52, *56*
Gliemann, G., 114, *119*
Goddard III, W. A., 39(18), 47(29), 54(49), *55, 56*
Goldstein, R., 195, *220*
Goscinski, O., 123(19), *154,* 161, *185*
Gould, S. H., 31(9), *55,* 244, *246*
Grasso, M. N., 123, *154*
Green, S., 64(23), *68*
Grein, F., 59, *67*
Grimaldi, F., 161, *185*
Gross, E. P., 115, *119*
Gruen, N., 54, *56*
Gupta, R. P., 189, *219*
Gutschick, V. P., 143, *154*
Guy, J., 120(3), *154*

H

Hall, G. G., 52(40), *56,* 61(16), *68,* 86(18), *118*
Hambro, L., 143, *154,* 216, 217, *220*
Handler, G. S., 27, *55*
Handy, N. C., 23(24), 24(37), *25*

Hansen, K. H.,114, *119*
Harriman, J. E., 124, *154*
Harris, F. E., 11(24), *25*
Harrison, M. C., 80(8), *118*
Hartree, D. R., 51(37), *56*
Hayes, E. F., 254(4), *259*
Haymaker, R. W., 216(33), 217(33), *220*
Hegyi, M. G., 23, *25*
Heinrichs, J., 257, *259*
Hellmann, H., 107(43), *119*
Herzberg, G., 9, *25*
Hibbert, A., 146(38), *154*
Hilbert, D., 8(11), *24*
Hirschfelder, J. O., 93(30), 96(33), 97(30), 98(33), 106(41), 107, 110(48), 115(33), 117(33), *118, 119,* 121, 124, 147, *154,* 194(11), *220,* 230(3), 231(3), *233,* 234(1), *243,* 259(9), *259*
Hunt, W. J., 47(29), 54, *56*
Hurley, A. C., 88, 109, 113(50), 114(51), 115(57), 117(57, 61), *118, 119*
Huron, B., 123, *154,* 156, 161, *185*
Hurst, R. P., 123, 144(36), *154,* 173(25), 176(25), *186,* 238(7), *243,* 256(5), *259*
Huzinga, S., 59, *67*
Hylleraas, E. A., 33(13), *55,* 110(46), *119,* 176(26), *186,* 189, *220*

I

Iafrate, G. J., 189, *219*

J

Jankowski, K., 21(32), *25*
Jaszuński, N., 234, *243*
Johnson, R. E., 234, 239, *243,* 256(5), *259*
Joy, H. W., 27, *55*
Jug, K., 54, *56*

K

Kane, E. O., 165, *186*
Kari, R., 14, *25*
Karplus, M., 238(7), *243,* 250(1), 256(1), *259*
Katriel, J., 27, *55,* 84, 97(34), 117(62), *118, 119*
Kaveeshwar, V. G., 256(5), *259*
Kelly, H. P., 61, *67*
Kestner, N. R., 234(4), 239(4, 5, 14), *243*
Kikuta, T. K., 121, *154*
Kincaid, J. F., 110(48), *119*
King, H. F., 59, *67*
Kirtman, B., 241(19), *243*
Klapisch, M., 60, *67*
Klopman, G., 54, *56*
Knight, R. E., 199(24), *220*
Kolos, W., 196, *220*
Koopmans, T., 46, *56*
Kostin, M. D., 121, *154*
Koutecký, J., 14, *25*
Krauss, M., 123, *154*
Kreiger, J. B., 241(19), *243*

L

Labarthe, J. J., 60, *67*
Laferriere, A., 30(4), *55*
Langhoff, P. W., 144, *154,* 173, 176, *186,* 238(7), *243,* 250, 256(1), *259*
LeFebvre, R., 59, *67*
Levy, B., 60, *67*
Linderberg, J., 183(36), *186*
Lindner, P., 195, *220*
Lippincott, E. R., 189(1), *219*
Lipscomb, W. N., 80, 91(26), *118,* 165(17), *186*
Liu, B., 69(2), *118*
Loeb, R. S., 69, *118*
Löwdin, P.-O., 8, *24,* 39(17), 40, 48(35), *55, 56,* 86(15), 110(47), 112(48), 115(49), *118, 119,* 195, 210(28), *220,* 252, *259*
Lunell, S., 48, *56,* 59, *67,* 86(14), *118*
Lyons, J. D., 144(36), *154,* 173(25), 176(25), *186*
Lyslo, G., 123(24), *154,* 156(1), 157, *185*

M

MacDonald, J. K. L., 33(13), *55*
MacDonald, N., 86(17), *118*
McElwain, D. L. S., 26(1), *55*
Macias, A., 10, *25,* 123, *154,* 165(15), *186*
McIver, Jr., J. W., 123, *154*
McKoy, V., 30(4), *55,* 143, *154,* 189, *219*
McLaughlin, D. R., 11(24), *25*
McLean, A. D., 110(48), *119,* 123, *154*
McWeeny, R., 41, *55,* 62, *68,* 82, 91, *118,* 165, *186*
McWilliams, D., 59, *67,* 86, *118*
Malrieu, J. P., 123(20, 23), *154,* 156(3,6), 161(3, 6), *185*

Mandan, R. N., 216, 217, *220*
Mandelstam, S., 1, *24*
Manne, R., 46, *56*
Manson, S. T., 59, *67*
Matcha, R. L., 69, *118*, 185, *186*
Matsen, F. A., 32(11), *55*
Mayer, I., 59, *67*
Meath, W. J., 106(41), 107, *119*, 198, *220*, 259(9), *259*
Melder, H. W., 46(26), *56*
Melius, C. F., 39(18), *55*
Mendelsohn, L. B., 189, *219*
Messiah, A., 32, *55*, 72, *118*, 263, *264*
Mestechkin, M. M., 165, *186*
Mezei, M., 23(33), *25*
Midtdal, J., 123(24), *154*, 156(1), 157(9), *185*
Mikhlin, S. G., 2, *24*, 31(6), *55*
Miller, W. H., 208(27), *220*
Mills, I. M., 64, *68*, 165, *186*
Mizuno, Y., 62, *68*
Moccia, R., 137, *154*, 165, *186*
Møller, C., 61(16), *68*
Mooney, J. R., 30, *55*
Morley, G. L., 115(55), *119*
Moskowitz, J. W., 80(8), *118*
Mukherjee, P. K., 252, *259*
Musher, J. I., 48, *56*, 87, *118*, 239(9), 241(9), *243*

N

Nazaroff, G. V., 146(37), *154*
Nelander, B., 108(45), *119*
Nesbet, R. K., 59, *67*

O

O'Leary, B. J., 54, *56*
Olive, J. P., 52, *56*
O'Malley, T. F., 210(28), *220*
Oppenheimer, M., 123(25), *154*, 156(2), *185*
Ostlund, N. S., 123, *154*

P

Pace, E., 54(45), *56*
Paldus, J., 48, *56*
Pan, K.-C., 59, *67*
Pandres, Jr., D., 32(11), *55*, 106(42), 117(42), *119*
Parr, R. G., 115(57), 117(57, 60), *119*, 183(32), *186*, 254(4), *259*
Pauncz, R., 44(23), 48(36), *55*, *56*, 86(16), *118*
Pekeris, C., 11, *25*
Percus, J., 239, 240, *243*
Perez, J. D., 46(26), *56*
Perkins, J. F., 36(15), 39(18), *55*, 245(2), *246*
Phillipson, P., 113(50), *119*
Pitzer, R. M., 91(26), *118*, 165(17), *186*
Pomraning, G. C., 121(9), *154*
Pople, J. A., 54, *56*, 91, *118*, 123, *154*
Prager, S., 194(11), *220*
Pritchard, H. O., 26(1), *55*

R

Raisel, Y., 69, *118*
Ramsey, N. F., 91(25), *118*
Rancurel, P., 123(23), *154*, 156(6), 161(6), *185*
Rao, B. K., 234(4), 239(4), *243*
Rebane, T. K., 144, *154*, 173, *186*
Rebello, I., 208(27), *220*
Reeken, M., 47(31), *56*
Riemenschneider, B. R., 234, 239, *243*
Roberts, H. G. Ff., 91, *118*, 239(10), 241(10), *243*
Roby, K. R., 54, *56*
Roos, B., 123, *154*, 156, 161, *185*
Roothaan, C. C. J., 47, 52(30), 54, *56*, 123, *154*
Rosenberg, L., 210(28), *220*
Rothstein, S. M., 23(36), *25*
Rudge, M. R. H., 8, *24*
Ruedenberg, K., 1, *24*, 44(22), *55*, 105, *118*

S

Sadlej, A. J., 197, *220*, 234, 238(7), *243*
Salem, L., 229(1), *233*
Salmon, L. S., 44(22), *55*
Sambe, H., 100(36), 101, 104(36), *118*
Sanders, F. C., 164, *186*, 199(24), *220*
Saxena, K. M. S., 64, *68*
Schaefer, H. F., III, 9, 11, *24*, *25*, 31(10), *55*, 69(2), 118
Scherr, C. W., 199, *220*

Schiffrer, G., 54(45), *56*
Schroeder, R., 189(1), *219*
Schrödinger, E., 1(3, 4), 8, *24*
Schulman, J. M., 147, *155,* 239(9), 241(9), *243*
Schutte, C. J. H., 105, *118,* 123, *154*
Schwartz, C., 241, *243*
Sen, S. K., 189, *219*
Sharma, C. S., 123, *154,* 183, *186,* 208(27), *220*
Shavitt, I., 123, *154,* 156, 161, *185*
Shirley, D. A., 46, *56*
Shorb, A. M., 189, *219*
Shustek, L. J., 241(19), *243*
Siegbahn, K., 46, *56*
Silverman, J. N., 124, 133(31), 144, *154*
Silverman, R. A., 120(1), *154*
Silverstone, H. J., 23(36), *25,* 54, *56,* 59, *67*
Simonetti, M., 52, *56*
Simons, J., 10, *25*
Sinanoglu, O., 208(26), *220*
Singh, T. R., 198, *220*
Slater, J. C., 41, 46, 51, *55, 56*
Sleaman, D. H., 54, *56*
Spruch, L., 210(28), *220*
Stanton, R. E., 54(51), *56*
Starace, A. F., 195, *220*
Steiner, E., 82, *118*
Sternheimer, R. M., 100(35), *118,* 147, *155,* 189, *219,* 259(8), *259*
Stevens, R. M., 91, *118,* 165, *186*
Stewart, A. L., 9, *25,* 120, 146(37), 147, *154, 155,* 192(10), *220*
Sugar, R., 210(28), *220*
Sutcliffe, B. T., 14, *25,* 80(8), *118*
Swanstrom, P., 123(18), *154*
Szondy, T., 23(33), *25*

T

Taylor, H. S., 33, *55*
Thomsen, K., 123(18), *154*
Thouless, D. T., 16(30), *25*
Tillieu, J., 120(3), *154*
Tobin, F. L., 147, *155*
Trickery, S. B., 115, *119*
Tuan, D. F.-T., 147, *155,* 234, 239(11), 241(11), 242, *243*
Turinsky, P. J., 121, *154*

U

Undheim, B., 33(13), *55*

V

Valkering, T. P., 157(7), *185*
Van Leuvan, J. L., 124, 133(31), 144, *154*
Vladimiroff, T., 196(19), *220*
Voitlander, J., 215, *220*
von Neumann, J., 157(8), *185*

W

Wahl, A. C., 11, *25*
Wallis, A., 26, *55*
Wang, P. S. C., 11, *25,* 212(30), *220*
Watson, R. E., 123, *154*
Weare, J. H., 183(32), *186*
Weber, T. A., 183, *186*
Weinhold, F., 10, *25,* 210(29), 212, 218, *220*
Weinstein, H., 44, *55*
Weiss, R. J., 241(17), *243*
Welch, J. E., 23(36), *25*
White, R. J., 72(4), *118*
Wigner, E., 157(8), *185*
Wilson, Jr., E. B., 229(1), *233*
Wilson, R. G., 183(34), *186*
Wilson, R. M., 9, *25*
Winter, N. W., 30, *55,* 189, *219*
Witriol, N. M., 115(55), *119*
Wolniewicz, L., 196, *220*
Wood, J. H., 51, *56*
Wrobel, H., 215, *220*
Wu, K. K., 234, *243*
Wyatt, R. E., 115(57), 117(57), *119*

Y

Yang, C. N., 62, *68*
Yaris, R., 230(2), *233*
Yde, P. D., 123, *154*
Yin, M.-L., 59, *67*
Yoshimine, M., 123, *154*
Young, R. H., 33(13), *55,* 104(37), *118,* 239(14), *243,* 256(6), *259,* 260, *264*
Yourgrau, W., 1, *24*

Z

Zubakov, V. A., 208(26), *220*

Subject Index

Because the text is divided into many rather precisely titled sections we have in this subject index very often not included the "obvious" pages on which certain subjects may be found. Thus the subject index should be used in conjunction with the table of contents.

A

"Accurate" Hamiltonian, 122, 141
Acronyms
  CI, 31
  HVM, 221
  RHF, 49
  SCF, 52
  SUHF, 49
  SUSCF, 165
  TRK, 205
  UHF, 38
  USCF, 52
  VVM, 221
"All," 14
Atomic units, 2, 4

B

Basis set, 22, 52
Bond angle, 75, 108
Bond length, 75, 108
Born–Oppenheimer approximation, 107
Born–Oppenheimer force, 107, 112
Brillouin's theorem, 238

C

Canonical spin orbitals (definition), 43
Charge current conservation, 255
Clamped nuclei Hamiltonian (definition), 2
Configuration interaction method, 31, 115, 156
Constrained invariance, 78, 224, 232-233
Correlation energy, 178
Coupled SCF, 166
Current conservation, 94, 95

D

Degenerate perturbation theory
  in Hylleraas variation method, 202
  in linear variation method, 165
  for natural spin orbitals, 63
  in USCF method, 171
Delta function molecule, 81, 82, 85
Density matrix, 41, 62
Diagonalize, 28
Dipole moment
  electric, 89, 96, 100, 101, 103, 151, 229
  frequency dependent, 164, 256
  magnetic, 90, 229
Dipole shielding factor, 100, 256
Dirichlet functional, 194–195

E

Ehrenfest theorem, 254
Electric dipole moment, 89, 100, 101, 103, 229
Electric field, 2, 89, 98–101, 102–104, 107, 176–177, 181, 229, 255–256, 258–259
Elliptic coordinates, 113
Energy variance, 10
Euler's theorem, 106, 108, 111
"Exact" Hamiltonian, 122
Extended Hartree–Fock, 146

F

First variation
  of energy, 13, 29
  of wave function, 12
Floating wave functions, 109

Force
  Born–Oppenheimer, 107, 112
  on electrons, 98, 261
  Hellmann–Feynman, 107
  on nuclei, 99, 100, 101, 255, 258
Force constant, 105, 122–123, 151
Frenkel variation method, 250
Frequency-dependent dipole polarizability, 164, 256, 258
Frozen spin orbitals, 45

G

Gauge invariance, 76, 94, 224–225, 232, 254
Gauge invariant atomic orbitals, 79, 89
Gauge origin, 77
Gauge transformation, 76, 94, 224–225, 254
Geometric approximation, 239, 242
Gram determinant, 212

H

H_2^+, 82
Hamiltonian
  "accurate," 122
  "exact," 122
  fixed nuclei, 2
  Hartree–Fock, 42, 59, 205
  model, 195
  projected, 27, 55
  Sternheimer, 198, 205
Harmonic oscillator, 189
Hartree–Fock Hamiltonian, 42, 59, 205
Heinrich's variational approximation, 257
Helium, 27, 47, 114
Hellmann–Feynman force, 107, 231
Hellmann–Feynman theorems, 87, 227
  for electric field, 89, 151, 229
  integral, 117
  for magnetic dipole moment, 90
  for nuclear charge, 184
  for nuclear coordinates, 107, 111, 113–114, 151, 230
  for perturbation parameters, 133, 151–153, 228
  time dependent, 254
  in various coordinate systems, 87, 113, 230
Hermitian, 3
Hilbert space, 26, 70
Homonuclear diatomic molecules, 97, 99
Hurley's theorem, 88
HVM, 221–233
Hydrogen atom, 180, 189
Hylleraas functional, 189, 190, 210, 212
Hypervirial theorem, 79, 226
  by constraint, 69, 93
  for coordinate, 96, 255, 260
  for gauge function, 94, 254
  for momentum, 98, 255
  for orbital angular momentum, 201, 222
  simultaneous, 117, 263
  time dependent, 254
  for velocity, 261

I

Improvable upper bounds, 9, 33, 137, 141, 190, 191
Inconsistent equations, 142, 217, 251
Interchange theorem, 181, 236
Intermediate normalization, 162
Invariance of sets of trial functions
  to complex conjugation, 71, 73, 83
  to gauge transformations, 76, 94, 224
  general remarks, 70
  to parameter change, 87, 89, 228
  to rotations, 75, 108
  to scaling, 105, 108, 110, 111
  to spatial translations, 76, 78–79, 98, 108, 112
  to time translation, 255
  to unitary transformations, 75, 83, 93, 222
Inversion symmetry, 97, 99

K

Koopman's theorem, 45, 46

L

Lagrange multipliers, 42, 43, 50, 69, 174, 240
Landau gauge transformation, 263
Lie algebra, 117
Linear independence, 26, 31, 40
Linear space, 18, 26–27, 30, 93–94, 115–117, 132
Linear variation method, 21, 66, 71, 72, 89, 93–94, 100–101, 115–117, 123, 136, 191, 231

Local potential, 205
Localized spin orbitals, 44
Long range forces, 230
Lower bounds, 11, 137, 218

M

Magnetic dipole moment, 90, 229
Magnetic field, 2, 82–83, 90, 176–177, 229
Magnetic shielding tensor, 90, 150
Model Hamiltonian, 195, 236
Moments, 21
Momentum space, 80

N

Natural spin orbitals, 63
No crossing theorem, 157
Nonlinear parameters, 37, 38, 52, 54, 115, 138, 165, 190, 217
ν^{2n+1} theorem, 120, 130, 172, 176, 188

O

One-electron excitation, 59, 67
Optimal trial function, 9
Order parameter, 150, 236
Orthogonality theorem, 116, 179, 231–232
Orthogonalization procedures, 18, 40
Oscillator strength, 165
Overall scale, 15
Overlap matrix, 27, 31, 53, 213

P

Pade approximant, 123, 239
Pauli principle, 7
Point transformation, 115
Polarizibility, 90, 137, 196
Positive definite Hermitian matrix, 4, 31
Projected Hamiltonian, 27, 55
Projection before variation, 48, 86
Projection of symmetry components, 48, 86
Projection operator, 8, 27, 32, 55

R

Rayleigh–Ritz method, 30
Rayleigh–Schrödinger perturbation theory, 57, 124, 133, 178, 188
  and the linear variation method, 159–165
Reality constraints, 19–21
Real wave functions, 97, 98
Reflection symmetry, 97, 99
RHF, 49, 51, 60–61, 71, 89, 100–101, 110
Ritz method, 30

S

Scalar product, 3
Scaling, 105–106, 263
Schmidt orthogonalization, 18, 40
Schwartz inequality, 212
Schwartz method, 241
Second differential of energy, 15, 134, 138, 140
  in linear variation method, 29, 32
  in USCF, 65
Second variation of trial function, 12
Secular determinant, 26
Secular equation, 26
Separation theorem, 35, 245
Shielding approximation, 182
Single-particle energies, 43
Slater determinant, 31, 38, 51
s limit, 27, 66
Spin, 47–49, 65, 84, 182
Spin–spin coupling constants, 215
Stark effect, 81
Sternheimer Hamiltonian, 198, 205, 241
SUHF, 49–51, 60–65, 71, 76, 89, 93, 94, 96, 98, 105, 115, 177–185, 260
Sum rules, 205–206
Supervariation principles, 121
Susceptibility, 90, 137

T

Tensor virial theorem, 106
Thomas–Reiche–Kuhn sum rule, 205–206
Time-dependent Hellmann–Feynman theorem, 254
Time-dependent hypervirial theorem, 254
Time-dependent linear variation method, 255
Time-dependent SUHF, 254
Time-dependent UHF, 252, 254
Time reversal, 72, 98
Torque on electrons, 108
Torque on nuclei, 108
Trial function, 9
Two perturbations, 122, 131–132, 147–153, 160, 164

U

UHF, 58, 59–65, 71, 76, 84, 89, 93, 94, 96, 98, 100–101, 105, 115, 136, 177–185, 221–231, 260
Uncoupled Hartree-Fock, 195, 238, 242
Uncoupled SCF, 167
Unitary transformations
  gauge, 76, 224–225, 254
  general operator, 72, 93, 222–227, 232, 254
  matrix, 28, 40–41, 43–44, 169–172
  rotation, 76, 223
  scaling, 105, 110
  translation, 76, 223
Upperbound, 6, 7, 10–12, 33
USCF, 65–66, 71, 89, 136, 176

V

Variation of trial function, 13–15
Variational parameter, 12
Vector potential, 2, 76–77, 82, 90, 260
Velocity operator, 260, 261
Velocity theorem, 260
Virial theorem, 108, 183, 262–263
Virtual spin orbitals, 65, 170

W

Wave function tests, 69, 100, 104, 105
Work energy theorem, 255

Z

$\mathscr{Z}^{-1}$ expansion, 116, 123, 146-147, 156, 177, 195, 231
$\mathscr{Z}^{-1}$ orthogonality theorem, 179
$\mathscr{Z}^{-1}$ theorem for expectation values, 63-65, 180

Physical Chemistry

A Series of Monographs

Ernest M. Loebl, Editor

Department of Chemistry

Polytechnic Institute of New York

Brooklyn, New York

1 W. JOST: Diffusion in Solids, Liquids, Gases, 1952
2 S. MIZUSHIMA: Structure of Molecules and Internal Rotation, 1954
3 H. H. G. JELLINEK: Degradation of Vinyl Polymers, 1955
4 M. E. L. MCBAIN and E. HUTCHINSON: Solubilization and Related Phenomena, 1955
5 C. H. BAMFORD, A. ELLIOTT, and W. E. HANBY: Synthetic Polypeptides, 1956
6 GEORGE J. JANZ: Thermodynamic Properties of Organic Compounds — Estimation Methods, Principles and Practice, Revised Edition, 1967
7 G. K. T. CONN and D. G. AVERY: Infrared Methods, 1960
8 C. B. MONK: Electrolytic Dissociation, 1961
9 P. LEIGHTON: Photochemistry of Air Pollution, 1961
10 P. J. HOLMES: Electrochemistry of Semiconductors, 1962
11 H. FUJITA: The Mathematical Theory of Sedimentation Analysis, 1962
12 K. SHINODA, T. NAKAGAWA, B. TAMAMUSHI, and T. ISEMURA: Colloidal Surfactants, 1963
13 J. E. WOLLRAB: Rotational Spectra and Molecular Structure, 1967
14 A. NELSON WRIGHT and C. A. WINKLER: Active Nitrogen, 1968
15 R. B. ANDERSON: Experimental Methods in Catalytic Research, 1968
16 MILTON KERKER: The Scattering of Light and Other Electromagnetic Radiation, 1969
17 OLEG V. KRYLOV: Catalysis by Nonmetals — Rules for Catalyst Selection, 1970
18 ALFRED CLARK: The Theory of Adsorption and Catalysis, 1970
19 ARNOLD REISMAN: Phase Equilibria: Basic Principles, Applications, Experimental Techniques, 1970
20 J. J. BIKERMAN: Physical Surfaces, 1970
21 R. T. SANDERSON: Chemical Bonds and Bond Energy, 1970
22 S. PETRUCCI, ED.: Ionic Interactions: From Dilute Solutions to Fused Salts (In Two Volumes), 1971
23 A. B. F. DUNCAN: Rydberg Series in Atoms and Molecules, 1971

24 J. R. Anderson: Chemisorption and Reactions on Metallic Films, 1971

25 E. A. Moelwyn-Hughes: Chemical Statics and Kinetics of Solution, 1971

26 Ivan Draganic and Zorica Draganic: The Radiation Chemistry of Water, 1971

27 M. B. Huglin: Light Scattering from Polymer Solutions, 1972

28 M. J. Blandamer: Introduction to Chemical Ultrasonics, 1973

29 A. I. Kitaigorodsky: Molecular Crystals and Molecules, 1973

30 Wendell Forst: Theory of Unimolecular Reactions, 1973

31 Jerry Goodisman: Diatomic Interaction Potential Theory. Volume 1, Fundamentals, 1973; Volume 2, Applications, 1973

32 Alfred Clark: The Chemisorptive Bond: Basic Concepts, 1974

33 Saul T. Epstein: The Variation Method in Quantum Chemistry, 1974

In Preparation

I. G. Kaplan: Symmetry of Many-Electron Systems

A 4
B 5
C 6
D 7
E 8
F 9
G 0
H 1
I 2
J 3